6th Roma International Conference on Astroparticle Physics (RICAP16)

EPJ Web of Conferences Volume 136 (2017)

Roma, Italy
21 - 24 June 2016

Editors:

Aldo Morselli
Antonio Capone
Gonzalo Rodriguez Fernandez

ISBN: 978-1-5108-3853-6

Printed from e-media with permission by:

Curran Associates, Inc.
57 Morehouse Lane
Red Hook, NY 12571

proceedings.com

Some format issues inherent in the e-media version may also appear in this print version.

Printed by Curran Associates, Inc. (2017)

For additional information, please contact EDP Sciences – Web of Conferences at the address below.

EDP Sciences – Web of Conferences
17, Avenue du Hoggar
Parc d'Activité de Courtabœuf
BP 112
F-91944 Les Ulis Cedex A
France

Phone: +33 (0) 1 69 18 75 75
Fax: +33 (0) 1 69 28 84 91

contact-edps@webofconferences.org

Additional copies of this publication are available from:

Curran Associates, Inc.
57 Morehouse Lane
Red Hook, NY 12571 USA
Phone: 845-758-0400
Fax: 845-758-2633
Email: curran@proceedings.com
Web: www.proceedings.com

TABLE OF CONTENTS

DARK MATTER AFTER LHC RUN I: CLUES TO UNIFICATION 1
Olive Keith A.
MODELS OF SUPERSYMMETRY FOR DARK MATTER 9
Muñoz Carlos
RARE EVENTS SEARCHES WITH CHERENKOV TELESCOPES 15
Doro Michele
GAMMA-RAY SIGNATURES OF DARK MATTER 21
Cirelli Marco
SEARCH FOR ANNIHILATING DARK MATTER TOWARDS DWARF GALAXIES WITH THE CHERENKOV TELESCOPE ARRAY 27
Morselli Aldo, Rodriguez Gonzalo
THE PLISA PROJECT IN ASTERICS 31
De Bonis Giulia, Bozza Cristiano
THE HIGH-ENERGY PARTICLE DETECTOR ON BOARD OF THE CSES MISSION 35
Vitale Vincenzo, Palma Francesco, Sotgiu Alessandro
THE LUNA EXPERIMENT 39
Broggini Carlo
UNDERGROUND STUDY OF BIG BANG NUCLEOSYNTHESIS IN THE PRECISION ERA OF COSMOLOGY 43
Gustavino Carlo
FEASIBILITY STUDY OF THE $^{13}C(\alpha, N)^{16}O$ REACTION AT LUNA 47
Kochanek I., Boeltzig A., Ciani G. F.
WHITE RABBIT FACILITY 51
Ammendola Roberto, Morselli Aldo, Rodriguez Gonzalo, Salina Gaetano, Verzi Valerio
THE MASS COMPOSITION OF COSMIC RAYS MEASURED WITH LOFAR 55
Hörandel Jörg R., Bonardi A., Buitink S., Corstanje A., Falcke H., Mitra P., Mulrey K., Nelles A., Rachen J.P., Rossetto L., Schellart P., Scholten O., ter Veen S., Thoudam S., Trinh T.N.G., Winchen T.
ASTROPHYSICAL INTERPRETATION OF PIERRE AUGER OBSERVATORY MEASUREMENTS OF THE UHECR ENERGY SPECTRUM AND MASS COMPOSITION 61
di Matteo Armando
THE PIERRE AUGER OBSERVATORY UPGRADE 65
Marsella Giovanni
THE AMS-02 DETECTOR ON THE INTERNATIONAL SPACE STATION - THE STATUS AFTER THE FIRST 5 YEARS ON ORBIT 69
Duranti Matteo
NUCLEI MEASUREMENTS WITH THE ALPHA MAGNETIC SPECTROMETER ON THE INTERNATIONAL SPACE STATION 75
Heil Melanie
THE PROTON AND HELIUM ANOMALIES IN THE LIGHT OF THE MYRIAD MODEL 79
Salati Pierre, Génolini Yoann, Serpico Pasquale, Taillet Richard
THE COSMIC RAY SPECTRUM IN THE ENERGY REGION BETWEEN 10^{12} AND 10^{16} EV MEASURED BY ARGO–YBJ 83
Montini Paolo
COSMIC-RAY ENERGY DENSITIES IN STAR-FORMING GALAXIES 87
Persic Massimo, Rephaeli Yoel
TWO-ZONE MODEL FOR THE BROADBAND CRAB NEBULA SPECTRUM: MICROSCOPIC INTERPRETATION 93
Fraschetti F., Pohl M.
THE DAMPE EXPERIMENT: FIRST DATA FROM SPACE 98
De Mitri Ivan
CALOCUBE: A NOVEL CALORIMETER FOR HIGH-ENERGY COSMIC RAYS IN SPACE 104
Rappoldi A., Cattaneo P.W., Adriani O., Agnesi A., Albergo S., Auditore L., Basti A., Berti E., Bigongiari G., Bonechi L., Bonechi S., Bongi M., Bonvicini V., Bottai S., Brogi P., Cappello G., Carotenuto G., Castellini G., D'Alessandro R., Detti S., Fasoli M., Finetti N., Italiano A., Lenzi P., Maestro P., Marrocchesi P.S., Miritello M., Mori N., Olmi M., Orzan G., Pacini L., Papini P., Pellegriti M.G., Pirzio F., Ricciarini S., Spillantini P., Starodubtsev O., Stolzi F., Suh J.E., Sulaj A., Tiberio A., Tricomi A., Trifirò A., Trimarchi M., Vannuccini E., Vedda A., Zampa G., Zampa N.

PRECISION STUDY OF RADIO EMISSION FROM AIR SHOWERS AT LOFAR 108
Scholten Olaf, Bonardi Antonio, Buitink Stijn, Corstanje Arthur, Ebert Ute, Falcke Heino, Hörandel Jörg, Mitra Pragati, Mulrey Katharine, Nelles Anna, Rachen Jörg, Rossetto Laura, Rutjes Casper, Schellart Pim, Thoudam Satyendra, Trinh Gia, ter Veen Sander, Winchen Tobias

EXPLOITING THE RADIO SIGNAL FROM AIR SHOWERS: THE AERA PROGRESS 112
Revenu B.

MEASUREMENT OF THE COSMIC RAY FLUX WITH THE ANITA EXPERIMENT 117
García-Fernández Daniel, Alvarez-Muñiz Jaime, Carvalho Washington R. Jr, Schoorlemmer Harm, Zas Enrique, Based on joint work with the ANITA collaboration

THE FAST PROJECT - A NEXT GENERATION UHECR OBSERVATORY 121
Fujii Toshihiro, Malacari Max, Bellido Jose A., Horvath Pavel, Hrabovsky Miroslav, Jiang Jiaqi, Mandat Dusan, Matalon Ariel, Matthews John N., Motloch Pavel, Palatka Miroslav, Pech Miroslav, Privitera Paolo, Schovanek Petr, Thomas Stan B., Travnicek Petr

RECENT DEVELOPMENT ON THE REALIZATION OF A 1-INCH VSIPMT PROTOTYPE 125
Barbato F.C.T., Barbarino G., Campajola L., Di Capua F., Mollo C.M., Valentini A., Vivolo D.

THE PIERRE AUGER OBSERVATORY STATUS AND LATEST RESULTS 130
Berat Corinne

H.E.S.S.-II - GAMMA RAY ASTRONOMY FROM 20 GEV TO HUNDREDS OF TEV'S 136
de Naurois Mathieu

A MAJOR UPGRADE OF THE H.E.S.S. CHERENKOV CAMERAS 142
Lypova Iryna, Giavitto Gianluca, Ashton Terry, Balzer Arnim, Berge David, Brun Francois, Chaminade Thomas, Delagnes Eric, Fontaine Gerard, Füßling Matthias, Giebels Berrie, Glicenstein Jean-Francois, Gräber Tobias, Hinton Jim, Jahnke Albert, Klepser Stefan, Kossatz Marko, Kretzschmann Axel, Lefranc Valentin, Leich Holger, Lüdecke Hartmut, Manigot Pascal, Marandon Vincent, Moulin Emmanuel, de Naurois Mathieu, Nayman Patrick, Ohm Stefan, Penno Marek, Ross Duncan, Salek David, Schade Markus, Schwab Thomas, Simoni Rachel, Stegmann Christian, Steppa Constantin, Thornhill Julian, Toussnel Francois

THE CRAB PULSAR AT VHE 146
Zanin Roberta

RECENT RESULTS IN γ-RAY ASTRONOMY WITH THE ARGO-YBJ DETECTOR 150
Di Girolamo Tristano

FIRST YEAR RESULTS FROM THE HAWC OBSERVATORY 154
Casanova Sabrina

HAWC HIGH ENERGY UPGRADE WITH A SPARSE ARRAY 160
Joshi V.

OBSERVATIONS OF THE GAMMA-RAY EMISSION FROM THE QUIESCENT SUN WITH FERMI LARGE AREA TELESCOPE DURING THE FIRST 7 YEARS IN ORBIT 164
Rainò S., Giglietto N., Moskalenko I., Orlando E., Strong A.W.

THE TAIGA TIMING ARRAY HISCORE - FIRST RESULTS 168
Tluczykont M., Budnev N., Astapov I., Barbashina N., Bogdanov A., Boreyko V., Brückner M., Chiavassa A., Chvalaev O., Gress O., Gress T., Grishin O., Dyachok A., Epimakhov S., Fedorov O., Gafarov A., Gorbunov N., Grebenyuk V., Grinuk A., Horns D., Kalinin A., Karpov N., Kalmykov N., Kazarina Y., Kiryuhin S., Kokoulin R., Kompaniets K., Konstantinov A., Korosteleva E., Kozhin V., Kravchenko E., Kunnas M., Kuzmichev L., Lemeshev Yu., Lubsandorzhiev B., Lubsandorzhiev N., Mirgazov R., Mirzoyan R., Monkhoev R., Nachtigall R., Osipova E., Pakhorukov A., Panasyuk M., Pankov L., Petrukhin A., Poleschuk V., Popova E., Porelli A., Postnikov E., Prosin V., Ptuskin V., Rubtsov G., Pushnin A., Samoliga V., Satunin P., Semeney Yu., Silaev A., Silaev A. (junior), Skurikhin A., Slunecka M., Sokolov A., Spiering C., Sveshnikova L., Tabolenko V., Tarashansky B., Tkachenko A., Tkachev L., Voronin D., Wischnewski R., Zagorodnikov A., Zurbanov V., Zhurov D., Yashin I.

THE FIRST FERMI-LAT SNR CATALOG AND COSMIC RAY IMPLICATIONS 172
de Palma F., Brandt T. J., Hewitt J.W.

THE FERMI-GBM GAMMA-RAY BURST CATALOGS: THE FIRST SIX YEARS 176
Bissaldi E., Bhat P. N., Yu H.-F.

FERMI BUBBLES: AN ELEPHANT IN THE GAMMA-RAY SKY 180
Malyshev Dmitry

SUPERNOVA REMNANTS WITH FERMI LARGE AREA TELESCOPE 184
Caragiulo M., Di Venere L.

LATTES: A NEW GAMMA-RAY DETECTOR CONCEPT FOR SOUTH AMERICA 191
Assis P., Barres de Almeida U., Blanco A., Conceição R., D'Ettoree Piazzoli B., De Angelis A., Doro M., Fonte P., Lopes L., Matthiae G., Pimenta M., Shellard R., Tomé B.

ANOMALIES IN THE GAMMA-RAY DIFFUSE EMISSION OF THE GALAXY AND IMPLICATIONS FOR THE INTERPRETATION OF ICECUBE RESULTS 196
Grasso D., Gaggero D., Marinelli A., Taoso M., Urbano A., Valli M.

GAMMA RAY ASTRONOMY ABOVE 30 TEV AND THE ICECUBE RESULTS 200
Vernetto Silvia, Lipari Paolo

MODELING THE GALACTIC CENTER EMISSION FROM GEV TO PEV ... 206
Marinelli A., Gaggero D., Grasso D., Taoso M., Urbano A., Valli M.
THE INNER 300 PARSECS OF THE MILKY WAY SEEN BY H.E.S.S.: A PEVATRON IN THE GALACTIC CENTRE ... 211
Moulin Emmanuel
LIMITS ON LORENTZ INVARIANCE VIOLATION AT THE PLANCK ENERGY SCALE FROM H.E.S.S. SPECTRAL ANALYSIS OF THE BLAZAR MRK 501 ... 215
Lorentz Matthias, Brun Pierre
PROSPECTS FOR GAMMA-RAY BURST DETECTION BY THE CHERENKOV TELESCOPE ARRAY ... 219
Bissaldi E., Di Girolamo T., Di Pierro F., Gasparetto T., Longo F., Vallania P., Vigorito C.
THE FERMI GBM AND LAT FOLLOW-UP OF GW150914 ... 223
Bissaldi E., Connaughton V., Omodei N., Burns E., Goldstein A., Vianello G.
MEASURING SPACE-TIME FUZZINESS WITH HIGH ENERGY γ-RAY DETECTORS ... 229
Cattaneo Paolo Walter, Rappoldi Andrea
TOWARDS THE DEVELOPMENT OF A SIPM-BASED CAMERA FOR THE CHERENKOV TELESCOPE ARRAY ... 233
Ambrosi G., Bissaldi E., Di Venere L., Fiandrini E., Giglietto N., Giordano F., Ionica M., Paoletti R., Simone D., Vagelli V.
STATUS AND PROSPECTS OF THE JUNO EXPERIMENT ... 237
Ranucci Gioacchino
RESULTS FROM THE ANTARES NEUTRINO TELESCOPE ... 242
Losa Agustín Sánchez
FOLLOW-UP OF GW150914 AND MULTI-MESSENGER STUDIES OF TRANSIENT ASTROPHYSICAL SOURCES WITH THE ANTARES NEUTRINO TELESCOPE ... 248
Coleiro Alexis
SEARCH FOR HIGH ENERGY NEUTRINOS FROM BRIGHT GRBS WITH ANTARES ... 252
Sanguineti M., Celli S., Turpin D.
SEARCHES FOR DIFFUSE FLUXES OF ASTROPHYSICAL NEUTRINOS WITH THE ANTARES TELESCOPE ... 256
Fusco Luigi Antonio, Hallmann Steffen, Navas Sergio, Versari Federico
A TIME-DEPENDENT SEARCH FOR HIGH-ENERGY NEUTRINOS FROM BRIGHT GRBS WITH ANTARES ... 261
Celli Silvia
BAIKAL-GVD ... 265
Avrorin A.D., Avrorin A.V., Aynutdinov V.M., Bannash R., Belolaptikov I.A., Brudanin V.B., Budnev N.M., Danilchenko I.A., Demidov S.V., Domogatsky G.V., Doroshenko A.A., Dvornicky R., Dyachok A.N., Dzhilkibaev Zh.-A.M., Fajt L., Fialkovsky S.V., Gafarov A.R., Gaponenko O.N., Golubkov K.V., Gress T.I., Honz Z., Kebkal K.G., Kebkal O.G., Konischev K.V., Korobchenko A.V., Koshechkin A.P., Koshel F.K., Kozhin A.V., Kulepov V.F., Kuleshov D.A., Milenin M.B., Mirgazov R.A., Osipova E.R., Panfilov A.I., Pan'kov L.V., Pliskovsky E.N., Rozanov M.I., Rjabov E.V., Shamakhov F.A., Shaybonov B.A., Sheifler A.A., Shelepov M.D., Simkovic F., Skurihin A.V., Smagina A.A., Stekl I., Suvorova O.V., Tabolenko V.A., Tarashansky B.A., Yakovlev S.A., Zagorodnikov A.V., Zurbanov V.L.
DESIGN AND PRODUCTION OF THE DIGITAL OPTICAL MODULE OF THE KM3NET PROJECT ... 270
Leonora Emanuele, Giordano Valentina
RESULTS AND SIMULATION OF THE PROTOTYPE DETECTION UNIT OF KM3NET-ARCA ... 274
Hugon C.M.F.
DAMA/LIBRA RESULTS AND PERSPECTIVES ... 278
Bernabei R., Belli P., Di Marco A., Montecchia F., d'Angelo A., Incicchitti A., Cappella F., Caracciolo V., Cerulli R., Dai C.J., He H.L., Kuang H.H., Ma X.H., Sheng X.D., Wang R.G., Ye Z.P.
$ZnWO_4$ ANISOTROPIC SCINTILLATOR FOR DARK MATTER INVESTIGATION WITH THE DIRECTIONALITY TECHNIQUE ... 284
Bernabei R., Belli P., Cappella F., Caracciolo V., Cerulli R., Danevich F.A., d'Angelo A., Di Marco A., Incicchitti A., Mokina V. M., Poda D.V., Polischuk O.G., Taruggi C., Tretyak V.I.
IMPACT OF NEXT FUTURE DIRECT DETECTION EXPERIMENTS ON DARK PORTALS AND BEYOND ... 289
Arcadi Giorgio
SEASONAL VARIATIONS OF THE RATE OF MULTIPLE-MUONS IN THE GRAN SASSO UNDERGROUND LABORATORY ... 294
Ronga F.

Presentation

RICAP-16 is the sixth edition of the RICAP Conference. The Conference, entirely dedicated to the study of high-energy cosmic rays, has been traditionally organized by the INFN sections and Physics Departments of the three public Universities of Roma (University "La Sapienza", University Roma "Tor Vergata" and University "Roma Tre"). The 2016 edition was held in Villa Tuscolana, Frascati, Roma from 21 June 2016 to 24 June 2016

Committees and List of reviewers

Local Organizing Committe

Aldo Morselli (Università Roma Tor Vergata & INFN, Roma) Chair
Antonio Capone (Università di Roma La Sapienza & INFN, Roma)
Giulia De Bonis (INFN, Roma)
Mario De Vincenzi (Università Roma Tre & INFN, Roma)
Anna Di Ciaccio (Università Roma Tor Vergata & INFN, Roma)
Marco Ciuchini (Università Roma Tre & INFN, Roma)
Giuseppe Di Sciascio (Università Roma Tor Vergata & INFN, Roma)
Irene Di Palma (Università di Roma La Sapienza & INFN, Roma)
Marcella Diemoz (Università di Roma La Sapienza & INFN, Roma)
Gonzalo Rodriguez (Università Roma Tor Vergata & INFN, Roma)
Roberta Sparvoli (Università Roma Tor Vergata & INFN, Roma)
Valerio Verzi (Università Roma Tor Vergata & INFN, Roma)

Organizing Committee

Felix Aharonian (Dublin Institute for Advanced Studies, Ireland)
Roberto Battiston (ASI Italian Space Agency)
Paschal Coyle (CPPM, Marseille, France)
Tom Gaisser (University of Delaware, USA)
Francis Halzen (University of Wisconsin, USA)
Juan Jose Hernandez (University of Valencia, Spain)
Paolo Lipari (INFN, Roma, Italy)
Szabolcs Marka (Columbia University, New York, USA)
Antonio Masiero (Università degli Studi di Padova & INFN, Padova)
Giorgio Matthiae (University "Tor Vergata" & INFN , Roma, Italy)
Carlos Muñoz (Universidad Autonoma de Madrid, Spain)
Piergiorgio Picozza (University "Tor Vergata" & INFN , Roma, Italy)
Christian Spiering (DESY Zeuthen, Germany)
Eli Waxman (The Weizmann Inst. of Science, Rehovot, Israel)
Juande Zornoza (Instituto de Física Corpuscular, Valencia, Spain)

Sponsors

SAPIENZA
UNIVERSITÀ DI ROMA

UNIVERSITA' degli STUDI di ROMA
TOR VERGATA

ROMA TRE
UNIVERSITÀ DEGLI STUDI
DIPARTIMENTO DI FISICA

INFN
Istituto Nazionale
di Fisica Nucleare

HAMAMATSU
PHOTON IS OUR BUSINESS

CAEN
Tools for Discovery

SITAEL

ARAP
Associazione Romana per le Astro-Particelle

edp sciences

EPJ Web of Conferences

Statement of Peer review

In submitting conference proceedings to EPJ Web of Conferences, the editors of the proceedings certify to the Publisher that they adhere to the **Policy on Publishing Integrity** of the journal in order to safeguard good scientific practice in publishing.

1. All articles have been subjected to peer review administered by the proceedings editors.
2. Reviews have been conducted by expert referees, who have been requested to provide unbiased and constructive comments aimed, whenever possible, at improving the work.
3. Proceedings editors have taken all reasonable steps to ensure the quality of the materials they publish and their decision to accept or reject a paper for publication has been based only on the merits of the work and the relevance to the journal.

Title, date and place of the conference: Ricap16, 6th Roma International Conference on Astroparticle Physics, Roma 21-24 June 2016

Proceedings editor(s): Aldo Morselli, Antonio Capone, Gonzalo Rodriguez Fernandez

Date and editor's signature: 02/03/ 2017

DOI: 10.1051/epjconf/201713601001

Dark Matter after LHC Run I: Clues to Unification

Keith A. Olive[1,a]

[1] *William I. Fine Theoretical Physics Institute, School of Physics and Astronomy, University of Minnesota, Minneapolis, MN 55455, USA*

Abstract. After the results of Run I, can we still 'guarantee' the discovery of supersymmetry at the LHC? It is shown that viable dark matter models in CMSSM-like models tend to lie in strips (co-annihilation, funnel, focus point). The role of grand unification in constructing supersymmetric models is discussed and it is argued that non-supersymmetric GUTs such as SO(10) may provide solutions to many of the standard problems addressed by supersymmetry.

1 Introduction

Among the motivations for supersymmetry is the theory's ability to provide gauge coupling unification [1] and address the gauge hierarchy problem [2]. It is well known that the additional fields predicted in minimal supersymmetric standard model (MSSM), if present at low energy, alter the running of the gauge couplings as shown in Figure 1. In order to ensure that the proton is long-lived, it is common to impose *R*-parity defined in terms of baryon number, lepton number and spin as $(-1)^{3B+L+2s}$. *R*-parity limits the models to include only those interactions which are direct supersymmetric analogues of Standard Model (SM) processes. As an additional consequence, supersymmetric models with *R*-parity predict the existence of a stable particle which can be a dark matter candidate [3].

Many models of supersymmetry breaking (such as gravity mediation [4]) predict universalities among the supersymmetry breaking parameters. For example, at some high energy input scale (usually taken to be the GUT scale), all gaugino masses take a common value, $m_{1/2}$, all scalar masses, m_0, and all trilinear mass terms, A_0. These three parameters, together with the ratio of the two Higgs expectation values, $\tan\beta$, defines the constrained MSSM (CMSSM). In the CMSSM, one uses the conditions derived by the minimization of the Higgs potential after radiative electroweak symmetry breaking to solve for the Higgs mixing mass, μ and the bilinear mass term B_0 (or equivalently μ and the Higgs pseudoscalar mass, m_A) for fixed $\tan\beta$.

As discussed later, non-supersymmetric SO(10) models can also naturally account for the presence of a dark matter candidate. If symmetry breaking to the SM occurs through an intermediate scale gauge group. The gauge couplings may be deflected at the intermediate scale and hence allow for gauge coupling unification as shown in Fig. 2 [5]. If the intermediate scale is broken via a **126** dimensional representation of SO(10), a Z_2 discrete symmetry (similar to *R*-parity) is preserved thus allowing for the existence of a stable dark matter candidate.

[a] e-mail: olive@umn.edu

DOI: 10.1051/epjconf/201713601001

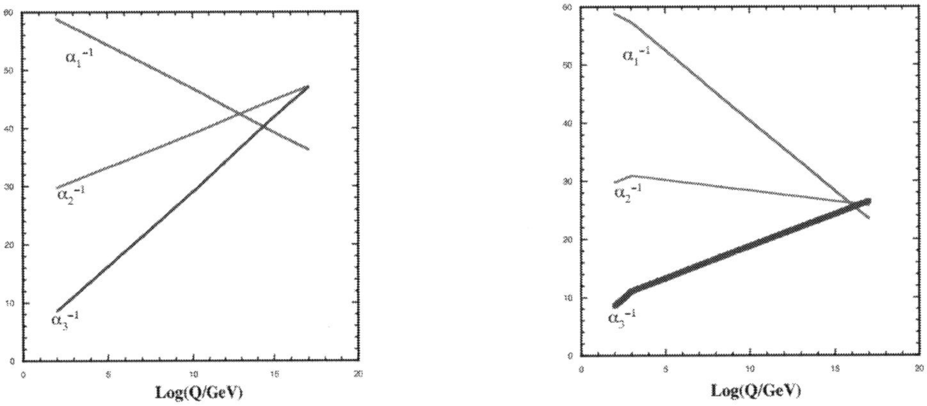

Figure 1. Running of the gauge couplings in the standard model (left) and in the MSSM (right). $\alpha_i = g_i^2/4\pi$ for each of the gauge couplings associated with $SU(3)_c$, $SU(2)_L$, and $U(1)_Y$. Unification of the couplings becomes possible in the MSSM.

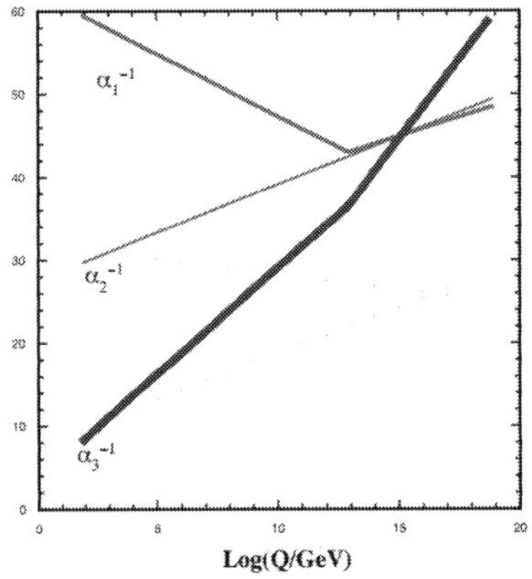

Figure 2. Running of the gauge couplings in a SO(10) model broken to $SU(4) \times SU(2)_L \times U(1)_R$ at the unification scale, $M_{GUT} = 1.3 \times 10^{15}$ GeV, which is subsequently broken to the Standard model at an intermediate scale $M_{int} = 7.8 \times 10^{12}$ GeV.

2 Pre-Run I

Before Run I at the LHC, there was much anticipation for the possibility of discovering supersymmetry as supersymmetric models such as the CMSSM provided definite improvements to low energy precision phenomenology and were well within the range of the LHC. The left panel of Fig. 3 shows the results of mastercode [6] - a frequentist Markov Chain Monte Carlo analysis of low energy experimental observables in the context of supersymmetry. At each point sampled in the CMSSM, mastercode computes a set of observables and compares that to experiment, thus establishing a χ^2 likelihood function across the parameter space. The figure shows the color coded values of $\Delta\chi^2$ relative to the best fit point shown by the white dot at low $m_{1/2}$ and low m_0. Marginalization over A_0 and $\tan\beta$ was performed to produce this $(m_0, m_{1/2})$ plane. The best-fit CMSSM point lies at $m_0 = 60$ GeV, $m_{1/2} = 310$ GeV, $A_0 = 130$ GeV, $\tan\beta = 11$ with $m_h = 114.2$ GeV. This was a pre-LHC prediction

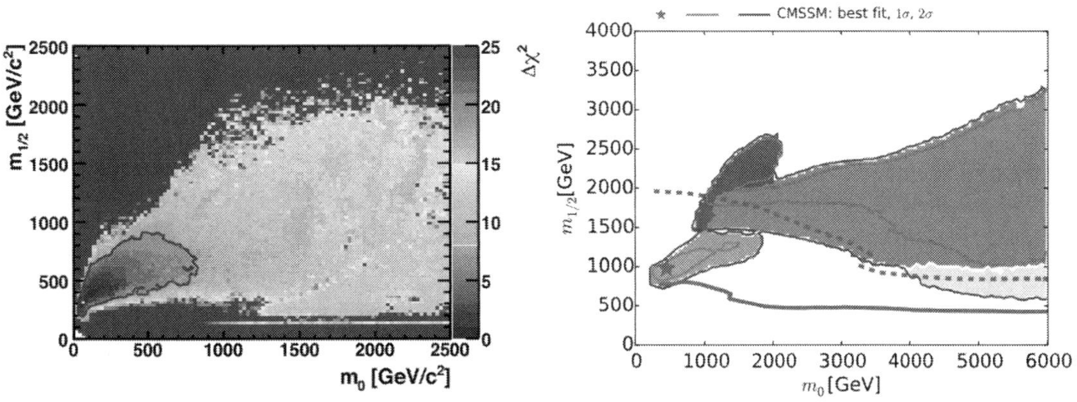

Figure 3. The $\Delta\chi^2$ functions in the $(m_0, m_{1/2})$ planes for the CMSSM from a mastercode frequentist analysis. The pre-LHC result is shown in the left panel [6]. Red and blue contours correspond to 68% and 95% CL contours and the best fit point is depicted by a white dot. The post-LHC result is shown in the right panel [7] using 8 TeV data at 20 fb^{-1}. Here the best fit point is shown by the filled star. The color of the shaded region indicates the dominant annihilation mechanism for obtained the correct relic density: stau coannihilation-pink; A/H funnel-blue; focus point-cyan; and a hybrid region of stau coannihilation and funnel-purple. The solid and dashed purple curves show the run I reach and the expected run II reach at 14 TeV at 3000 fb^{-1} respectively. The latter corresponds approximately to the 95% CL exclusion sensitivity with 300/fb at 14 TeV.

and used no LHC data. The low energy observables used included $(g_\mu - 2)$, M_W, $B \to \tau\nu$, $b \to s\gamma$, the LEP limit on the Higgs mass, forward-backward asymmetries among others (for a full list of observables used see [6]). The relatively low value of m_h was a common prediction of MSSM models [8]. A dedicated scan for the distribution of Higgs masses in the CMSSM was made in [9]. It was found that when all phenomenological constraints (with or without $(g_\mu - 2)$) are included, all models yielded $m_h \le 128$ GeV. When $(g_\mu - 2)$ is included, only models with $m_h < 126$ GeV were found. Note that the scan sampled scalar and gaugino masses only out to 2 TeV.

There was equal optimism for discovering supersymmetric dark matter in direct detection experiments. The left panel of Fig. 4 displays the pre-LHC preferred range of the spin-independent DM scattering cross section $\sigma_p^{\rm SI}$ (calculated here assuming an optimistic π-N scattering term $\Sigma_N = 64$ MeV) as a function of m_χ [6]. The expected range of $\sigma_p^{\rm SI}$ lied just below the then present experimental upper limits (solid lines) [10, 11]. As one can see from the successive lower upper limits from later experiments [12–14] shown by the bands, the pre-LHC values for the elastic scattering cross section was well within reach of current experiments.

3 Post Run I

After run I at the LHC, the prospects for discovering supersymmetry looked very different. In the right panel of Fig. 3, the post-Run I likelihood contours in the $(m_0, m_{1/2})$ plane [7] are shown using 8 TeV results at 20 fb^{-1} [18]. The best fit point based on the 8 TeV data is shown by the filled star at (420,970) GeV with $A_0 = 3000$ GeV and $\tan\beta = 14$, though the likelihood function is quite flat and the exact position of the best point is not very well defined.

The right panel of Fig. 4 shows that there is still hope for direct detection experiments though the new best fit point implies a cross section of $\sim 10^{-47}$ cm^2, nearly two orders of magnitude below the current upper bound. The likelihood function, however, is rather flat between 10^{-47} cm$^2 \lesssim \sigma_p^{\rm SI} \lesssim$

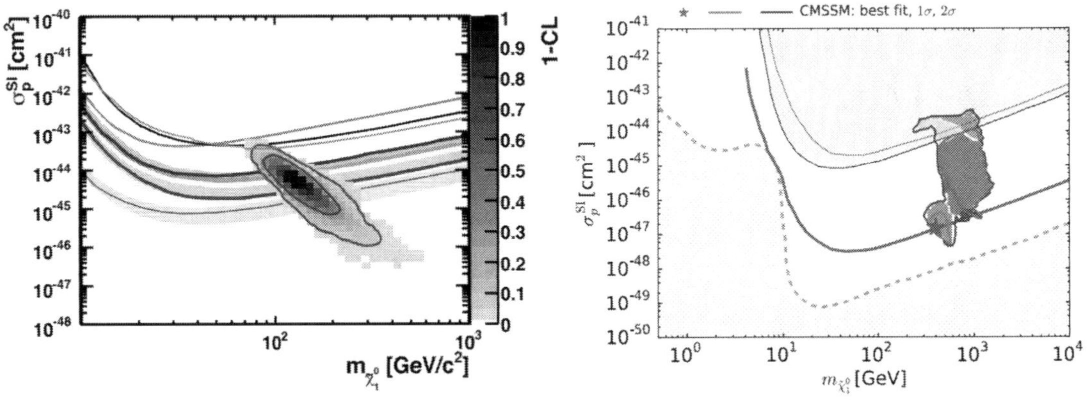

Figure 4. (left) The pre-LHC prediction for the spin-independent DM scattering cross section, σ_p^{SI}, versus m_χ in the CMSSM [6]. The solid lines are the pre-LHC experimental upper limits from CDMS [10] and XENON10[11], while the bands are the more recent limits from XENON100 [12, 13] and LUX [14]. (right) The post-run I likelihood contours for σ_p^{SI} [7]. Shading within the likelihood contours is the same as in Fig. 3, though here we also see a region where chargino coannihilations are dominant (green). The green and black lines show the sensitivities of the XENON100 [13] and LUX [14] experiments, respectively, and the solid purple lines show the projected 95% exclusion sensitivity of the LUX-Zeplin (LZ) experiment [15]. The dashed orange line shows the astrophysical neutrino 'floor' [16, 17], below which astrophysical neutrino backgrounds dominate (yellow region).

10^{-45} cm^2. Note that in this case, a lower value of $\Sigma_{\pi N} = 50 \pm 7$ MeV was used. In addition to the 90% CL upper limits on σ_p^{SI} given by the XENON100 and LUX experiments [13, 14], the expected reach from LZ [15] is also displayed. The level of the atmospheric neutrino background [16, 17] is shown by the shaded region at small cross sections.

As a result of the constraints imposed by the LHC searches, parameter regions compatible with the Planck determination of the relic density, are largely found in 'strips' of the parameter space, due to necessary relations in the sparticle mass spectrum. For parameters where the stau and LSP are nearly degenerate, we obtain the stau co-annihilation strip [19], or when the stop and LSP are nearly degenerate at large A_0/m_0, we find a stop co-annihilation strip [20]. At large $\tan\beta$, the funnel strips [21] appear when $2m_\chi \simeq m_{H,A}$, where $m_{H,A}$ are the masses of the heavy Higgs scalar and pseudo-scalar. Finally, there is also the possibility of a focus point region [22], where the value of the μ term becomes relatively small near the edge of where radiative electroweak symmetry breaking is possible.

An example showing the stau and stop coannihilation strips is found in left panel of Fig. 5 which shows the $(m_0, m_{1/2})$ plane for fixed $\tan\beta = 20$ and $A_0 = 2.3m_0$ [20]. In the dark red shaded region at small m_0, the lighter stau is the LSP and that region is excluded. Along the border of that region, the stau and lightest neutralino are degenerate. The stau coannihilation strip tracks that boundary up to roughly $m_{1/2} = 1$ TeV and is shown as a barely visible blue shaded strip. Along the strip, the Higgs mass (shown by the red dot dashed curves) does not exceed 124 GeV. The stop coannihilation strip is also seen in the left panel of Fig. 5. The stop strip corresponds to the thin blue line which tracks the dark red wedge in the lower right of the panel. This strip extends past $m_0 = 10$ TeV. The current and future reach of the LHC is shown by the solid black, blue, green and purple lines which are particle exclusion reaches for $\not{E}_T$ searches with 20/fb at 8 TeV, 300 and 3000/fb at 14 TeV, and 3000/fb at a prospective HE-LHC at 33 TeV, respectively [20]. Unlike the stau strip, it is unlikely that the entire strip will be fully probed as it is seen to extend beyond the reach of a future 33 TeV LHC upgrade.

DOI: 10.1051/epjconf/201713601001

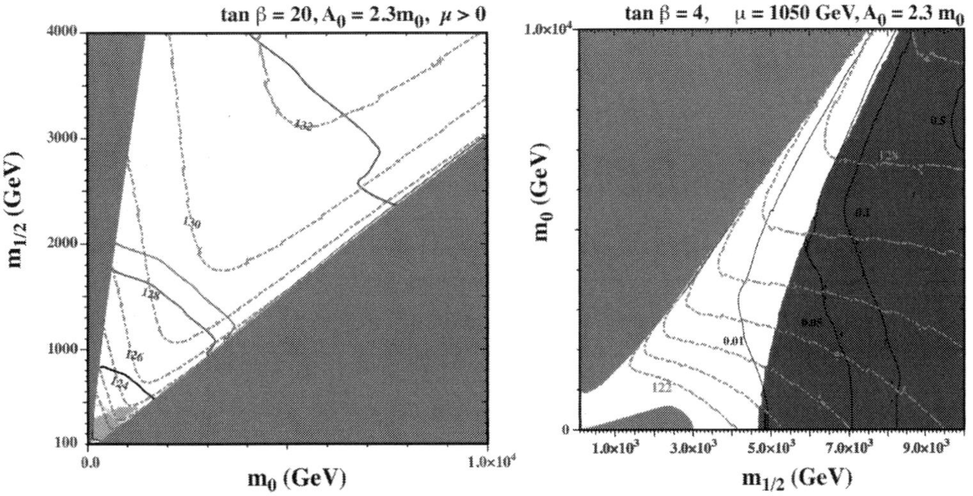

Figure 5. (left) The $(m_0, m_{1/2})$ plane for fixed $\tan\beta = 20$ and $A_0 = 2.3m_0$ in the CMSSM. (right) The $(m_{1/2}, m_0)$ plane for fixed $\tan\beta = 4$, $\mu = 1050$ GeV, and $A_0 = 2.3m_0$. The dark red shaded regions are excluded because of a charged LSP and/or a tachyon, and the green region are excluded by $b \to s\gamma$ decay. In the dark blue strips and region (in the right panel) the relic LSP density lies within the range allowed by cosmology, and the dashed red lines are contours of m_h. The solid black, blue, green and purple lines in each panel are particle exclusion reaches for $\not{E}_T$ searches with the LHC at 8 TeV, 300 and 3000/fb with LHC at 14 TeV, and 3000/fb with HE-LHC at 33 TeV, respectively. The solid black curves in the right panel correspond to contours of constant proton lifetime in units of 10^{35} yrs.

If one goes beyond the CMSSM, dark matter regions are no longer confined to strips as seen in the right panel of Fig. 5 which shows the $(m_{1/2}, m_0)$ plane for fixed $\tan\beta = 4$, $\mu = 1050$ GeV, and $A_0 = 2.3m_0$ [23]. Here, the two Higgs soft masses are no longer constrained to equal the sfermion masses, m_0, though they are constrained to equal each other, thus creating a one-parameter extension of the CMSSM, known as NUHM1 [24]. In this model, one may choose μ as the free parameter in lieu of the Higgs soft masses. Here, we see a large region where relic density is in agreement with Planck. The LSP is bino-like to the left of the blue region and the relic density is too high except for the extremely thin strips along the stop (higher m_0) and stau (lower m_0) LSP areas. In this case, since μ is fixed, as the gaugino mass is increased, the LSP becomes more Higgsino-like, however asymptotically, at large $m_{1/2}$, the Higgsino mass tends toward $\simeq$ 1.1 TeV where Higgsinos provide the correct dark matter density (this is of course tied to the choice of $\mu = 1050$ GeV) [25]. Indeed, this region would extend infinitely far to the right. As one can see, there is a significant area where the Higgsino provides the correct relic density with an acceptable Higgs mass. Also shown in this panel are contours of constant proton lifetime (in units of 10^{35} yrs) assuming a minimal SU(5) GUT [23]. The experimental limit would exclude points to the left of the curve labelled 0.05 or require a non-minimal GUT for which the calculated lifetime satisfies the bound.

4 SO(10) GUT Dark Matter

In addition to gauge coupling unification, the stability of the gauge hierarchy, and the presence of a dark matter candidate as noted earlier, supersymmetry is often motivated by the stability of the electroweak vacuum [26] and radiative electroweak symmetry breaking [27] . With the exception of the hierarchy problem, non-supersymmetric SO(10) GUT [28, 29] models may contain equivalences of all of these desirable features. As discussed above, in models with an intermediate scale between the electroweak scale and the GUT scale, gauge coupling unification becomes possible [29, 30] when the intermediate scale is determined by the unification conditions given a field content below the GUT scale as seen in Fig. 2.

To construct an SO(10) dark matter model, we should first pick an intermediate scale gauge group and a representation for the Higgs field, R_1 which breaks SO(10). For example, to obtain $SU(4)_C \otimes SU(2)_L \otimes SU(2)_R$, we should take $R_1 = \mathbf{210}$. If the intermediate scale is broken by a Higgs in a **126** representation, a residual Z_2 discrete symmetry survives enabling the possibility of a stable dark matter candidate [5, 31–33]. Furthermore, in models with gauge coupling unification and a stable dark matter candidate, it is also possible to stabilize the electroweak vacuum while at the same time radiatively break the electroweak symmetry [34]. The coupling of the **126** to SM matter fields embedded in a **16** representation of SO(10) naturally gives rise to a majorana mass mass to the ν_R component of the **16** of order $\langle \mathbf{126} \rangle \sim M_{int}$ which when combined with the Dirac mass arising from the vev of the SM Higgs (now residing in a 10-plet of SO(10)) gives rise to the seesaw mechanism for light neutrino masses [35].

The dark matter in SO(10) models may be either fermionic or bosonic. A fermionic DM candidate should be parity even and belong to a **10**, **45**, **54**, **120**, **126**, **210** or **210′** representation, while scalar DM is parity odd and belongs to a **16** or **144** representation.

5 Summary

It is becoming apparent that recent LHC searches for supersymmetry have pushed CMSSM into corners of the parameter space which rely on the near degeneracy between the LSP and the next lightest superpartner, thus allowing coannihilations to reign in the relic density. While the stau coannihilation strip is nearly ruled out by LHC searches, possibilities remain for the stop strip and if there are non-universal gaugino masses, gluino coannihilation. It is also possible that m_0 is large near the focus point strip so that the LSP is mostly Higgsino-like. There are several variants of the CMSSM which still permit neutralino dark matter. These include models with non-universal Higgs scalar masses (NUHM), models where the input universality scale is below the GUT scale (subGUT models), or pure gravity mediated models with either wino or Higgsino dark matter.

While supersymmetry has many motivations beyond dark matter, with the exception of the hierarchy problem, almost of these motivating factors can be resolved in non-supersymmetric versions of SO(10) grand unification. The real challenge lies in the detection of dark matter and our ability to discriminate between the various models.

References

[1] John R. Ellis, S. Kelley and Dimitri V. Nanopoulos, *Phys. Lett.* B249, 441, 1990; John R. Ellis, S. Kelley and Dimitri V. Nanopoulos, *Phys. Lett.* B260, 131, 1991; Ugo Amaldi, Wim de Boer, and Hermann Furstenau. *Phys. Lett.*, B260, 447, 1991; Paul Langacker and Ming-xing Luo, *Phys. Rev.* D44, 817, 1991; C. Giunti, C. W. Kim and U. W. Lee, *Mod. Phys. Lett.* A6, 1745, 1991.

[2] L. Maiani, in Proceedings, Gif-sur-Yvette Summer School On Particle Physics, 1979, 1-52; Gerard 't Hooft and others (eds.), *Recent Developments in Gauge Theories, Proceedings of the Nato Advanced Study Institute, Cargese, France, August 26 - September 8, 1979*, Plenum press, New York, USA, 1980, Nato Advanced Study Institutes Series: Series B, Physics, 59.; Edward Witten, *Phys. Lett.* B105, 267, 1981.

[3] H. Goldberg, Phys. Rev. Lett. **50** (1983) 1419; J. Ellis, J. Hagelin, D. Nanopoulos, K. Olive and M. Srednicki, Nucl. Phys. B **238** (1984) 453.

[4] R. Barbieri, S. Ferrara and C. A. Savoy, Phys. Lett. B **119**, 343 (1982); J. R. Ellis, K. A. Olive, Y. Santoso and V. C. Spanos, Phys. Lett. B **573** (2003) 162 [arXiv:hep-ph/0305212], and Phys. Rev. D **70** (2004) 055005 [arXiv:hep-ph/0405110].

[5] Y. Mambrini, K. A. Olive, J. Quevillon and B. Zaldivar, Phys. Rev. Lett. **110**, no. 24, 241306 (2013) [arXiv:1302.4438 [hep-ph]].

[6] O. Buchmueller *et al.*, Eur. Phys. J. C **64**, 391 (2009) [arXiv:0907.5568 [hep-ph]].

[7] O. Buchmueller *et al.*, Eur. Phys. J. C **74** (2014) 2922 [arXiv:1312.5250 [hep-ph]]; O. Buchmueller *et al.*, Eur. Phys. J. C **74** (2014) 12, 3212 [arXiv:1408.4060 [hep-ph]]; E. A. Bagnaschi *et al.*, Eur. Phys. J. C **75**, 500 (2015) [arXiv:1508.01173 [hep-ph]].

[8] John R. Ellis, Giovanni Ridolfi and Fabio Zwirner, *Phys. Lett.* B**257**, 83, 1991; John R. Ellis, Giovanni Ridolfi and Fabio Zwirner, *Phys. Lett.* B**262**, 477, 1991; Y. Okada, Masahiro Yamaguchi and T. Yanagida, *Phys. Lett.* B**262**, 54, 1991; Yasuhiro Okada, Masahiro Yamaguchi and Tsutomu Yanagida, *Prog. Theor. Phys.* 85, 1, 1991; Howard E. Haber and Ralf Hempfling, *Phys. Rev. Lett.* **66**, 1815, 1991.

[9] J. R. Ellis, D. V. Nanopoulos, K. A. Olive and Y. Santoso, Phys. Lett. B **633**, 583 (2006) [arXiv:hep-ph/0509331].

[10] Z. Ahmed *et al.* [CDMS Collaboration], Phys. Rev. Lett. **102**, 011301 (2009) [arXiv:0802.3530 [astro-ph]].

[11] J. Angle *et al.* [XENON Collaboration], Phys. Rev. Lett. **100**, 021303 (2008) [arXiv:0706.0039 [astro-ph]].

[12] E. Aprile *et al.* [XENON100 Collaboration], Phys. Rev. Lett. **107**, 131302 (2011) [arXiv:1104.2549 [astro-ph.CO]].

[13] E. Aprile *et al.* [XENON100 Collaboration], Phys. Rev. Lett. **109**, 181301 (2012) [arXiv:1207.5988 [astro-ph.CO]].

[14] D. S. Akerib *et al.* [LUX Collaboration], Phys. Rev. Lett. **112**, 091303 (2014) [arXiv:1310.8214 [astro-ph.CO]].

[15] D. C. Malling, D. S. Akerib, H. M. Araujo, X. Bai, S. Bedikian, E. Bernard, A. Bernstein and A. Bradley *et al.*, arXiv:1110.0103 [astro-ph.IM].

[16] J. Billard, L. Strigari and E. Figueroa-Feliciano, Phys. Rev. D **89**, no. 2, 023524 (2014) [arXiv:1307.5458 [hep-ph]];

[17] P. Cushman *et al.*, arXiv:1310.8327 [hep-ex].

[18] G. Aad *et al.* [ATLAS Collaboration], JHEP **1409** (2014) 176 [arXiv:1405.7875 [hep-ex]]; JHEP **1510**, 054 (2015) [arXiv:1507.05525 [hep-ex]].

[19] M. Citron, J. Ellis, F. Luo, J. Marrouche, K. A. Olive and K. J. de Vries, Phys. Rev. D **87**, no. 3, 036012 (2013) [arXiv:1212.2886 [hep-ph]], and references therein.

[20] J. Ellis, K. A. Olive and J. Zheng, Eur. Phys. J. C **74** (2014) 2947 [arXiv:1404.5571 [hep-ph]], and references therein; O. Buchmueller, M. Citron, J. Ellis, S. Guha, J. Marrouche, K. A. Olive, K. de Vries and J. Zheng, Eur. Phys. J. C **75**, no. 10, 469 (2015) [arXiv:1505.04702 [hep-ph]].

[21] M. Drees and M. M. Nojiri, Phys. Rev. D **47** (1993) 376 [arXiv:hep-ph/9207234]; H. Baer and M. Brhlik, Phys. Rev. D **53** (1996) 597 [arXiv:hep-ph/9508321]; Phys. Rev. D **57** (1998) 567 [arXiv:hep-ph/9706509]; H. Baer, M. Brhlik, M. A. Diaz, J. Ferrandis, P. Mercadante, P. Quintana and X. Tata, Phys. Rev. D **63** (2001) 015007 [arXiv:hep-ph/0005027]; J. R. Ellis, T. Falk, G. Ganis, K. A. Olive and M. Srednicki, Phys. Lett. B **510** (2001) 236 [arXiv:hep-ph/0102098].

[22] J. L. Feng, K. T. Matchev and D. Sanford, Phys. Rev. D **85**, 075007 (2012) [arXiv:1112.3021 [hep-ph]]; P. Draper, J. Feng, P. Kant, S. Profumo and D. Sanford, Phys. Rev. D **88**, 015025 (2013) [arXiv:1304.1159 [hep-ph]].

[23] J. Ellis, J. L. Evans, F. Luo, N. Nagata, K. A. Olive and P. Sandick, Eur. Phys. J. C **76**, no. 1, 8 (2016) [arXiv:1509.08838 [hep-ph]].

[24] H. Baer, A. Mustafayev, S. Profumo, A. Belyaev and X. Tata, Phys. Rev. D **71**, 095008 (2005) [arXiv:hep-ph/0412059]; H. Baer, A. Mustafayev, S. Profumo, A. Belyaev and X. Tata, *JHEP* **0507** (2005) 065, hep-ph/0504001; J. R. Ellis, K. A. Olive and P. Sandick, Phys. Rev. D **78**, 075012 (2008) [arXiv:0805.2343 [hep-ph]].

[25] K. A. Olive and M. Srednicki, Phys. Lett. B **230**, 78 (1989); K. A. Olive and M. Srednicki, Nucl. Phys. B **355**, 208 (1991).

[26] John R. Ellis and Douglas Ross, *Phys. Lett.* B **506**, 331, 2001, hep-ph/0012067.

[27] L. E. Ibanez and G. G. Ross, Phys. Lett. B **110**, 215 (1982); K. Inoue, A. Kakuto, H. Komatsu and S. Takeshita, Prog. Theor. Phys. **68**, 927 (1982) [Erratum-ibid. **70**, 330 (1983)] [Prog. Theor. Phys. **70**, 330 (1983)]; L. E. Ibanez, Phys. Lett. B **118**, 73 (1982); J. R. Ellis, D. V. Nanopoulos and K. Tamvakis, Phys. Lett. B **121**, 123 (1983); J. R. Ellis, J. S. Hagelin, D. V. Nanopoulos and K. Tamvakis, Phys. Lett. B **125**, 275 (1983); L. Alvarez-Gaume, J. Polchinski and M. B. Wise, Nucl. Phys. B **221**, 495 (1983).

[28] H. Georgi, AIP Conf. Proc. **23**, 575 (1975); H. Fritzsch and P. Minkowski, Annals Phys. **93**, 193 (1975); M. S. Chanowitz, J. R. Ellis and M. K. Gaillard, Nucl. Phys. B **128**, 506 (1977); H. Georgi and D. V. Nanopoulos, Nucl. Phys. B **155**, 52 (1979).

[29] H. Georgi and D. V. Nanopoulos, Nucl. Phys. B **159**, 16 (1979); C. E. Vayonakis, Phys. Lett. B **82**, 224 (1979) [Phys. Lett. **83B**, 421 (1979)].

[30] A. Masiero, Phys. Lett. B **93**, 295 (1980); Q. Shafi, M. Sondermann and C. Wetterich, Phys. Lett. B **92**, 304 (1980); F. del Aguila and L. E. Ibanez, Nucl. Phys. B **177**, 60 (1981); R. N. Mohapatra and G. Senjanovic, Phys. Rev. D **27**, 1601 (1983); M. Fukugita and T. Yanagida, In *Fukugita, M. (ed.), Suzuki, A. (ed.): Physics and astrophysics of neutrinos* 1-248. and Kyoto Univ. - YITP-K-1050 (93/12,rec.Feb.94) 248 p. C.

[31] M. Kadastik, K. Kannike and M. Raidal, Phys. Rev. D **80** (2009) 085020 [Erratum-ibid. D **81** (2010) 029903] [arXiv:0907.1894 [hep-ph]]; M. Kadastik, K. Kannike and M. Raidal, Phys. Rev. D **81**, 015002 (2010) [arXiv:0903.2475 [hep-ph]]; M. Frigerio and T. Hambye, Phys. Rev. D **81** (2010) 075002 [arXiv:0912.1545 [hep-ph]].

[32] Y. Mambrini, N. Nagata, K. A. Olive, J. Quevillon and J. Zheng, Phys. Rev. D **91**, no. 9, 095010 (2015) [arXiv:1502.06929 [hep-ph]].

[33] N. Nagata, K. A. Olive and J. Zheng, JHEP **1510**, 193 (2015) [arXiv:1509.00809 [hep-ph]].

[34] Y. Mambrini, N. Nagata, K. A. Olive and J. Zheng, arXiv:1602.05583 [hep-ph].

[35] P. Minkowski, Phys. Lett. B **67** (1977) 421; M. Gell-Mann, P. Ramond and R. Slansky, in *Supergravity*, eds. D. Freedman and P. Van Nieuwenhuizen (North Holland, Amsterdam, 1979), pp. 315-321. ISBN 044485438x; T. Yanagida, in *Proceedings of the Workshop on the Unified Theory and The Baryon Number of the Universe*, eds O. Sawada and S. Sugamoto. KEK79-18 (1979); R. N. Mohapatra and G. Senjanovic, Phys. Rev. Lett. **44**, 912 (1980); J. Schechter and J. W. F. Valle, Phys. Rev. D **22** (1980) 2227; J. Schechter and J. W. F. Valle, Phys. Rev. D **25** (1982) 774.

Models of Supersymmetry for Dark Matter

Carlos Muñoz[1,2,a]

[1] *Departamento de Física Teórica, Universidad Autónoma de Madrid, Cantoblanco, E-28049 Madrid, Spain*
[2] *Instituto de Física Teórica UAM-CSIC, Campus de Cantoblanco, E-28049 Madrid, Spain*

Abstract. A brief review of supersymmetric models and their candidates for dark matter is carried out. The neutralino is a WIMP candidate in the MSSM where R-parity is conserved, but this model has the μ problem. There are natural solutions to this problem that necessarily introduce new structure beyond the MSSM, including new candidates for dark matter. In particular, in an extension of the NMSSM, the right-handed sneutrino can be used for this job. In R-parity violating models such as the $\mu\nu$SSM, the gravitino can be the dark matter, and could be detected by its decay products in gamma-ray experiments.

1 Introduction

The Higgs particle in the standard model is intriguing, being the only elementary scalar in the spectrum, and introducing the hierarchy problem in the theory. In supersymmetry (SUSY), the presence of the Higgs is more natural: scalar particles exist by construction, the hierarchy problem can be solved, and the models predict that the Higgs mass must be $\lesssim$ 140 GeV if perturbativity of the relevant couplings up to high-energy scales is imposed. In a sense, the latter has been confirmed by the detection of a scalar particle with a mass of about 125 GeV. However, in SUSY at least two Higgs doublets are necessary, and as a consequence new neutral and charged scalars should be detected in the future to confirm the theory. Not only that, as is well known, the spectrum of elementary particles is in fact doubled with masses of about 1 TeV, and therefore even the simplest SUSY model, the Minimal Supersymmetric Standard Model (MSSM, see Ref. [1] for a review), predicts a rich phenomenology, including interesting candidates for dark matter (DM) such as the neutralino and sneutrino in R-parity conserving (R_p) models and the gravitino in R parity violating ($\not R_p$) models. However, the LHC started operations several years ago and, with Run 1 already finished, SUSY has not been discovered yet. Because of this, it has been raised the question of whether SUSY is still alive. In our opinion the answer is yes, and we think that there are several arguments in favor of it:

• The lower bounds on SUSY particle (sparticle) masses are smaller or about 1 TeV, depending on the sparticle analyzed. Thus SUSY masses are still reasonable, and in that sense we can remember that the Higgs particle was discovered with a mass close to its (SUSY) perturbative upper bound.

• Because of the complicated parameter space of SUSY, experimentalists use in their analyses simplified models that do not cover the full MSSM. For example, branching-ratio variations are not considered in much detail, and other assumptions are also made.

[a] e-mail: c.munoz@uam.es

• Run 2 is going on, and for the moment with a low luminosity of about 20 fb^{-1}. Therefore, to (be prepared) wait for results with higher luminosity seems to be a sensible strategy, since 100 fb^{-1} are expected for the end of Run 2.

• Most searches at the LHC assume R_p, with the lightest supersymmetric particle (LSP) stable, requiring therefore missing energy in the final state to claim for detection. However, in the case of $R\!\!\!/_p$, sparticles can decay to standard model particles, and the bounds on their masses become weaker.

Nevertheless, despite all these arguments in favour of SUSY, it is honest to recognize that it has its own theoretical problems in the low-energy formulation. By construction, the MSSM produces too fast proton decay. In particular, the simultaneous presence of the couplings $\lambda'_{ijk} L_i Q_j d^c_k$ and $\lambda''_{ijk} u^c_i d^c_j d^c_k$ violating lepton (L) and baryon (B) number respectively, as well as R_p, would produce this effect. The usual assumption in the literature of invoking R_p to avoid the problem, forbidding all $R\!\!\!/_p$ couplings, is perhaps too stringent, since forbidding only one of the above couplings would have been sufficient. We will come back to this point in Section 4. So, once eliminated (all) B and L number violating operators, we are left with the superpotential of the MSSM:

$$W = Y^e_{ij}\, \hat{H}_d\, \hat{L}_i\, \hat{e}^c_j + Y^d_{ij}\, \hat{H}_d\, \hat{Q}_i\, \hat{d}^c_j - Y^u_{ij}\, \hat{H}_u\, \hat{Q}_i\, \hat{u}^c_j + \mu\, \hat{H}_u \hat{H}_d\,, \tag{1}$$

where $i, j = 1, 2, 3$ are family indexes, and our convention for the contraction of two $SU(2)$ doublets is e.g. $\hat{H}_u\, \hat{H}_d \equiv \epsilon_{ab} \hat{H}^a_u\, \hat{H}^b_d$, with ϵ_{ab} the totally antisymmetric tensor $\epsilon_{12} = 1$.

In superpotential (1), the μ term is necessary e.g. to generate Higgsino masses, given the current experimental lower bound of about 100 GeV on chargino masses. Here we find another problem of SUSY models, the so-called μ problem [2]. In the presence of a high-energy theory like a grand unified theory (GUT) or a string theory, with a typical scale of the order of 10^{16} GeV or larger, and/or a gravitational theory at the Planck scale, one should be able to explain how to obtain a SUSY mass parameter in the superpotential of the order of the electroweak (EW) scale. The MSSM does not solve the μ problem. One takes for granted that the μ term is there, of the order of the EW scale, and that's it. In this sense, the MSSM is a kind of effective theory. Nevertheless, there are natural solutions to this problem that necessarily introduce new structure beyond the MSSM at low energies. Several of these solutions, and the associated SUSY models, will be discussed below.

The work is organized as follows. In Section 2, we will briefly review the popular neutralino DM in the MSSM. We will also see that the left-handed sneutrino is excluded as candidate for DM from experimental constraints. In Section 3, the Next-to-Minimal Supersymmetric Standard Model (NMSSM, see Ref. [3] for a review) is introduced as a solution to the μ problem, and neutralino DM discussed. In an extension of the NMSSM, we will also see that right-handed sneutrino DM is possible. Finally, in Section 4 we will argue that models with $R\!\!\!/_p$ are viable, solving also the μ problem. In particular, the 'μ from ν' Supersymmetric Standard Model ($\mu\nu$SSM [4, 5], see Refs. [6, 7] for reviews), solves the μ-problem through the presence of right-handed neutrino superfields, while simultaneously explains the origin of neutrino masses, i.e. in addition it solves the ν problem. Let us emphasize in this sense that in the MSSM, by construction, neutrinos are massless. Of course, the typical sparticle candidates for DM, the neutralino or the right-handed sneutrino, have very short lifetimes in $R\!\!\!/_p$ models, and can no longer be used as DM. Nevertheless, the gravitino can be the DM, and we will discuss its feasibility in the $\mu\nu$SSM, as well as its possible detection in gamma-ray satellite experiments such as the Fermi Large Area Telescope (LAT).

2 Neutralino DM in the MSSM

As mentioned in the Introduction, the MSSM superpotential in Eq. (1) conserves by construction R_p. This is a discrete symmetry which assigns quantum number +1 for particles and -1 for sparticles.

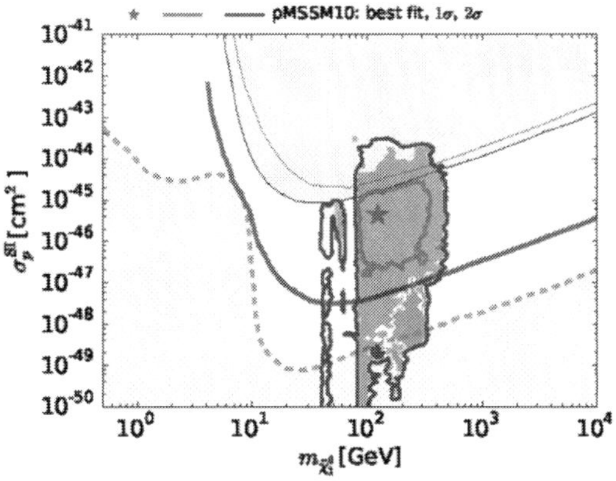

Figure 1. The $(m_{\tilde{\chi}_1^0}, \sigma_p^{SI})$ plane in the pMSSM10. The green and black lines show the current sensitivities of the XENON100 and LUX experiments, respectively. The solid purple line show the projected 95% exclusion sensitivity of the LUX-Zeplin (LZ) experiment, and the dashed orange line show the astrophysical neutrino 'floor', below which astrophysical neutrino backgrounds dominate (yellow region). Figure from [10].

As a consequence, SUSY particles are produced or destroyed only in pairs, and the LSP has to be stable. This implies that the LSP is a possible candidate for DM. In this sense, it is remarkable that in many regions of the parameter space of the MSSM the LSP is the lightest neutralino, a physical superposition of the Bino, and neutral Wino and Higgsinos:

$$\tilde{\chi}_1^o = N_{11}\tilde{B}^0 + N_{12}\tilde{W}^0 + N_{13}\tilde{H}_d^0 + N_{14}\tilde{H}_u^0 . \qquad (2)$$

Since the neutralino is an electrically neutral particle, it avoids the problem of charged particles as DM: they would bind to nuclei and would be excluded from unsuccessful searches for exotic heavy isotopes (see e.g. Ref. [8] and references therein). Besides, the neutralino is a weakly interacting massive particle (WIMP) and therefore is able to reproduce naturally the amount of relic density that is observed in the Universe, $\Omega_{DM}h^2 \sim 1$. We can then conclude that, in the MSSM, the lightest neutralino is a very good DM candidate (see Ref. [9] for a review).

The LHC could produce a neutralino with a mass of the order of the GeV-TeV. Such a production and detection would be of course a great success, but not a complete test of the DM theory. Even if we are able to measure the mass and interactions of the new particle, checking whether the amount of relic density is correct, we would never be able to test if the candidate is stable on cosmological scales. A complete confirmation can only arise from experiments where the DM particle is detected as part of the galactic halo or extragalactic structures. This can come from direct and indirect DM searches. Actually, there has been an impressive progress on this issue in recent years, with significant improvements in the precision and sensitivity of experiments. The combination of LHC data with those provided by direct and indirect searches can be a crucial tool for the identification of the DM.

The neutralino WIMP DM could be detected directly in underground laboratories through its elastic interaction with nuclei inside detectors. In view of the LHC1 constraints on SUSY, Higgs data, and flavour physics observables, in Fig. 1 the current constraints on the parameter space of neutralino DM are shown for the phenomenological MSSM, in which 10 of the effective Lagrangian parameters are treated as independent inputs specified at the EW scale (pMSSM10). Indirect DM searches of WIMPs are carried out in neutrino and Cherenkov telescopes, and satellites, through the analysis of the DM annihilation or decay products in the Sun, galactic center, galactic halo or extragalactic structures. Such products can be neutrinos, gamma rays and antimatter, and their non-observation put constraints on neutralino DM as well (see Ref. [11] for a review).

2.1 Sneutrino DM in the MSSM?

We might wonder whether are there other candidates for DM in the MSSM. In principle, the left-handed sneutrino fulfils the three interesting properties to become a DM candidate [12, 13]. It is a neutral particle, it is stable when it becomes the LSP, and it is a WIMP. However, at the end of the day, it turns out not to be viable for DM. Given its sizable coupling to the Z boson, left-handed sneutrinos either annihilate too rapidly, resulting in a very small relic abundance, or give rise to a large scattering cross section and are excluded by direct DM searches.

3 Is there life beyond MSSM/neutralino DM?

3.1 Neutralino DM in the NMSSM

The NMSSM provides an elegant solution to the μ problem of the MSSM via the introduction of a singlet superfield $\hat{S}$ under the standard model gauge group. Substituting now the μ-term in (1) by

$$W = \lambda\, \hat{S}\hat{H}_u\hat{H}_d \; + k\, \hat{S}\hat{S}\hat{S}\,, \tag{3}$$

when the scalar component of the superfield $\hat{S}$, denoted by S, acquires a vacuum expectation value (VEV) of order the SUSY breaking scale, an effective interaction $\mu\hat{H}_1\hat{H}_2$ is generated through the first term in (3), with $\mu \equiv \lambda\langle S\rangle$. This effective coupling is naturally of order the EW scale if the SUSY breaking scale is not too large compared with M_W, as expected. In fact, in the NMSSM the EW scale exclusively originates from the SUSY-breaking scale. The second term in (3) is allowed by the gauge symmetry, and avoids, as the μ-term in the MSSM, the existence of a Goldstone boson.

Due to the presence of the superfield $\hat{S}$, in addition to the MSSM fields, the NMSSM contains an extra scalar and pseudoscalar in the Higgs sector, as well as an additional singlino/neutralino. These new fields mix with the corresponding MSSM ones, giving rise to a richer and more complex phenomenology. For example, the results concerning the possible detection of neutralino DM turn out to be modified with respect to those of the MSSM in regions of the parameter space.

3.2 Sneutrino DM in an extended NMSSM

An interesting extension of the NMSSM can help to explain the origin of neutrino masses. Since experiments induce us to introduce right-handed neutrino superfields, superpotential (3) can be extended with [14]:

$$\delta W = Y^{\nu}_{ij}\, \hat{H}_u\, \hat{L}_i\, \hat{\nu}^c_j + \kappa_{ij}\, \hat{S}\hat{\nu}^c_i\hat{\nu}^c_j\,. \tag{4}$$

Majorana masses for right-handed neutrinos of the order of the EW scale are generated dynamically through the VEV of the singlet S, $M_\nu = \kappa\langle S\rangle$. This is an example of a seesaw at the EW scale. Light masses are then obtained with a value $m_\nu \simeq Y_\nu^2 v_u^2/M_\nu$, which implies Yukawa couplings $Y_\nu \lesssim 10^{-6}$, i.e. of the same order as the electron Yukawa.

As discussed above in the context of the MSSM, the left-handed sneutrino cannot be used as a DM candidate. Actually, a purely right-handed sneutrino either in a natural way, because of its very weak couplings with the rest of the matter implying a scattering cross section too small (supressed by Y^ν), and a relic density too large. However, now, through its direct coupling to the singlet in (4), the right-handed sneutrino can be not only a thermal relic DM, but also have a large enough scattering cross section with nuclei as to be detected (see [15] and [16, 17], and references therein).

4 Is there life beyond R_p/neutralino-sneutrino DM

As discussed in the Introduction, to impose R_p in SUSY models may be too stringent, since the $\not{R}_p$ couplings which are harmless for proton decay would also be forbidden. A less drastic solution, taking into account that the choice of R_p is *ad hoc*, is to use other Z_N discrete symmetries to forbid only λ''_{ijk}. This is the case e.g. of Z_3 Baryon-parity [18] which also prohibits dimension-5 proton decay operators, unlike R_p. In addition, this strategy seems reasonable if one expects all discrete symmetries to arise from the breaking of gauge symmetries of the underlying unified theory, because Baryon-parity and R_p are the only two generalized parities which are 'discrete gauge' anomaly free. Actually, this can occur in string compactifications where the matter superfields can have several extra $U(1)$ charges broken spontaneously at high energy, and as a consequence residual Z_N symmetries are left in the low-energy theory. The same result can be obtained by the complementary mechanism pointed out in Ref. [5], that stringy selection rules can naturally forbid the λ''_{ijk} couplings since matter superfields are located in general in different sectors of the compact space.

The gravitino turns out to be an interesting candidate for DM in $\not{R}_p$ models. It has an interaction term in the supergravity Lagrangian with the photon and the photino. Since the photino and the neutrinos are mixed in the neutral fermion mass matrix due to the $\not{R}_p$, the gravitino will be able to decay into a photon and a neutrino. Nevertheless, this decay is suppressed both by the gravitational interaction (the gravitino is a superWIMP) and by the small $\not{R}_p$ coupling, making the gravitino lifetime much longer than the age of the Universe [19]. Adjusting the reheating temperature one can also reproduce the correct relic density.

4.1 Gravitino DM in the $\mu\nu$SSM

Right-handed neutrinos are likely to exist in order to generate neutrino masses. Then, given the fact that sneutrinos are allowed to get VEVs, we may wonder why not to use $\not{R}_p$ terms of the type $\hat{\nu}^c \hat{H}_u \hat{H}_d$ to produce an effective μ term. This would allow us to solve the μ problem of the MSSM, without having to introduce an extra singlet superfield as in case of the NMSSM. This is the basic idea of the $\mu\nu$SSM [4, 5]: natural particle content without the μ problem. Thus, in addition to the MSSM Yukawa couplings for quarks and charged leptons, the $\mu\nu$SSM superpotential contains:

$$W = -Y^{\nu}_{ij}\,\hat{H}_u\,\hat{L}_i\,\hat{\nu}^c_j + \lambda_i\,\hat{H}_u\hat{H}_d\,\hat{\nu}^c_i + \frac{1}{3}\kappa_{ijk}\hat{\nu}^c_i\hat{\nu}^c_j\hat{\nu}^c_k\,. \quad (5)$$

When the scalar components of the superfields $\hat{\nu}^c_i$, denoted by $\tilde{\nu}^*_{iR}$, acquire VEVs of order the EW scale, an effective interaction $\mu\hat{H}_u\hat{H}_d$ is generated through the second term in (5), with $\mu \equiv \lambda_i\langle\tilde{\nu}_{iR}\rangle^*$. The third term in (5) is allowed by all symmetries, and avoids the presence of a Goldstone boson associated to a global $U(1)$ symmetry, similarly to the case of the NMSSM. In addition, it contributes to generate effective Majorana masses for neutrinos at the EW scale $M_{ij} = \sqrt{2}\kappa_{ijk}\langle\tilde{\nu}_{kR}\rangle^*$, which together with the Dirac masses generated by the first term, produce correct neutrino masses. Thus, the $\mu\nu$SSM solves the μ and the ν problems, by simply introducing right-handed neutrinos.

Since the gravitino decays producing a monochromatic photon with an energy half of the gravitino mass, the prospects for detecting these γ rays in satellite experiments can be very interesting, and therefore it seems important to know $\mu\nu$SSM predictions concerning gravitino DM detection, first studied in Ref [20]. In recent works [21, 22], a complete analysis of the detection of $\mu\nu$SSM gravitino dark matter through γ-ray observations was carried out. In addition to the two-body decay producing an anisotropic sharp line, the three-body decays producing a smooth spectral signature were included in the analysis. First, a deep exploration of the low-energy parameter space of the $\mu\nu$SSM was performed, taking into account that neutrino data must be reproduced. Then, the γ-ray fluxes predicted by the model were compared with *Fermi*-LAT observations. In particular, with the 95% CL upper limits on the total diffuse extragalactic γ-ray background using 50 months of data, together with the upper limits on line emission from an updated analysis using 69.9 months of data. For standard values of bino and wino masses, gravitinos with masses larger than 4 GeV, or lifetimes smaller than 10^{28} s, produce too large fluxes and are excluded as DM candidates. However, when limiting scenarios with large and close values of the gaugino masses are considered, the constraints turn out to be less stringent, excluding masses larger than 17 GeV and lifetimes smaller than 4×10^{25} s.

Acknowledgments
This work was supported by grants FPA2015-65929-P MINECO/FEDER UE, Consolider-Ingenio 2010 MultiDark CSD2009-00064, and SEV-2012-0249 'Centro de Excelencia Severo Ochoa'.

References

[1] S.P. Martin, Adv. Ser. Direct. High Energy Phys. 18 (1998) 1 (1997), hep-ph/9709356
[2] J.E. Kim, H.P. Nilles, Phys. Lett. **B138**, 150 (1984)
[3] U. Ellwanger, C. Hugonie, A.M. Teixeira, Phys. Rept. **496**, 1 (2010), 0910.1785
[4] D.E. López-Fogliani, C. Muñoz, Phys. Rev. Lett. **97**, 041801 (2006), hep-ph/0508297
[5] N. Escudero, D.E. López-Fogliani, C. Muñoz, R.R. de Austri, JHEP **12**, 099 (2008), 0810.1507
[6] C. Muñoz, in *AIP Conf. Proc.* (2010), Vol. 1200, p. 413, 0909.5140
[7] C. Muñoz, in *PoS* (2016), 1608.07912
[8] A. Kudo, M. Yamaguchi, Phys. Lett. **B516**, 151 (2001), hep-ph/0103272
[9] C. Muñoz, Int. J. Mod. Phys. **A19**, 3093 (2004), hep-ph/0309346
[10] E.A. Bagnaschi et al., Eur. Phys. J. **C75**, 500 (2015), 1508.01173
[11] C. Muñoz, Nucl. Instrum. Meth. **A692**, 13 (2012), 1203.0678
[12] L.E. Ibañez, Phys. Lett. **B137**, 160 (1984)
[13] J.S. Hagelin, G.L. Kane, S. Raby, Nucl. Phys. **B241**, 638 (1984)
[14] R. Kitano, K.y. Oda, Phys. Rev. **D61**, 113001 (2000), hep-ph/9911327
[15] D.G. Cerdeño, C. Muñoz, O. Seto, Phys. Rev. **D79**, 023510 (2009), 0807.3029
[16] D.G. Cerdeño, M. Peiró, S. Robles, JCAP **1408**, 005 (2014), 1404.2572
[17] D.G. Cerdeño, M. Peiró, S. Robles, Phys. Rev. **D91**, 123530 (2015), 1501.01296
[18] L.E. Ibanez, G.G. Ross, Nucl. Phys. **B368**, 3 (1992)
[19] F. Takayama, M. Yamaguchi, Phys. Lett. **B485**, 388 (2000), hep-ph/0005214
[20] K.Y. Choi, D.E. López-Fogliani, C. Muñoz, R.R. de Austri, JCAP **03**, 028 (2010), 0906.3681
[21] A. Albert, G. Gomez-Vargas, M. Grefe, C. Muñoz, C. Weniger, E. Bloom, E. Charles, M. Mazziotta, A. Morselli, JCAP **10**, 023 (2014), 1406.3430
[22] G.A. Gomez-Vargas, D.E. López-Fogliani, C. Muñoz, A.D. Pérez, R. Ruiz de Austri (2016), 1608.08640

Rare Events searches with Cherenkov Telescopes

Michele Doro[1,a]

[1] *University and INFN Padova, via Marzolo 8, I-35131 Padova (Italy)*

Abstract. Ground-based Imaging Cherenkov Telescope Arrays observe the Cherenkov radiation emitted in extended atmospheric showers generated by cosmic gamma rays in the TeV regime. The rate of these events is normally overwhelmed by 2–3 orders of magnitude more abundant cosmic rays induced showers. A large fraction of these "background" events is vetoed at the on-line trigger level, but a substantial fraction still goes through data acquisition system and is saved for the off-line reconstruction. What kind of information those events carry, normally rejected in the analysis? Is there the possibility that an exotic signature is hidden in those data? In the contribution, some science cases, and the problems related to the event reconstruction for the current and future generation of these telescopes will be discussed.

1 Introduction

Gamma-ray Astronomy is the branch of science that observes the cosmic radiation beyond the keV. Below some tens of GeV, such observation is done mostly through pair-production instruments (e.g. the Fermi-LAT instrument[1]) or Compton-scattering instruments, mounted on satellites. Above few tens of GeV and below several tens of TeV, observations are mostly done with Imaging Atmospheric Cherenkov Telescope Arrays (IACTA) that observe indirectly gamma rays through the Cherenkov light produced by atmospheric particle shower initiated in the high Earth atmosphere by cosmic gamma rays. Despite this technique has only 3 decades now, it has already reached a mature stage [1, 2], with about 150 sources detected, and a world-wide installation soon to be deployed, under the name of CTA (Cherenkov Telescope Array [3]).

There are currently three major installations of IACTA: H.E.S.S., MAGIC and VERITAS[2] that are under operations for about a decade now. These instruments perform stereoscopic observations of the same event with multiple telescopes: the Cherenkov radiation from the atmospheric shower, generates, on the cameras of the telescopes, an ellipse-like shape, whose image treatment allow inferring the direction and energy of the corresponding primary cosmic gamma ray. To image an event, IACTA cameras are constituted by more than a thousand pixel each (the individual pixel is typically a photomultiplier tube of typically 0.1 deg aperture). In such instruments, there are several layers of triggers and selection of events, some acting online, some offline. The first levels need to exclude the noise events caused by the Light of the Night Sky, due to starlight, zodiacal light and airglow. This is

[a] e-mail: michele.doro@pd.infn.it
[1] fermi.gsfc.nasa.gov/
[2] H.E.S.S.: www.mpi-hd.mpg.de/hfm/HESS/, MAGIC: wwwmagic.mppmu.mpg.de, VERITAS: veritas.sao.arizona.edu

done online. A rate of about 200 Hz of events passing this selection is typically stored on disk[3]. However, most of these events do not correspond to gamma rays, but instead are comprised of atmospheric shower events initiated by cosmic rays (mostly protons, with traces of heavier nuclei). The hadronic background at this stage outnumbers the gamma-ray events by more than a factor of hundred. Later on during the data reconstruction, these hadronic events are rejected by further image cleaning and selection. However, not all background can be rejected, specially at the lowest energies, where the images are more dim.

In this contribution, we briefly discuss the possibility that some of the background events can have actually a different origin, in some cases even hiding signatures of exotic and fundamental physics. We argue that one can develop special reconstruction and analysis treatment to extract these events. We are motivated to discuss this issue by gathering together different phenomena, for two reasons: from one side, the search of hidden signals in the background data of IACTA share similarities (special image cleaning, special data selection, whole data sample access, blind signal searches), and from the other side, it could be timely to consider fast selection filter for the CTA instrument. The reason is that, while current IACTA can manage to save data on disk because the space occupation is limited (about 1 TB of data/day for, e.g., MAGIC), for CTA the situation will be more dramatic, with expected 100 TB data/day or even more. In order to reduce the occupancy, CTA is planning to preselect and delete some information on the events. If this will not be done efficiently, CTA will risk to throw away possible extremely interesting events in its data haystack.

A search for such needles in the haystack would require several dedicated steps in the reconstruction and analysis:

1. A dedicated Monte Carlo. All events of IACTA are determined by comparing the image in the multipixel camera with the corresponding Monte Carlo simulation. For gamma rays and hadrons, this is done using the Corsika code. For peculiar events, one should additionally develop a code for the interaction of the cosmic particle with the atmosphere. It is clear that in some cases, when an exotic particle is under scrutiny, such Monte Carlo will be not only complex to develop, but will rely on theoretical ansatz;
2. A dedicated image cleaning. The standard image cleaning (although different techniques were proposed in the past) relies on the extrapolation of the event image by "cleaning out" those pixels whose signal is very likely caused by the Light of the Night Sky. The procedure is optimized for ellipse-like shapes (like those coming from gamma rays) through the so-called Hillas parameterization [4]. Some rare events could have instead very peculiar images (small bright spots, multiple images, etc, – see below). A dedicated procedure should thus be prepared;
3. A dedicated parameterization of the event and extraction of primary information (direction, energy);
4. A dedicated high-level analysis.

It is clear that the finding of one event will very likely not be sufficient to infer a detection. All rare events should happen with sufficient statistics to be visible above an unresolvable background.

2 Rare Events in the Background sample

The first class of rare phenomena that will be discussed is composed of events that have passed the first on-telescope trigger criteria, have been rejected by the standard analysis, and are stored on the disks. Of these events, some could have a classic nature, some could belong to more exotic explanations.

[3]For the ten years of operation of MAGIC, considering an average datataking of 5 h per night, this rate corresponds to about 12 GEvents saved on disk

2.1 Heavy Nuclei

In 2007, H.E.S.S. reported the measurement of the spectrum of cosmic iron nuclei from 13 TeV to 200 TeV [5] with five spectral points. Their data nicely overlap previous measurements taken with balloon experiments. Events from iron nuclei are two orders of magnitude less frequent than proton-induced shower, and for this reasons are harder to detect. However, the signatures in IACTA from heavy-nuclei-initiated showers have different features than those of proton or gamma rays. The nucleus is charged and proceeding with relativistic speed. Therefore, a small but intense burst of Cherenkov radiation is *directly* produced by the nucleus itself in the high atmosphere. As soon as it travels down, the nucleus has interactions with the denser atmospheres initiating an hadronic shower, rather similar to that of the protons. Therefore, in the camera, an iron event is composed of two spots: a bright spot toward the center of the camera (high in the atmosphere) arriving earlier, followed later in time by the classical ellipsoidal shape of the atmospheric showers and aligned with the main shower axis. The analysis is not straightforward, but proven possible. Besides H.E.S.S., no other IACTA has tested this method.

One could ask whether other heavy ions can be seen in the cosmic ray spectrum, whose abundances per element are measured at lower energies with balloons (see, e.g. [6]). Particles like CNO or Si are not only rarer because of lower fluxes, but also would provide less photon yield (that goes as Z^2). However, specially with future generation of telescopes like CTA, with better sensitivity and larger energy range, such searches will be possible.

2.2 Tau-Neutrino searches

Several classes of astronomical targets including massive black holes at the center of active galaxies or gamma-ray bursts, are expected to produce significant radiation of neutrinos. Irrespective of the family of neutrino at the production place, for extragalactic distances, the mixing foresees that the neutrino families at the earth should arrive in equal fraction, and thus that cosmic tau-neutrinos should be observable at Earth. These have not yet been discovered in cosmic neutrino detectors, however, they may be observable with IACTA through a phenomenon called Earth-skimming taus [7, 8]. Shortly, if a tau-neutrino crosses the right amount of ground (the Earth crust, or water), of the order of few tens of km, tau-leptons can be generated through deep inelastic scattering processes like $\nu_\tau + N \rightarrow W^+ \rightarrow X + \tau^-$ If the tau-lepton later on emerges from the medium, it creates an atmospheric shower. Suppose now that a telescope is located at the right distance from the exit point of the tau-lepton, it could detect the emerging atmospheric shower. From such directions, a shower could be not explained by other mechanisms. Searches like this were performed by MAGIC looking at the right direction toward the Canarian sea, reporting for now only results on the feasibility of the technique, but still no detection [9]. The expectation on the flux are extremely low: the diffuse neutrino flux can provide few events per decade. However, in case of strong or flaring astrophysical sources, the neutrino flux could be enhanced, thus providing still dim, but detectable signals. When one then compares the sensitivity of e.g. CTA compared to other instruments like Auger or IceCube, one can see that for "low-energy" PeV neutrinos, the CTA sensitivity could be larger than the others, thus providing sufficient ground for a careful search [10]. MAGIC developed the selection criteria for these events, showing that tau-neutrino induced events are in principle observable in the data.

2.3 Magnetic Monopoles

Magnetic monopoles were predicted back in 1930 by Dirac to explain the electric quantization. Later on during the century, it was found that magnetic monopoles appear naturally in Grand Unification models [11, 12]. In particular, some theories predict that they are formed during the QCD phase transition in the early Universe, and, being stable, they could still be present in the actual Universe. When a magnetic monopole crosses the Earth atmosphere, it will produce a huge number of Cherenkov photons, about 4700 times than those produced by a gamma ray [13]. In addition, the Cherenkov photons from magnetic monopoles will be produced throughout the full length of the atmosphere, and not from a limited path as when originated by atmospheric showers. This event would be observable by an IACTA as extremely bright spots or short lines, and not like ellipse-like shapes. The search for magnetic monopoles events has already been accomplished by H.E.S.S. [14] and the expectation for CTA were discussed in [15]. However, other instruments like Auger or IceCube seems to have higher sensitivity [16, 17]. One should also mention that IACTA would be sensitive only to ultrarelativistic magnetic monopoles, while other instruments have wider capabilities [18].

2.4 Antiquark Matter

In order to explain the matter-antimatter density inequality in the present Universe, some theories predict that during baryogenesis, the antimatter content was confined into very high dense states of quark plasma by the formation and subsequent collapse of domain walls in the existing quark-gluon plasma [19, 20]. Such aggregation would be composed of a huge number of antiquarks (or quarks), in the order of 10^{25} – 10^{35}, and have survived until present times in the intergalactic medium. These aggregation are called "quark nuggets" and share similarities with the strangelets [21]. They can be considered as viable dark matter candidate, at least comprising a fraction of the total density. The quark nuggets would be dressed with leptons to be globally neutral. In several works of K. Lawson, and specially [22], the direct and indirect detection techniques for quark nuggets are described. In particular, the quark nuggets are expected to emit charged particles and high-energy radiation when crossing the Earth atmosphere, thus initiating an extended atmospheric shower. The main difference with respect to standard cosmic showers would stem from the fact that the nugget will not decay in the atmosphere, and that its velocity is much lower than that of cosmic rays, typically of the order of the galactic velocities. The passage will then be seen as a "stripe" on the camera of the telescope, developing slowly from one side of the camera to the other, considering the nugget velocity, and increasing in brightness toward the ground, where the nuggets interactions with the denser atmosphere would increase.

No dedicated search for these exotic states has been performed with IACTs so far. However, the search would share similarities with the case of magnetic monopoles, as discussed in [18].

3 Rare Events in the Field of View

Not only one can have peculiar events in the background data haystack, with specific signatures in duration, time evolution, shape, etc., as described in the previous section, but additional rare events could occur serendipitously within the field of view, passing undetected, unless a specific analysis is developed. It is clear that a steady source or very brilliant flaring source in the field of view is recognized through the standard analysis. Here we are discussing examples of very brief events, lasting seconds or less, that would not appear when integrating over larger time windows.

3.1 Primordial Black Hole Evaporation

There are several mechanisms that allow the creation of primordial black holes (PBHs) in the Early Universe, besides those of astrophysical origin. Depending on the Universe average density at a given time after the Big Bang, these PBHs could have a specific mass. However, the range of possible masses is large: from the Planck mass to 10^5 $M_\odot$ [23]. As the time passes, a PBH increase its temperature and radiate energy, toward the final phase when Hawking radiation is emitted, and the BH evaporates. The life expectancy of a BH can be computed and depends solely on the BH mass. As the evaporation time approaches, the BH radiates more on more. This means that, at present times, we could be seeing the evaporation of all the PBHs of a *given* mass. A description of the lightcurve and gamma-ray spectrum of emission from an exploding PBH can be found in [24]. Shortly, the gamma-ray emission would be stable for most of the time, while an exponential increase in the last minutes to seconds to the evaporation is predicted.

PBH evaporation could be therefore appear as short bursts of emission randomly in the FOV of an IACTA regular observation. Bright and steady sources in the FOV are in principle easily seen in these data. However, in this case the emission would be more subtle to find and its observation would require a dedicated analysis: the emission could be dim, and specially short in time, and therefore washed out by integration over large duration. The PBH search should be performed over the whole data sample of an IACTA. This requires some non-standard data handling. The Whipple gamma-ray telescope pioneered this search [25], however, much better sensitivity can be expected with CTA.

3.2 Fast Radio Bursts

Fast Radio Bursts are very short ($1 - 10$ ms) bursts of radiation discovered in archival radio data few years ago [26]. Besides their short duration, the main characteristics is that the radio emission shows a large wavelength dispersion, which hints to extragalactic origin ($z \sim 1$). They could be originated out of neutron stars or magnetars formation or merger events [27]. These peculiar extreme events recently raised attention in the astrophysics community, however, their true nature is still to be clarified. When computing the intensity, it is possible that these events are accompanied by the emission of gamma-ray radiation at TeV energies. Very similarly to the PBH case discussed above, they would therefore appear as very short and intense spots in IACTA skymaps. The light curve should be different from that of PBH, so they could be discriminated. No result is published yet from IACTA in search of these targets.

4 Discussion and Conclusions

In this contribution, we have briefly discussed few possibilities to search for rare events among the data gathered by ground-based gamma-ray detectors of the IACTA class. These rare events share some features: they would probably go undetected by standard reconstruction and analysis techniques, thus they would require dedicated simulations, data selection, image treatments and so on. We grouped these events into two classes: events that would be mostly tagged as background in the IACTA standard reconstruction, and events that would appear serendipitously in the field of view of the instrument, and could go unnoticed because of short duration or faintness.

Past and current instruments have performed searches and published results around some of these topics, some instead are not investigated yet. These projects share complexity in terms of data handling and often suffer from incomplete theoretical mapping. However, in most cases, these investigations would not require allocation of instrumental time. They would mostly imply a careful treatment the large dataset of archival data gathered by the current instruments, which is now comprised of about a decade of data.

Because the current instruments generate a large but not huge amount of data per night, basically all data that triggered the telescopes are safely stored on disks, including a large fraction of background (cosmic ray) data which is normally partly unused in the advanced steps of the analysis. However, the future large installation of CTA will produce a very large amount of data per night, which would demand an effort to reduce consistently the full information, e.g. by excluding some pixels from the image event, or reducing the amount of background-tagged events stored on disk. For gamma-ray searches, this is an optimal solution, but for the search of rare events proposed here, this could be a killer factor. It is therefore envisaged to develop robust and fast routines that could tag interesting not-standard events and save them for further analysis. It is clear that such routines should be developed well on time before CTA starts operation, which is expected soon after 2020.

Acknowledgements

I want to thank A. Morselli for the invitation at the conference, M. Martinez for having pushed me to investigate these topics some years ago and collect some ideas, A. De Angelis, D. Gora and J. Rico and the anonymous referee for comments on the text.

References

[1] E. Lorenz, R. Wagner, R. Wagner, Eur. Phys. J. **H37**, 459 (2012), 1207.6003
[2] A.M. Hillas, Astropart. Phys. **43**, 19 (2013)
[3] M. Actis et al. (CTA Consortium), Exper. Astron. **32**, 193 (2011), 1008.3703
[4] Hillas, A.M., Procs. of the 19th International Cosmic Ray Conference (????)
[5] F. Aharonian et al. (H.E.S.S.), Phys. Rev. **D75**, 042004 (2007), astro-ph/0701766
[6] H. Hu (2009), 0911.3034
[7] Y. Asaoka, M. Sasaki, Astroparticle Physics **41**, 7 (2013)
[8] D. Fargion et al., Nucl. Instrum. Meth. **A588**, 146 (2008), 0710.3805
[9] M. Gaug et al., Procs of the 30th International Cosmic Ray Conference (2007)
[10] D. Gora, E. Bernardini, Astropart. Phys. **82**, 77 (2016), 1606.01676
[11] G. 't Hooft, Nucl. Phys. B **79** (1974)
[12] A. Polyakov, JETP Lett. **20** (1974)
[13] D. Tompkins, Phys. Rev. **138** (1965)
[14] G. Spengler, U. Schwanke, Procs. of the 32nd ICRC (2011)
[15] M. Doro et al. (CTA Consortium), Astropart. Phys. **43**, 189 (2013), 1208.5356
[16] T. Fujii, PoS **ICRC2015**, 319 (2016)
[17] M.G. Aartsen et al. (IceCube), Eur. Phys. J. **C76**, 133 (2016), 1511.01350
[18] M.G. Aartsen et al. (IceCube), Eur. Phys. J. **C74**, 2938 (2014), 1402.3460
[19] A.R. Zhitnitsky, JCAP **0310**, 010 (2003), hep-ph/0202161
[20] D.H. Oaknin, A. Zhitnitsky, Phys. Rev. **D71**, 023519 (2005), hep-ph/0309086
[21] E. Farhi, R.L. Jaffe, Phys. Rev. D **30**, 2379 (1984)
[22] K. Lawson, EPJ Web Conf. **99**, 12005 (2015)
[23] B.J. Carr et al., Phys. Rev. **D81**, 104019 (2010), 0912.5297
[24] J.H. MacGibbon et al., Procs. of the 5th International Fermi Symposium (2015)
[25] E.T. Linton et al., JCAP **0601**, 013 (2006)
[26] D.R. Lorimer et al., Science **318**, 777 (2007), 0709.4301
[27] K. Murase et al., Mon. Not. Roy. Astron. Soc. **461**, 1498 (2016), 1603.08875

DOI: 10.1051/epjconf/201713601004

Gamma-ray signatures of Dark Matter

Marco CIRELLI[1,a]

[1] *Laboratoire de Physique Théorique et Hautes Energies (LPTHE), UMR 7589 CNRS & UPMC, 4 Place Jussieu, F-75252, Paris, France*

Abstract. I review the ways in which Dark Matter can produce gamma-rays (and lower energy photons) and I apply the considerations to three specific examples: the tests of multi-TeV pure-WIMP candidates based on high energy gamma-rays from the Galactic Center or dwarfs, the importance of bremsstrahlung emission for light DM, the updated constraints on DM from all-sky radio surveys of the Milky Way.

1 Introduction

Cosmology and astrophysics provide several convincing evidences of the existence of Dark Matter (DM). The observation that some mass is missing to explain the internal dynamics of galaxy clusters and the rotations of galaxies dates back respectively to the '30s and the '70s. The observations from weak lensing, for instance in the spectacular case of the so-called 'bullet cluster', provide evidence that there is mass where nothing is optically seen. More generally, global fits to a number of cosmological datasets (Cosmic Microwave Background, Large Scale Structure and also Type Ia Supernovae) allow to determine very precisely the amount of DM in the global energy-matter content of the Universe at $\Omega_{\rm DM} h^2 = 0.1188 \pm 0.0010$ [1][1].

All these signals pertain to the gravitational effects of Dark Matter at the cosmological and extra-galactical scale. Searches for explicit manifestation of the DM particles that are supposed to constitute the halo of our own galaxy (and the large scale structures beyond it) have instead so far been giving negative results, but this might be on the point of changing.

Indirect searches for DM are of particular interest. These searches aim at detecting the signatures of the annihilations or decays of DM particles in the fluxes of Cosmic Rays (CRs), intended in a broad sense: charged particles (electrons and positrons, antiprotons, antideuterium), photons (gamma rays, X-rays, synchrotron radiation), neutrinos. In this contribution I am going to focus in particular on a few examples involving photon signals, from the high energy (multi-TeV gamma rays) to the low energy ones (radio-waves). Before discussing the examples, though, let's review generically how Dark Matter produces photons.

[a] e-mail: marco.cirelli@gmail.com

[1] Here $\Omega_{\rm DM} = \rho_{\rm DM}/\rho_c$ is defined as usual as the energy density in Dark Matter with respect to the critical energy density of the Universe $\rho_c = 3H_0^2/8\pi G_N$, where H_0 is the present Hubble parameter. h is its reduced value $h = H_0/100$ km s^{-1}Mpc^{-1}.

2 Indirect DM detection via gamma-rays

In general, DM annihilation (or decay) can produce photon fluxes in many ways, among which:

I) 'Prompt' gamma-rays: produced directly by DM annihilations themselves. In turn, however, these gamma-rays can originate from different stages of the annihilation process:

Ia) From the bremsstrahlung of charged particles and the fragmentation of hadrons, e.g. π^0, in the final states of the annihilations. These processes generically give origin to a *continuum* of γ-rays which peaks at energies somewhat smaller than the DM mass $m_{\rm DM}$, i.e. typically in the range of tens of GeV to multi-TeV. The spectra can be computed in a model independent way (see e.g. [2]), since all one needs to know is the pair of primary SM particles.

Ib) From the *bremsstrahlung* from one of the *internal* particles in the annihilation diagram. This typically gives rise to a sharp feature that peaks at an energy corresponding to the DM mass. The process is in general subdominant with respect to the continuum, but it can be particularly important in cases in which the continuum itself is suppressed by some mechanism, e.g. helicity constraints, which are lifted by the internal radiation. The spectrum from this contribution cannot be computed without knowing the details of the annihilation model.

Ic) From an annihilation directly into a pair of gamma-rays, which gives rise to a *line* spectrum at the energy corresponding to the mass of the DM. Since DM is neutral, this annihilation has to proceed via some intermediation (typically a loop of charged particles) and it is therefore suppressed by (typically) 2 to 4 orders of magnitude.

In any case, since these γ-rays originate directly from the annihilations themselves, their spatial distribution follows closely the distribution of DM.

II) Secondary radiation, emitted by the e^+e^-, which have been produced by the annihilation process, when they interact with the environment:

IIa) ICS gamma-rays: produced by the Inverse Compton Scattering (ICS) of the energetic electrons and positrons, created in the DM annihilation, onto the low energy photons of the CMB, the galactic star-light and the infrared-light, which are thus upscattered in energy. Typically, they cover a wider range of energies than prompt gamma rays, from energies of a fraction of the DM mass to almost up to the DM mass itself. Their spatial distribution traces the distribution of $e^\pm$, which originate from DM but then diffuse out in the whole galactic halo (as seen above).

IIb) Bremsstrahlung gamma-rays: produced by the same energetic electrons and positrons onto the gas in the interstellar medium. Typically, they are of lower energy than ICS rays; their spatial distribution traces the $e^\pm$ but also the density of target gas, so it is typically concentrated in the galactic disk.

IIc) Synchrotron emission: consisting in the radiation emitted in the magnetic field of the Galaxy by the $e^\pm$ produced by DM annihilations. For an intensity of the magnetic field of $\mathcal{O}(\mu$Gauss), like in the case of the Milky Way halo, and for $e^\pm$ of GeV-TeV energies, the synchrotron emission falls in the MHz-GHz range, i.e. in the radio band. For large magnetic fields and large DM masses it can reach up to EHz, i.e. the X-ray band. Their region of origin is necessarily concentrated where the magnetic field is highest; in particular the GC is the usual target of choice. However, it has been recently suggested that the galactic halo at large, or even the extragalactic ones, can be interesting sources, as we will discuss below.

Individuating the best targets to search for these annihilation signals is one of the main games in the field. Not very surprisingly, the preferred targets have to be **(i)** regions with high DM densities *and/or*

(ii) regions where the astrophysical 'background' is reduced and therefore the signal/noise ratio is favorable. The distinction between (i) and (ii) is of course not clear-cut, and of course there are specific cases in which other environmental reasons make a region more suitable than another (such as in the case of synchrotron radiation which needs a region with a strong magnetic field). Moreover, new promising targets keep being individuated. However, for the sake of schematizing, one can list the following targets at which the experiments look:

- The Milky Way Galactic Center (GC) – **(i)**.
- Small regions around or just outside the GC, such as the Galactic Center Halo (GCH, an annulus of about 1° around the GC, excluding the Galactic Plane) etc – **(i)** + **(ii)**.
- Wide regions of the Galactic Halo (GH) itself (such as observational windows centered at the GC and several tens of degrees wide in latitude and longitude, from which a diffuse flux of gamma-rays is expected, including the one due to the ICS emission from the diffused population of $e^{\pm}$ from DM annihilations – **(ii)**.
- Globular clusters (GloC), which are dense agglomerates of stars, embedded in the Milky Way galactic halo. They are a peculiar kind of target since they are not supposed to be DM dominated, quite the opposite, as they are rich of stars. The interest in them arises from two facts: that they may have formed inside a primordial DM subhalo and some of the DM may have remained trapped; that the density of baryonic matter may create by attraction a DM spike and thus enhance the annihilation flux– **(i)**.
- Subhalos of the galactic DM halo, whose position, however, is of course unknown a priori.
- Satellite galaxies of the Milky Way, often of the dwarf spheroidal (dSph) class, such as Sagittarius, Segue1, Draco and several others, which are star-deprived and believed to be DM dominated – **(i)** + **(ii)**. The 'problem' with dSphs is precisely that the determination of their DM content and distribution relies on stars as kinematical tracers and therefore suffers from rather large uncertainties.
- Large scale structures in the relatively nearby Universe, such as galaxy clusters (e.g. the Virgo, Coma, Fornax, Perseus clusters, and several others with catalog names that are less pleasant to write) – **(i)** + **(ii)**
- The Universe at large, meaning looking at the *isotropic* flux of (redshifted) γ-rays that come to us from DM annihilation in all halos and all along the recent history of the Universe. Often this flux is called 'extragalactic' or 'cosmological' – **(ii)**

Let us now move to discuss a few examples of how photon searches can play a relevant role in the Dark Matter parameter space, in very different contexts.

3 Testing pure WIMP multi-TeV DM models with high energy gamma rays

In this section we summarize the findings of [3]: we start by reconsidering the model of Minimal Dark Matter (a fermionic, hypercharge-less quintuplet of the EW interactions) and compute its gamma ray signatures. We compare them with a number of gamma ray probes. Those that turn out to be most significant are the galactic halo diffuse measurements, the galactic center line searches and recent dwarf galaxies observations. We find that the original minimal model, whose mass is fixed at 9.4 TeV by the relic abundance requirement, is constrained by the line searches from the Galactic Center: it is ruled out if the Milky Way possesses a cuspy profile such as NFW but it is still allowed if it has a cored one. This is illustrated by fig. 1, right panel. Observations of dwarf spheroidal galaxies are also relevant, and ongoing astrophysical progresses on these systems have the potential to eventually rule

 DOI: 10.1051/epjconf/201713601004

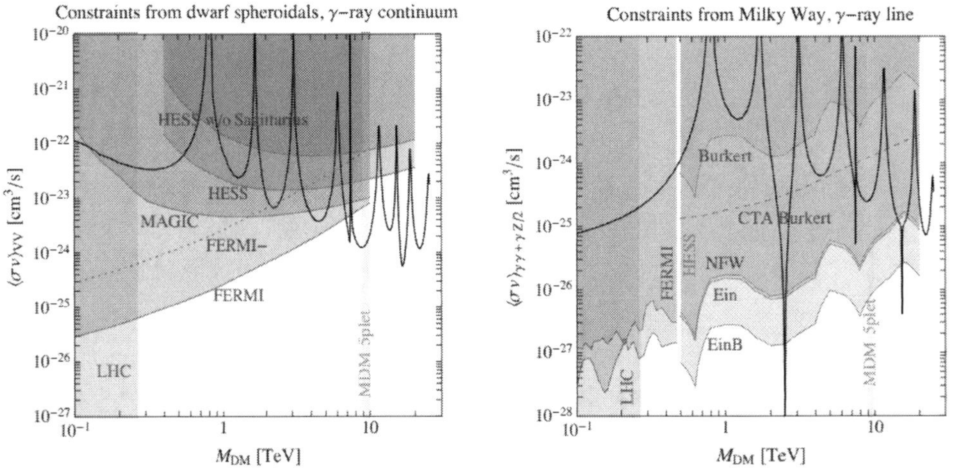

Figure 1. Constraints on the DM annihilation cross section into gauge bosons from dwarf spheroidal galaxies (left) and the Milky Way (right). Figure from [3].

out the model. As the figures show, we also explore a wider mass range, which applies to the case in which the relic abundance requirement is relaxed. Our constraints can therefore be applied to generic DM candidates with a mass in the multi-TeV range which annihilate into gauge bosons.

4 Bremsstrahlung gamma-rays from low mass DM

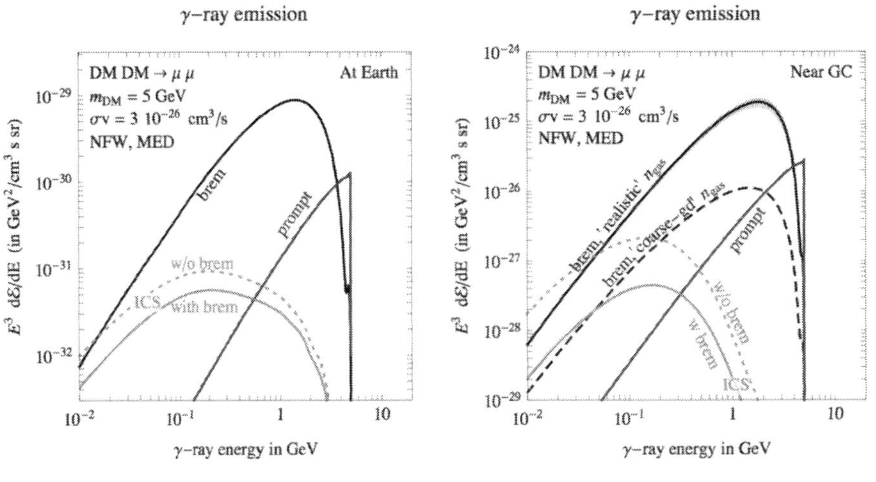

Figure 2. Full gamma-ray emission from light DM annihilations, including the bremsstrahlung contribution (and the modification to the Inverse Compton contribution due to taking into account bremsstrahlung), for the conditions typical of the location of the Earth (left) or of the Galactic Center (right). Figure from [4].

This section is based on the results in [4]. We stress the often-neglected role of bremsstrahlung processes on the interstellar gas in computing indirect signatures of DM annihilation in the Galaxy, particularly for light DM candidates in the phenomenologically interesting $\mathcal{O}(10)$ GeV mass range. Especially from directions close to the Galactic Plane, the gamma-ray spectrum is altered via two effects: directly, by the photons emitted in the bremsstrahlung process by energetic electrons which are among the DM annihilation byproducts; indirectly, by the modification of the same electron spectrum, due to the additional energy loss process in the diffusion-loss equation (e.g. the resulting inverse Compton emission is altered). Fig. 2 illustrates this point by plotting the total gamma-ray emission from light DM annihilations in conditions that are typical of the region close to the Earth or of the Galactic Center. So, for computing precise spectra in the (sub-)GeV range, it is important to obtain a reliable description of the Galaxy gas distribution as well as to compute self-consistently the gamma-ray emission and the solution to the diffusion-loss equation.

5 Constraining DM with galactic radio surveys

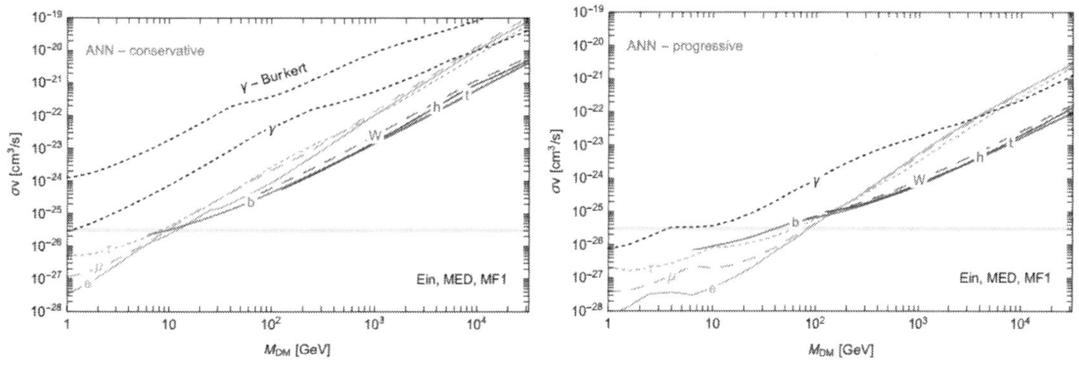

Figure 3. Constraints on DM annihilation from radio surveys, adopting a conservative (left) or progressive (right) approach (see text). Figure from [5].

This section is based on the study in [5], where we perform a detailed analysis of the synchrotron signals produced by dark matter annihilations (and decays). We consider different set-ups for the propagation of electrons and positrons, the galactic magnetic field and dark matter properties. We then confront these signals with radio and microwave maps, including Planck measurements, from a frequency of 22 MHz up to 70 GHz. We derive two sets of constraints: conservative and progressive. The former is obtained with no assumptions on the astrophysical background and just requires that the DM signal does not exceed the measured intensity. The latter is instead based on a modeling of the astrophysical emission: reducing the room avaiilable for DM, it leads to more stringent constraints. The results are depicted in fig. 3, for different annihilation channels. In general, radio and microwave constraints are complementary to those obtained with other indirect detection methods, especially for dark matter annihilating into leptonic channels.

6 Conclusions

In this contribution I have tried to review the different ways in which galactic gamma rays and radio surveys can help in the search for Dark Matter signatures. I first considered the searches of high energy gamma-rays, by Fermi, Hess and the other Cherenkov telescopes, as a test for models featuring pure-WIMP multi-TeV DM candidates, such as for instance the Minimal Dark Matter model. Then I stressed the importance of including bremsstrahlung as a process for the energy losses suffered by the electrons and positrons produced by DM in the galaxy, especially close to the galactic plane where the density of gas is sizable, and therefore its importance for the computation of DM secondary radiation. Finally, I presented the results of a new evaluation of the constraints on DM from all-sky radio surveys of the Milky Way.

As a general rule, DM indirect searches, and a fortiori gamma-ray searches, are complementary to one another, for the different annihilation channels, different chosen targets and different energy regimes. Fig. 4 collects some of the most relevant constraints, for the case of the $b\bar{b}$ channel.

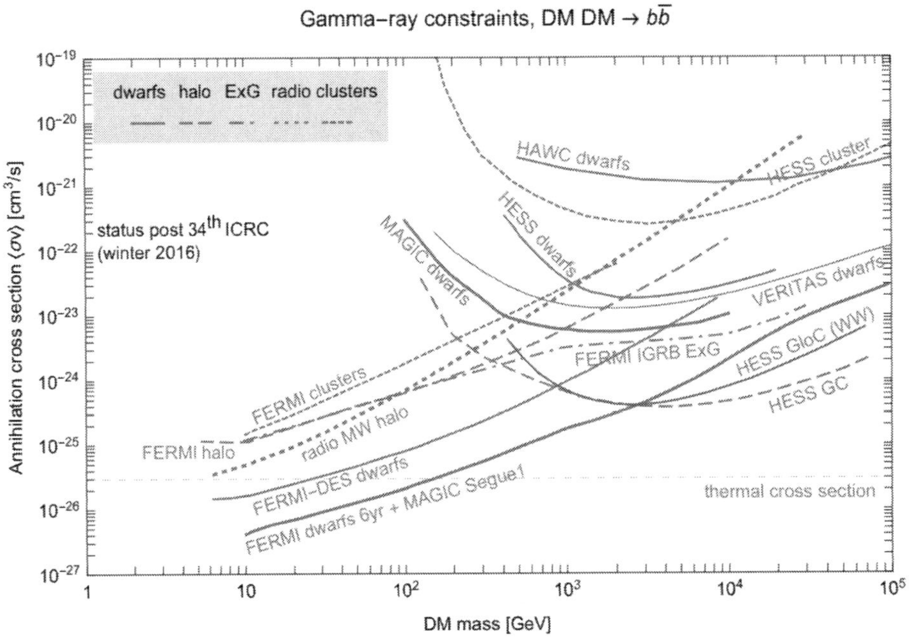

Figure 4. Bounds on DM annihilation imposed by different gamma-ray (and radio) observations. Figure from [6].

References

[1] P. A. R. Ade *et al.* [Planck Coll.], arXiv:1502.01589 [astro-ph.CO].

[2] M. Cirelli *et al.*, JCAP **1103** (2011) 051 [JCAP **1210** (2012) E01] [arXiv:1012.4515 [hep-ph]].

[3] M. Cirelli, T. Hambye, P. Panci, F. Sala and M. Taoso, JCAP **1510** (2015) no.10, 026 doi:10.1088/1475-7516/2015/10/026 [arXiv:1507.05519 [hep-ph]].

[4] M. Cirelli, P. D. Serpico and G. Zaharijas, JCAP **1311** (2013) 035 doi:10.1088/1475-7516/2013/11/035 [arXiv:1307.7152 [astro-ph.HE]].

[5] M. Cirelli and M. Taoso, JCAP **1607** (2016) no.07, 041 doi:10.1088/1475-7516/2016/07/041 [arXiv:1604.06267 [hep-ph]].

[6] M. Cirelli, arXiv:1511.02031 [astro-ph.HE].

Search for annihilating Dark Matter towards dwarf galaxies with the Cherenkov Telescope Array

Aldo Morselli[1] and Gonzalo Rodríguez[1,a] for the CTA Consortium[2]

[1] *Sezione INFN di Tor Vergata, Roma, Italy*
[2] *See www.cta-observatory.org for full author and affiliation list*

Abstract. The standard model of cosmology indicates that approximately 27% of the energy density of the Universe is in the form of dark matter. The nature of dark matter is an open question in modern physics. Indirect dark matter searches with imaging atmospheric Cherenkov telescopes are playing a crucial role in constraining the nature of the dark matter particle through the study of their potential annihilation that could produce very high energy gamma rays from different astrophysical structures. The Cherenkov Telescope Array will provide an unprecedented sensitivity over a range of dark matter mass from ~100 GeV to ~30 TeV. In this contribution we review the status of indirect dark matter searches at dwarf spheroidal galaxies.

1 Introduction

The existence of dark matter (DM) in our Universe (27% of the total energy) is well established, but its nature is still unknown. Among the most promising particle candidates are Weakly Interacting Massive Particles (WIMPs), which typically can self-annihilate and generate gamma rays [1, 2]. If WIMPs particles are produces thermally in the early Universe then the self-annihilation cross-section has a natural value of approximately $3\times10^{-26}\mathrm{cm}^3\mathrm{s}^{-1}$ [3]. WIMPs models, such as the supersymmetric neutralino give prediction the for gamma-ray energy spectra from the annihilations, which are crucial inputs, together with the DM distribution in the observed target, to achieve prospects for the sensitivity of indirect searches [4]. The goal of the present study is to provide preliminary comparative expectations on indirect DM searches with the Cherenkov Telescope Array [5], taking into account continuum gamma-ray signatures coming from typical DM annihilation channels.

2 Cherenkov Telescope Array performance

The Cherenkov Telescope Array (CTA) is a next-generation imaging atmospheric Cherenkov telescopes (IACTs) which is currently in the planning and prototyping stage and scheduled to finish construction in 2020. In order to provide full sky coverage, the CTA observatory will consist of two arrays, one in the southern hemisphere close to Paranal Observatory (Chile), and a second one in the northern hemisphere in the Observatorio del Roque de Los Muchachos on the island of La Palma

[a] e-mail: grferna@roma2.infn.it

DOI: 10.1051/epjconf/201713601005

(Spain). An improved sensitivity over the full energy range as well as improved energy and angular resolution with respect to current generation IACTs, will be achieved by combining many telescopes of three different sizes distributed over a large area: few large size telescopes (LST), several medium size telescopes (MST), and many small size telescopes (SST). The baseline northern array will consist of 4 LSTs and 15 MSTs and the southern one will be composed of 4 LSTs, 25 MSTs and 70 SSTs. An extension for the southern observatory is envisaged based on a fourth, medium-sized Schwarzschild Couder telescope type. The expected performance of CTA has been evaluated with very detailed Monte Carlo simulations for many different layouts and sites [6]. The expected angular resolution of CTA is compared with that of some current and future gamma-ray detectors in the left panel of figure 1. The angular resolution of CTA will be better than that of any current IACT array over the full energy range. The CTA differential flux sensitivity for both southern and northern arrays for an observation time of 50 hours is compared in the right panel of figure 1 with the sensitivities of MAGIC-II and VERITAS for the same observation time. HAWC sensitivities for an observation time of one year and five years are shown too for comparison. The flux sensitivity of the CTA northern array will be less than that of the southern one over the entire energy range due to the lower number of telescopes and different site altitude.

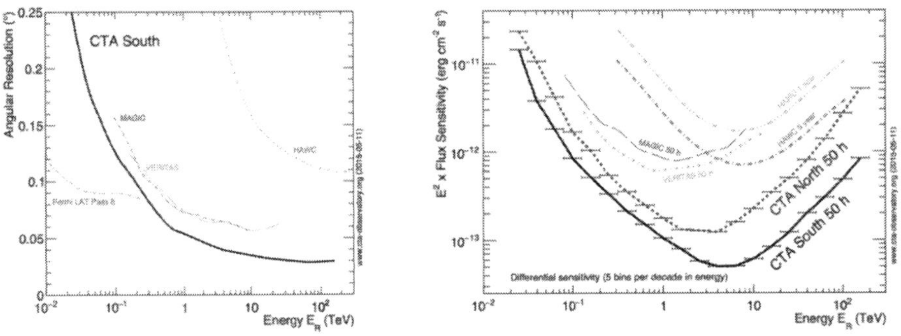

Figure 1: Left panel: Angular resolution (68% containment radius of the gamma-ray PSF) as a function of energy for CTA southern array (black line). Right Panel: Expected differential sensitivity for CTA southern array (black line) and CTA northern array (blue line). Figure from [7]

3 Indirect Dark Matter searches

The indirect detection experiments aim at searching for a flux of annihilation products created in astrophysical environments where DM annihilation may be occurring at an appreciable rate [8]. In particular, the indirect DM searches carried out by IACTs look for photons from WIMPs in the mass range ~100 GeV up to ~100 TeV. The flux expected from a DM-dominated region depends respectively on the so-called particle physics and astrophysical (or J) factors:

$$\Phi_s(\Delta\Omega) = \frac{1}{4\pi}\frac{<\sigma v>}{2m_{DM}^2}\int_{E_{min}}^{E_{max}}\frac{dN_\gamma}{dE_\gamma}dE_\gamma \times J(\Delta\Omega), \qquad (1)$$

where $<\sigma v>$ is thermally averaged self-annihilation cross-section times velocity, m_{DM} is the dark matter particle mass, E_{min} and E_{max} are the energy limits for the measurement and $\frac{dN_\gamma}{dE_\gamma}$ is the energy spectrum of the gammas produced in the annihilation (as, e.g., from [9]). The J-factor is the integral along the line of sight of the squared DM density profile of the given target integrated within an aperture angle, $\int_{\Delta\Omega} d\Omega \int_{l.o.s.} \rho^2_{DM}(\mathbf{r}) dl d\Omega'$. The products of DM annihilation are thought to come from decay and/or hadronization of the primary Standard Model (SM) particles: quark-antiquark, lepton and boson, and each channel is expected to have its own branching ratio.

4 Dwarf Spheroidal Galaxies

The indirect dark matter search with CTA has several possible astrophysical targets each with its own inherent advantages and disadvantages. The Galactic Center (GC) and the Dwarf Spheroidal Galaxies (dSphs) are the most promising targets. The large amount of DM and its proximity makes the GC a very interesting target to search for DM annihilation signals. However, the expected high background in very high energy gamma rays makes it hard to separate the DM signal from the gamma-rays of astrophysical origin. On the other hand, the dSphs are believed to be the smallest and faintest astronomical objects whose dynamics is dominated by DM, and they are free from gamma-ray astrophysical backgrounds [10]. Furthermore, ongoing optical surveys (such as DES, PanSTARSS, VLT, ATLAS), devoted to explore large areas of the sky, are increasing the pool of new dSphs in the Local Group and the knowledge needed to better constrain the values of the J-factors. On the basis of the Dark Energy Survey (DES) two years data [8], some of the newly discovered satellites in the southern hemisphere could already be considered among the most promising targets to be observed for the first time by the next generation of IACTs. The final choice of the most promising dwarf galaxy targets for CTA observations will be made at the start of array operation based on all available information at that time. In the left panel of figure 2 we show the sensitivity for different annihilation channels for the classical dSphs Sculptor [11]. On the right panel of the same figure we compare the sensitivity for Sculptor, Draco, Coma Berenices and Segue 1, showing the $\pm 1\sigma$ limits on the J-factor uncertainty.

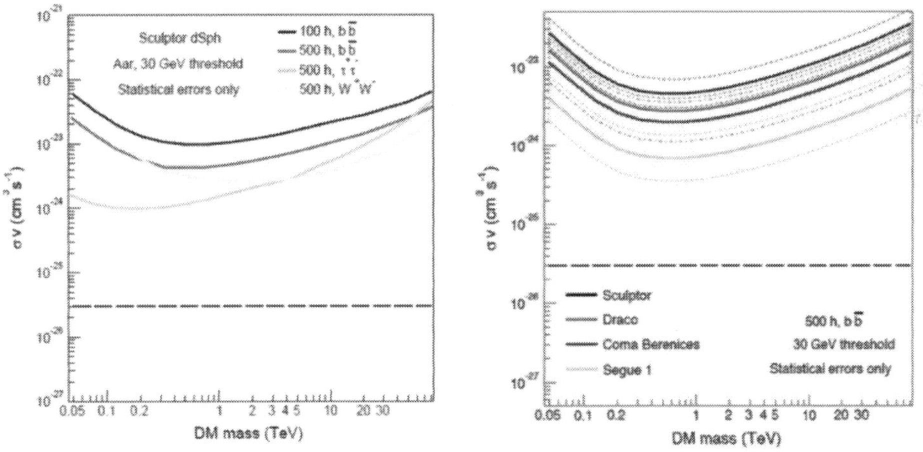

Figure 2: Left panel: Sensitivity for σv from observation of the classical dwarf galaxy Sculptor for different annihilation modes as indicated. Right panel: Sensitivity for 500 h observation of the classical dSphs Draco and Sculptor, and the ultra-faint dwarf galaxies Segue 1 and Coma Berenices as indicated. Dashed lines correspond to $\pm 1\sigma$ on the J-factors. Sensitivity is computed assuming the $b\bar{b}$ annihilation channel.

5 Conclusions

The Cherenkov Telescope Array will improve the sensitivity to DM annihilation for a range of interesting DM masses. The expected improvement in the angular resolution means that some dwarf spheroidal galaxies could be resolvable; the typical point-source assumption will no longer be valid, and the density profile could be constrained using gamma-ray measurements. Observations of dwarf spheroidal galaxies with CTA will be valuable for providing robust legacy constraints (in case of no detection). There is also a good complementary in the WIMPs mass range with other experiments like Fermi, MAGIC [12] and ANTARES [13]. New ultra-faint dwarf spheroidal galaxies are being and will be discovered with the next generation of sky surveys. When CTA starts to operate the best constrained/most promising known dwarf spheroidal galaxies will be chosen for observation time.

References

[1] L. Bergstrom, Non-Baryonic Dark Matter - Observational Evidence and Detection Method, Rep. Prog. Phys. 63, 793 (2000).

[2] G. Bertone, D. Hooper and J .Silk, Particle Dark Matter: Evidence, Candidates and Constraints, Phys. Rep. 405, 279 (2005).

[3] Steigman G. et al, Precise relic WIMP abundance and its impact on searches for dark matter annihilation, Phys. Rev. D 86, 2, 023506 (2012).

[4] S. P. Martin, A Supersymmetry Primer, arXiv:hep-ph/970935.

[5] Acharya B. S. et al, (CTA Consortium), Introducing the CTA concept, Astro. Phys. 43, 3 (2013).

[6] K.Bernlohr et al.(CTAConsortium),Monte Carlo design studies for the Cherenkov Telescope Array, Astroparticle Physics 43 (2015) 171-188.

[7] CTA Construction Project Science Technical Design Report, OBS-TDR/141106

[8] M. Cirelli et al, PPPC 4 DM ID: A Poor Particle Physicist Cookbook for Dark Matter Indirect Detection, , JCAP 1103 (2011) 051, erratum: JCAP 1210 (2012) E01, arXiv: 1012.4515, http://www.marcocirelli.net/PPPC4DMID.html.

[9] Drlica-Wagner A. et al. (2015), arXiv:1508.03622

[10] Bergstrom L. (2012), Annalen der Physik, vol 524 issue 9-10, pp479-496

[11] Carr et al. arXiv:1508.06128

[12] M. L. Ahnen et al. arXiv:1601.06590

[13] A. Albert et al. arXiv:1612.04595

The pLISA project in ASTERICS

Giulia De Bonis[1,a] and Cristiano Bozza[2,3,b]

[1] *Istituto Nazionale di Fisica Nucleare (INFN), Sezione di Roma*
[2] *Dipartimento di Fisica "E. R. Caianiello", Università di Salerno*
[3] *Istituto Nazionale di Fisica Nucleare (INFN), Gruppo Collegato di Salerno*

Abstract. In the framework of Horizon 2020, the European Commission approved the ASTERICS initiative (ASTronomy ESFRI and Research Infrastructure CluSter) to collect knowledge and experiences from astronomy, astrophysics and particle physics and foster synergies among existing research infrastructures and scientific communities, hence paving the way for future ones. ASTERICS aims at producing a common set of tools and strategies to be applied in Astronomy ESFRI facilities. In particular, it will target the so-called multi-messenger approach to combine information from optical and radio telescopes, photon counters and neutrino telescopes. pLISA is a software tool under development in ASTERICS to help and promote machine learning as a unified approach to multivariate analysis of astrophysical data and signals. The library will offer a collection of classification parameters, estimators, classes and methods to be linked and used in reconstruction programs (and possibly also extended), to characterize events in terms of particle identification and energy. The pLISA library aims at offering the software infrastructure for applications developed inside different experiments and has been designed with an effort to extrapolate general, physics-related estimators from the specific features of the data model related to each particular experiment. pLISA is oriented towards parallel computing architectures, with awareness of the opportunity of using GPUs as accelerators demanding specifically optimized algorithms and to reduce the costs of processing hardware requested for the reconstruction tasks. Indeed, a fast (ideally, real-time) reconstruction can open the way for the development or improvement of alert systems, typically required by multi-messenger search programmes among the different experimental facilities involved in ASTERICS.

1 Introduction

1.1 ASTERICS

ASTERICS (ASTronomy ESFRI and Research Infrastructure CluSter) [1] is a Research Infrastructure funded by the European Commission's Horizon 2020 framework. It collects experiences from astronomy, astrophysics and particle physics and aims at producing a common set of tools and strategies to be applied in the Astronomy ESFRI facilities[1], pushing for creating synergies among different

[a] e-mail: giulia.debonis@roma1.infn.it
[b] e-mail: cbozza@unisa.it

[1] See [2] for a description of the mission and the objectives of the European Strategy Forum on Research Infrastructures (ESFRI) and see [3] for a detailed report of the current roadmap.

experiments and scientific communities and for carrying out a multi-wavelength and multi-messenger approach in data analysis[2]. The ESFRI facilities in ASTERICS fully match the request for multi-wavelength and multi-messenger: SKA (radio) [5], E-ELT (infrared/optical) [6], CTA (gamma-rays) [7] and KM3NeT (neutrinos) [8]; complementary facilities (LOFAR, Euclid, VIRGO, LIGO, HESS, MAGIC, ANTARES, IceCube...) complete the framework.

1.2 OBELICS

Activities in ASTERICS are organised in five Working Packages (Figure 1-a). WP3 is named OBELICS (OBservatory E-environments Linked by common ChallengeS) and focuses on interoperability and software re-use for data generation, integration and analysis (Figure 1-b). Specific tasks of action, aimed at promoting multi-wavelength/multi-messenger data analyses, are the establishment of open standards and software libraries, the development of common solutions for data processing and extremely large databases, the study of advanced analysis algorithms and strategies.

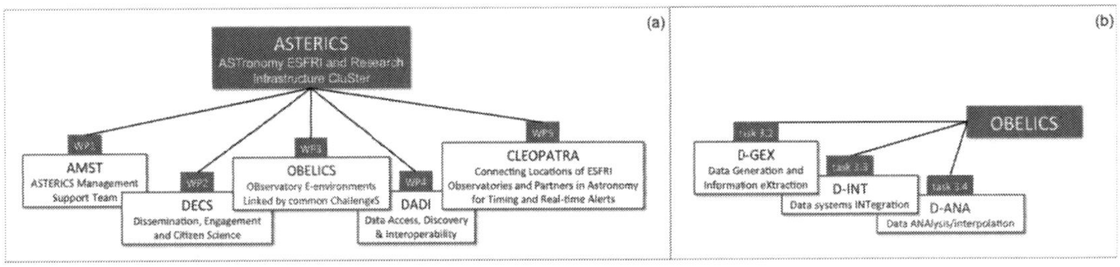

Figure 1. (a) ASTERICS and its Working Packages (WPs). (b) Sub-tasks in OBELICS.

2 pLISA

2.1 General Features

The pLISA project is inserted in sub-task 3.4 (D-ANA) and fulfils the mission and the objectives of ASTERICS and OBELICS, in particular for what concerns interoperability. The name of the project stands for *parallel Library for the Identification and Study of Astroparticles* and each word in this definition recalls important features of the ASTERICS/OBELICS initiative and distinctive elements of the project itself.

The term *parallel* refers to parallel programming and parallel computing architectures and reflects the pLISA plans of using Graphical Processing Units (GPUs) as computing accelerators, consistent with OBELICS about the development of new computing technologies. The acceleration feature is a key element when addressing the issue of real-time reconstruction, essential for the development of alert systems between experiments in a multi-messenger perspective. We have focused in particular on NVIDIA boards and CUDA [9] [10], and the implementations of the code have been devised as explicitly parallel, for running on GPUs since the very beginning.

The term *Library* qualifies pLISA as a toolbox, aiming at flexibility, interoperability, open standards and common solutions, independently of specific implementations operated in individual infrastructures, in order to adapt the code structure for the data analysis of a generic event-based experiment.

[2]See [4] for an example of the multi-messenger approach in neutrino astronomy.

That said, the study for pLISA has been initiated in the framework of KM3NeT and the peculiarities of KM3NeT (and ANTARES [11]) data taking, trigger systems and reconstruction algorithms have been taken into account and have been the starting point for the definition of the main features of pLISA.

The approach followed in pLISA for what concerns *Study and Identification* is through Multi-Variate Analysis (MVA) [12]: neural networks, boosted decision trees and similar machine-learning techniques have been investigated, in agreement with the claim of advanced algorithms and strategies professed by OBELICS; both regression and classification problems have been considered when working out the features of the library.

Finally, *Astroparticles* refers to the scientific target of the ASTERICS enterprise.

2.2 Implementation

The programming language used for pLISA is C++[3]; the current implementation is the "skeleton" of an header file. Classes and constants are very strongly typed, and extensive use of *namespaces* has been made, for minimising the chances of name clashes when pLISA is used alongside other libraries and for helping developers and users in producing bug-free code by enforcing type compliance. In addition, it is convenient to follow the structuring in namespaces to get an overview of the code (Figure 3). Classes are interfaces (i.e. purely abstract classes containing only pure virtual methods); furthermore, concerning the data storage pLISA puts requirements only on the information to be provided (*Features*, Figure 3) not on the way it is stored. This approach has been adopted to meet the objectives of ASTERICS and OBELICS for what concerns flexibility and interoperability, and also because GPU-based implementations require data structures in a format that differs from those in CPU-based implementations. Indeed, every experiment has its own data model and choosing one might lead to incompatibilities with others; with the solution under study in pLISA, non-implemented properties for data that are non-existing or meaningless in a specific dataset do not take memory/disk space; in addition the transient/persistent data model of user code need not to be changed, provided "reader" classes are produced by users. Optimised memory access (Figure 2) is obtained allowing that data in the GPU memory can be read just on-demand, and proper seamless caching mechanisms can be implemented (e.g. "lazy retrieval", i.e. retrieval of a full memory block only after a certain number of accesses are performed); moreover, memory transfer is optimised by adding flags that describe the internal encoding (namespace *Devices*, Figure 3).

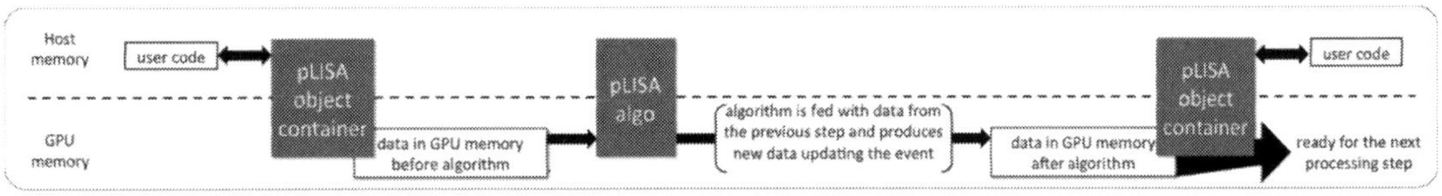

Figure 2. pLISA operates a connection between the Host memory (CPU) and the Device memory (GPU). The user code resides in the CPU; the processing is completed in the GPU.

Acknowledgement

ASTERICS is a project supported by the European Commission Framework Programme Horizon 2020, Research and Innovation action, under grant agreement n. 653477.

[3]Successfully compiled with *Visual C++* and *g++* (GCC) 4.8.5.

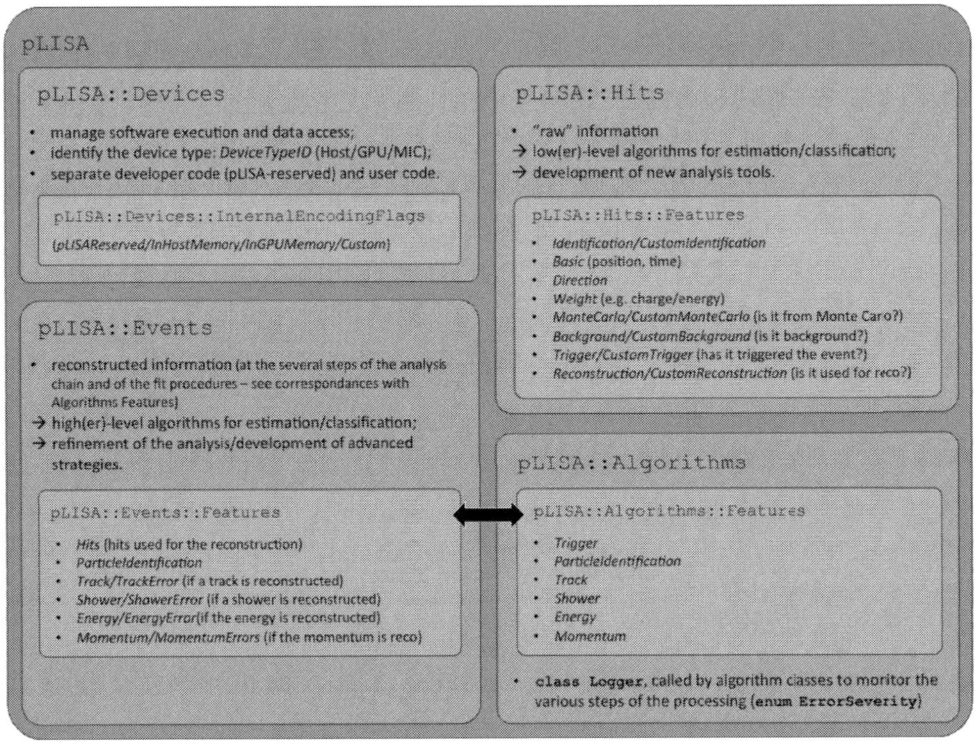

Figure 3. The structure of pLISA, in terms of namespaces. The outermost "container" is the namespace pLISA.

References

[1] ASTERICS Web Page, `https://www.asterics2020.eu`

[2] ESFRI Web Page, `http://ec.europa.eu/research/infrastructures/index_en.cfm?pg=esfri`

[3] *The ESFRI Roadmap 2016. Strategy Report on Research Infrastructures*, `http://ec.europa.eu/research/infrastructures/pdf/esfri/esfri_roadmap/esfri_roadmap_2016.pdf`

[4] De Bonis, G. (ANTARES), EPJ Web of Conferences **121**, 05014 (2016)

[5] SKA Web Page, `https://www.skatelescope.org`

[6] E-ELT Web Page, `http://www.eso.org/public/teles-instr/e-elt/`

[7] CTA Web Page, `https://portal.cta-observatory.org/Pages/Home.aspx`

[8] KM3NeT Web Page, `http://www.km3net.org`

[9] NVIDIA CUDA Web Page, `https://developer.nvidia.com/cuda-zone`

[10] J. Sanders, E. Kandrot, *CUDA by Examples.* (Addison-Wesley, 2011)

[11] ANTARES Web Page, `http://antares.in2p3.fr`

[12] T. Hastie, R. Tibshirani, J. Friedman, *The Elements of Statistical Learning.* (Springer, 2009)

The High-Energy Particle Detector on board of the CSES mission

Vincenzo Vitale[1,2,a], Francesco Palma[1,3] and Alessandro Sotgiu[1,4], for the CSES/HEPD collaboration

[1] *INFN sez. Roma Tor Vergata, Via della Ricerca Scientifica 1 - 00133 Rome (Italy)*
[2] *Agenzia Spaziale Italiana (ASI) Science Data Center, Via del Politecnico snc, 00133 Rome (Italy)*
[3] *University of Trento, Via Sommarive, 14, 38123 Povo, Trento (Italy)*
[4] *University of Rome Tor Vergata, Via della Ricerca Scientifica 1 - 00133 Rome (Italy)*

Abstract.
The High-Energy Particle Detector (HEPD) is a range-calorimeter for the near-Earth measurement of electrons, protons and light nuclei fluxes up to few hundreds of MeV. HEPD will fly on board of the China Seismo-Electromagnetic Satellite (CSES), scheduled to be launched in July/August 2017. This mission will investigate possible correlations between geomagnetic properties (electromagnetic wave emissions, plasma properties and particle fluxes in the ionosphere and magnetosphere) and seismic events. The study of the solar-terrestrial environment, Coronal Mass Ejections (CMEs), Solar Energetic Particles (SEPS) events and low-energy cosmic rays are also within the scientific objectives of this mission. A detailed description of the HEPD and its characteristics will be reported.

1 Introduction

The China Seismo-Electromagnetic Satellite (CSES) [1] is a space mission dedicated to the monitoring of the electromagnetic (e.m.) perturbations in atmosphere, ionosphere, magnetosphere and Van Allen belts, and to investigate possible correlations with seismic events. The satellite is based on the Chinese CAST2000 platform. It is a 3-axis attitude stabilized satellite and will be placed in a 98 degrees inclination Sun-synchronous circular orbit, at an altitude of 500 km. The launch is scheduled for July/August 2017 and the expected lifetime is 5 years.

There will be several instrument systems on board of CSES: (i) a Tri-Band Beacon and GNSS Occultation Receiver for the study of profile disturbance of plasma; (ii) a Langmuir Probe and Plasma Analyzer for measurements of local plasma disturbances; (iii) a High-Precision Magnetometer, an Electric Field Detector and a Search-Coil Magnetometer for the measurement of electromagnetic field; (iv) a High-Energy Particle Package, which has three components (H, L and X) and measures electrons (few keV to 50 MeV) and protons (20 MeV to 200 MeV) with acceptance of some tens of cm^2sr; (v) the High-Energy Particle Detector (HEPD), which is discussed in this article.

Then CSES science program can cover several fields and topics. In the field of the e.m. wave emissions, we list the monitoring of the e.m near-Earth space environment, of the e.m. man-made

[a] e-mail: vvitale@roma2.infn.it

effects at low Earth orbit altitude and the observations of e.m. transients caused by tropospheric activity, as also to search for direct correlation of e.m. wave emission and earthquakes. For the Ionosphere and plasma-sphere studies, the measurement of plasma perturbations is considered, while for the energetic charged particles, there are the measurements of electron and proton fluxes, the study of short time scale variability and precipitation of inner Van Allen belt particles. When at large latitudes, the HEPD detector can also perform measurements of not-trapped low-energy charged particles up to few hundreds of MeV. This allows the study of several topics related to the solar-terrestrial environment and the lower end of the cosmic-ray spectrum.

1.1 Van Allen Belts and Cosmic Rays

At low and intermediate magnetic latitudes the HEPD detector can access particles, which are trapped or significantly influenced by the geomagnetic field. Trapped charged particles constitute two (and sometimes more [2]) radiation belts. Different species (electrons and protons) have maximum intensities at different values of the McIlwain L-parameter (L defines a specific set of geomagnetic field lines). In the energy range of interest, particle fluxes within the belts are orders of magnitude larger than outside. A detailed characterization of the charged particle fluxes within the geomagnetic field is difficult for several reasons: - the departure of the geomagnetic field structure from a pure dipole and the presence of anomalies such as the South Atlantic Anomaly (SAA); - the time variation of the particle fluxes on several time scales; - the existence of few particle populations (particle albedo, likely produced with collisions of cosmic rays with the atmosphere gas, quasi-trapped component with limited lifetime and low flux); - the "penumbra" region, where both cosmic rays and trapped charged particles are found. Inner and outer belts have a different degree of flux stability, and likely different particle sources. Protons within the inner belt are thought to be produced via the CRAND process [3] by the decay of albedo neutrons. At large magnetic latitudes the solar particles and the low-energy component of cosmic ray might be measured. Coronal mass ejections (CME), emitted by the Sun, propagate through solar magnetic field and inter-planetary plasma. When directed along the proper trajectory CMEs can reach the Earth, and cause geomagnetic storms and other important perturbations of the Earth's magnetosphere. CMEs, as also powerful solar flares, can accelerate Solar energetic particles (SEPs), which carry unique information on the impulsive phenomena in the Sun atmosphere. An intense SEP event can also determine significant changes in the fluxes of trapped particles, The effects of intense SEP events can even reach the planetary ground as shown by the increase of the neutron flux measured by ground-based neutron monitors during the so-called ground-level enhancements (GLEs). A wide investigation program on low-energy cosmic rays and solar particles was pursued with the PAMELA satellite experiment. This included the study of trapped, albedo and solar energetic particles [4], of SEP events [5] and also the discovery of anomalously high fluxes of trapped anti-protons [6] in the kinetic energy range between 60 and 750 MeV. At lower energies, the measurements cited above can be complemented with the HEPD.

2 The HEPD Detector

HEPD, developed and built by the Italian members of the CSES mission, is designed for detecting electrons in the energy range between 3 and 100 MeV, protons between 30 and 300 MeV, and light nuclei. Being a space detector the HEPD mass budget is limited to 45 kg, while power to 43 W, and despite the small dimensions (20x20x40) cm^3 it reaches a relatively large geometrical acceptance of hundreds of cm^2sr. The HEPD includes four sub-detectors: - the **tracker** (TRK), made of two planes of double-sided silicon micro-strip sensors. Each plane includes 3 ladders made of 2 modules, each

DOI: 10.1051/epjconf/201713601007

module with an area of (10.7x7.2)cm^2, and read out by twelve VLSI chips; - the **trigger system** (TS), consisting of 6 plastic scintillator paddles, each with dimensions (20x3x0.5) cm^3 and read out by two Hamamatsu PhotoMultiplier Tubes (PMTs). The paddles fully cover the entrance of the detector and are placed below the tracker and above the calorimeter; - the **range-calorimeter** (CALO), divided in two sections. The top part consists in 16 plastic scintillator counters, each of which with dimensions (15x15x1) cm^3 and read out by two PMTs. The PMTs are placed at the ends of a counter diagonal and the instrumented diagonal alternates for even and odd counters. The lower calorimeter section is made of 9 LYSO crystals,which are arranged in a layer. A crystal has a height of 4 cm and a side of 5 cm and is read out by a PMT; - the **veto system** (VETO), made of five plastic scintillator counters, 5 mm thick, read out by two PMTs. Four counters are at the sides of the CALO, while the fifth one is below the LYSO layer.

The HEPD associated electronics (sub-system ELS) is inside a box placed at one side of the detector. The ELS includes: the CPU, the silicon detector DAQ, the PMT trigger and acquisition, the LV Control Board, the High Voltage Power Supply (HVPS), the Main power DC/DC converter (LVPS) and 6 mechanical modules. A scheme of the HEPD sub-systems can be seen in Fig. 1. The HEPD

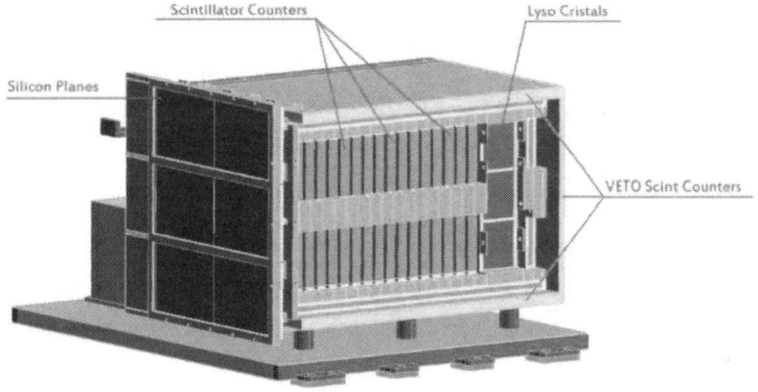

Figure 1. A scheme of the HEPD:- in blue, the silicon tracker, labeled as "Silicon Planes"; - below the tracker there is the trigger plane (not seen here), composed of 6 paddles; - in light blue 16 plastic scintillators which compose the upper part of the calorimeter, labeled as "Scintillation Counters"; - on the right, the LYSO cubes (3 of them are visible), which compose the lower part of the calorimeter, labeled as "LYSO Crystals"; - above, below and on the right of the calorimeter 3 counters of the VETO are visible, labeled as "VETO Scint Counters" ; - in green the power supply and electronics.

trigger is formed with the coincidence of signals from one of the TS paddles and from the upper CALO counter (P1). In order to vary the acceptance, when needed, other trigger combinations can be formed, using the other CALO counters, or requiring only a specific paddle to be hit. The dE/dX and incidence direction of the charged particle can be obtained with the TRK measurements, and a further dE/dX measurement is obtained with the TS paddles. Charged particles with trajectories fully inside the calorimeter can be selected by asking no signals from the lateral and the bottom VETO counters. In such a case the total energy can be measured with the CALO. For particles stopping in the upper CALO a range measurement can also be obtained. The lower CALO (LYSO layer) is designed to provide a larger matter thickness and to increase the operational energy range. Within the energy range of interest electrons are always relativistic, while protons are slower and not relativistic.

Then for fully contained particles, the electron/proton separation can be achieved with the dE/dX vs E_{tot} method. Two examples of electrons, which are stopped within the CALO, are in Fig. 2.

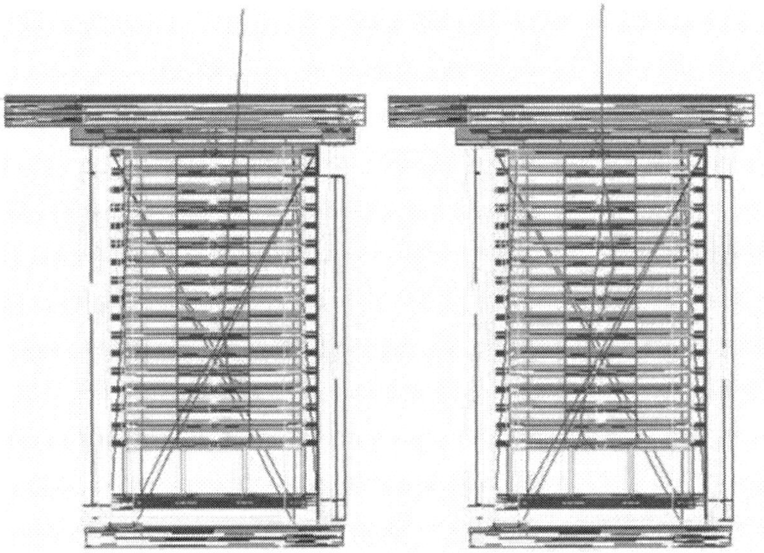

Figure 2. Monte Carlo simulations of electron measurement with the HEPD apparatus. On the right a 60 MeV electron, on the left a 30 MeV one, incident on HEPD from above. In red are the electrons trajectories, the apparatus elements are drawn as solids in color. While the 30 MeV electron is contained in the upper plastic scintillator CALO, the 60 MeV electron reaches the LYSO layer. These simulations are based on Geant4 [7].

2.1 Status of the project

Four HEPD versions were built, as required by Chinese space procedures: the Electrical Model (EM), the Structural and Thermal Model (STM), the Qualification Model (QM), the Flight Model (FM). During Spring-Summer 2016 a test and qualification campaign was performed on the QM, to assess the compliance of the HEPD to the space operation requirements, and to study the detector performances. The campaign included: -a beam test, carried on at Beam Test Facility (BTF) of the "Laboratori Nazionali di Frascati" of INFN, with electrons in the 30-150 MeV energy range; -a full set of thermo-vacuum-vibrational tests, performed at SERMS laboratory in Terni. At the end of Summer the FM model was also integrated, and at the beginning of Fall a test campaign started on the FM.

References

[1] Shen,X. et al, Earthq.Sci. 28(4),303–310 (2015)
[2] Shprits,Y.Y. et al, NatPh, 9, 699S
[3] Farley,T.A. and Walt,M., J.Geophys. Res., 76,(1971)
[4] Adriani,O. et al, ApJ, 799L, 4A (2015)
[5] Adriani,O. et al, ApJ, 801L, 3A (2015)
[6] Adriani,O. et al, ApJL, 737, L29, (2011)
[7] Agostinelli, S. et al, NIM A 506 (2003) 250-303,

DOI: 10.1051/epjconf/201713601008

The Luna experiment

Carlo Broggini[1,a] for the LUNA Collaboration

[1] *INFN Sezione di Padova, via Marzolo 8, I-35131*

Abstract. One of the main ingredients of nuclear astrophysics is the cross section of the thermonuclear reactions which power the stars and synthesize the chemical elements in the Universe. Deep underground in the Gran Sasso Laboratory the cross section of the key reactions of the proton-proton chain and of the Carbon-Nitrogen-Oxygen (CNO) cycle have been measured right down to the energies of astrophysical interest. The main results obtained during the solar phase of LUNA are reviewed before describing the current LUNA program devoted to the study of the nucleosynthesis of the light elements in AGB stars and Classical Novae. Finally, the future of LUNA with the new 3.5 MV accelerator devoted to the study of helium and carbon burning is discussed.

1 Introduction

Only hydrogen, helium and lithium were synthesized in the first minutes after the big-bang. All the other elements in the Universe are produced by the thermonuclear reactions taking place inside the cauldrons active in the cosmos, i.e. the stars. Nuclear astrophysics studies the reactions which transmute the chemical elements and provide energy to the stars. The reactions occur in the hot plasma of a star, with temperatures in the range of tens to hundreds of millions degrees, inside an energy window, the Gamow peak, which is far below the Coulomb energy arising from the repulsion between nuclei. In this region the cross section is given by:

$$\sigma(E) = \frac{S(E)}{E} exp(-2\pi\eta), \tag{1}$$

where $S(E)$ is the astrophysical factor (which contains the nuclear physics information) and η is given by $2\pi\eta = 31.29 Z_1 Z_2(\mu/E)^{1/2}$. Z_1 and Z_2 are the nuclear charges of the interacting particles, μ is the reduced mass (in units of amu), and E is the center of mass energy (in units of keV).

At low energies the cross sections are extremely small. Such smallness makes the star life-time of the length we observe, but it also makes the direct measurement impossible in the laboratory. LUNA, Laboratory for Underground Nuclear Astrophysics, started twenty five years ago to run nuclear physics experiments in an extremely low-background environment, the Gran Sasso Laboratory (LNGS), to reproduce in the laboratory what Nature makes inside the stars [1, 2].

[a] e-mail: broggini@pd.infn.it

2 Accelerators under Gran Sasso

Two electrostatic accelerators able to deliver hydrogen or helium beams have been installed in LUNA: first a compact 50 kV home made machine and then, in the year 2000, a commercial 400 kV one. Common features of the two accelerators are the high beam current, the long term stability and the precise beam energy determination. In particular, the 400 kV accelerator can provide up to 0.5 mA of hydrogen and 0.25 mA of helium at the target stations, with 0.3 keV accuracy on the beam energy, 100 eV energy spread, and 5 eV per hour long-term stability.

The dolomite rock of Gran Sasso provides a natural shielding equivalent to at least 3800 meters of water which reduces the muon and neutron fluxes by a factor 10^6 and 10^3, respectively. The neutron flux underground is mainly due to (α,n) reactions in the rock, with the alpha particles coming from the ^{238}U and ^{232}Th decay chains. Finally, the activity due to Radon from the rock is suppressed down to the level of few tens of Bequerel/m^3 thanks to frequent air volume exchanges.

3 LUNA and the Sun

The first phase of LUNA, the solar phase, has been driven by the solar neutrino problem [3]. In particular, the initial activity has been focused on the ^{3}He(^{3}He,2p)^{4}He cross section measurement within the solar Gamow peak (15-27 keV). Such a reaction is a key one of the hydrogen burning proton-proton chain, which is responsible for more than 99% of the solar luminosity. Fig.1 shows the results from LUNA [4] together with higher energy measurements [5–7]. For the first time a nuclear reaction has been measured in the laboratory at the energy occurring in a star. In particular, at the lowest energy of 16.5 keV the cross section is 0.02 pbarn, which corresponds to a rate of about 2 events/month, rather low even for the "silent" experiments of underground physics. No narrow resonance has been found and, as a consequence, the astrophysical solution of the ^{8}B and ^{7}Be solar neutrino problem based on its existence has been definitely ruled out. ^{3}He(α,γ)^{7}Be, the competing reaction for ^{3}He burning, has also been measured by LUNA both by detecting the prompt γ rays and by counting the ^{7}Be nuclei from their decay.

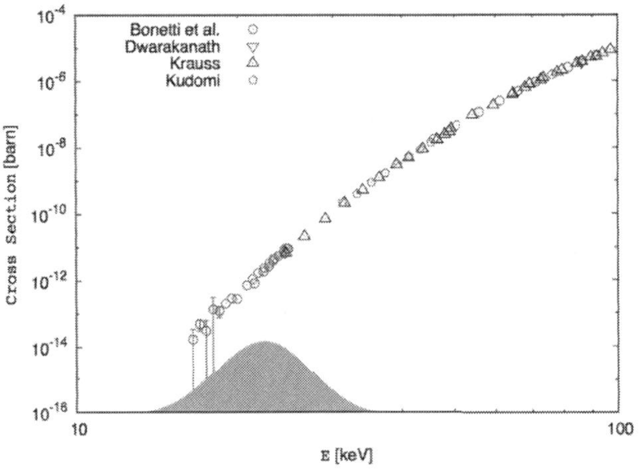

Figure 1. *The cross section of ^{3}He(^{3}He,2p)^{4}He as function of energy.*

The CNO cycle is responsible for the remaining 1% of the solar luminosity. $^{14}N(p,\gamma)^{15}O$ is the slowest reaction of the cycle. In particular, it is the key reaction to predict the ^{13}N and ^{15}O solar neutrino flux, which depends almost linearly on its cross section. The results obtained first with the germanium detector [8], then with the BGO set-up [9] and, finally, with a composite germanium detector [10] were about a factor two lower than the existing extrapolation from previous data at very low energy (with a total error now reduced to 8%). Because of this reduction the CNO neutrino yield in the Sun is decreased by about a factor of two. Thanks to the relatively small error on the cross section, it will be possible in the near future to measure the carbon and nitrogen content of the Sun core by comparing the predicted CNO neutrino flux with the measured one.

4 AGB stars and Classical Novae

A rich program of nuclear astrophysics mainly devoted to CNO, Ne-Na and Mg-Al cycles started a few years ago after the solar phase of LUNA. Of particular interest are those bridge reactions which are connecting one cycle to the next, as $^{15}N(p,\gamma)^{16}O$ [11] and $^{17}O(p,\gamma)^{18}F$ [12], the latter competing with $^{17}O(p,\alpha)^{14}N$ [13], or which are key ingredients of gamma astronomy, as $^{25}Mg(p,\gamma)^{26}Al$ [14].

Due to the higher Coulomb barrier of the reactions involved, the cycles become important at temperatures higher than the one of our Sun, i.e. during hydrogen burning in the shell of AGB stars and during the thermonuclear runaway of classical Novae (about 30-100, and 100-400 million degrees, respectively). Relatively unimportant for energy generation, these cycles are essential for the cooking of the light nuclei up to ^{27}Al. In particular, LUNA is now measuring $^{22}Ne(p,\gamma)^{23}Na$ [15], the reaction of the Ne-Na cycle with the highest uncertainty, and $^{23}Na(p,\gamma)^{24}Mg$, the reaction connecting the Ne-Na cycle to the Mg-Al one.

5 What next: helium and carbon burning

After hydrogen burning the natural evolution of LUNA is the study of the next steps in the fusion chain towards ^{56}Fe: helium and carbon burning. In particular, $^{12}C(\alpha,\gamma)^{16}O$ determines the abundance ratio between carbon and oxygen, the two key elements to the development of life, and it shapes the nucleosynthesis in massive stars affecting the outcomes of both thermonuclear and core-collapse supernovae.

Equally important are $^{13}C(\alpha,n)^{16}O$ and $^{22}Ne(\alpha,n)^{25}Mg$, the sources of the neutrons which synthesize half of the trans-iron elements through the S-process: neutron capture followed by β decay. Finally, the $^{12}C+^{12}C$ fusion reaction is switching on the carbon burning. Its rate determines the evolution of a massive star up to a slowly cooling white dwarf or up to a core-collapse supernova. It also affects the ignition conditions and time scales of thermonuclear supernovae, the standard candles of Cosmology.

This program requires a new 3.5 MV accelerator which is going to be installed underground in the north side of Hall B of Gran Sasso in spring 2018. The accelerator room (fig.2) will have 80cm thick concrete walls and ceiling working as neutron shielding. The room will host a single-ended electrostatic accelerator, with two beam lines (only one of them fed at a time) able to deliver H, He^+, C^+ and C^{2+} beams at high current (from 1 mA of H to 100 eμA of C^{2+}). After the underground mounting of the accelerator and its commissioning, the first physics run will start at the beginning of 2019.

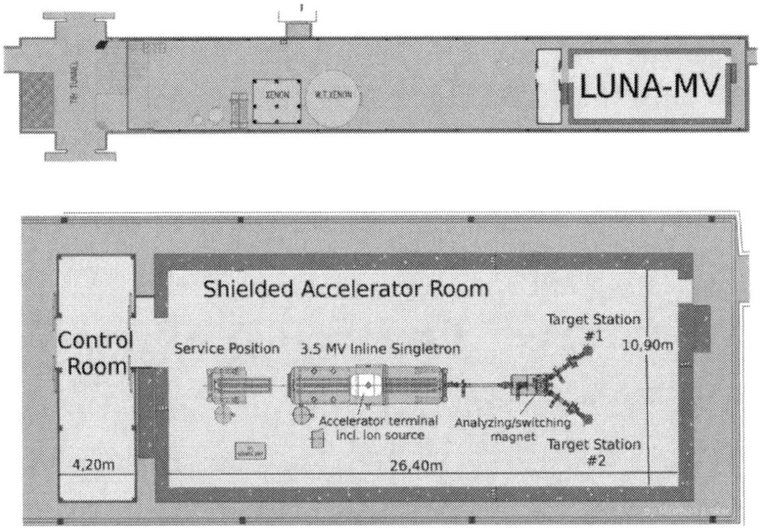

Figure 2. *The LUNA-MV facility in the north side of Hall B at LNGS and the 3.5 MV accelerator.*

6 Conclusions

Underground nuclear astrophysics was born twenty five years ago in the core of Gran Sasso, with the aim of measuring cross sections at very low energy. After the solar phase, LUNA is now studying the hydrogen burning reactions which are responsible for the cooking of the light elements in AGB stars and classical Novae. The future of LUNA is going to start with the installation of a new 3.5 MV accelerator in spring 2018 and it will be focused on the study of helium and carbon burning in stars.

References

[1] H. Costantini *et al*, Rep. Prog. Phys. **72**, 086301 (2009).
[2] C. Broggini *et al*, Annu. Rev. Nucl. Part. Sci. **60**, 53 (2010).
[3] G. Bellini, C. Broggini, A. Guglielmetti, Eur. Phys. J. A **52**, 88 (2016).
[4] R. Bonetti *et al*, Phys. Rev. Lett. **82**, 5205 (1999).
[5] A. Krauss *et al*, Nucl. Phys. A**467**, 273 (1987).
[6] M. Dwarakanath and H. Winkler, Phys. Rev. C **4**, 1532 (1971).
[7] N. Kudomi *et al*, Phys. Rev. C **69**, 015802 (2004).
[8] C. Formicola *et al*, Phys. Lett. B **591**, 61 (2004).
[9] A. Lemut *et al*, Phys. Lett. B **634**, 483 (2006).
[10] M. Marta *et al*, Phys. Rev. C **78**, 022802 (2008).
[11] A. Caciolli *et al*, Astronomy and Astrophysics **533**, A66 (2011).
[12] D.A. Scott *et al*, Phys. Rev. Lett. **109**, 202501 (2012).
[13] C.G. Bruno *et al*, Phys. Rev. Lett. **117**, 142502 (2016).
[14] O. Straniero *et al*, Astrophysical Journal **763**, 100 (2013).
[15] F. Cavanna *et al*, Phys. Rev. Lett. **115**, 252501 (2015).

Underground Study of Big Bang Nucleosynthesis in the Precision Era of Cosmology

Carlo Gustavino[1,a]

[1] *INFN, Sezione di Roma, Piazzale A. Moro 2, 00185 Roma, Italy*

Abstract. Big Bang Nucleosinthesis (BBN) theory provides definite predictions for the abundance of light elements produced in the early universe, as far as the knowledge of the relevant nuclear processes of the BBN chain is accurate. At BBN energies ($30 \lesssim E_{cm} \lesssim 300$ MeV) the cross section of many BBN processes is very low because of the Coulomb repulsion between the interacting nuclei. For this reason it is convenient to perform the measurements deep underground. Presently the world's only facility operating underground is LUNA (Laboratory for Undergound Nuclear astrophysics) at LNGS ("Laboratorio Nazionale del Gran Sasso", Italy). In this presentation the BBN measurements of LUNA are briefly reviewed and discussed. It will be shown that the ongoing study of the $D(p,\gamma)^3He$ reaction is of primary importance to derive the baryon density of universe Ω_b with high accuracy. Moreover, this study allows to constrain the existence of the so called "dark radiation", composed by undiscovered relativistic species permeating the universe, such as sterile neutrinos.

1 Big Bang Nucleosynthesis

Big Bang Nucleosynthesis (BBN) occurs in the first minutes of cosmic time, as result of the competition between the expansion of the Universe and the nuclear reaction rates. The expansion rate of the universe is governed by the Freidmann equation:

$$H^2 = \frac{8\pi}{3} G\rho \tag{1}$$

Were H is the Hubble parameter, G is the Newton's gravitational constant and ρ is the energy density which, in the early Universe, is dominated by the *"radiation"*, i.e. the contributions from massless or extremely relativistic particles. The radiation density is often expressed as follows:

$$\rho = \rho_\gamma \left(1 + \frac{7}{8} \left(\frac{4}{11} \right)^{4/3} N_{eff} \right) \tag{2}$$

In this formula ρ_γ is the photon density and N_{eff} is the contribution of other relativistic species. Using this formula $N_{eff} = 3.046$ if only the three known neutrino families are considered.
Constraints on cosmology and particle physics can be obtained by comparing the result of direct

[a] e-mail: carlo.gustavino@roma1.infn.it

observations of light isotopes (^2H, ^3He, ^4He, ^6Li, ^7Li) in primordial astrophysical sites with BBN calculated abundances. Therefore, are very important both the accuracy of astronomical observations and the precise determination of nuclear cross sections of BBN leading processes.

2 Nuclear Astrophysics

At energies of interest in astroparticle physics ($0.01 \div 1\ MeV$) the cross-section $\sigma(E)$ drops almost exponentially with decreasing energy E, due to the repulsion of charged nuclei. For this reason, in nuclear astrophysics the nuclear cross section $\sigma(E)$ is often factorized as follows:

$$\sigma(E) = \frac{S(E)e^{-2\pi\eta}}{E} \tag{3}$$

In this formula, the exponential term takes into account the Coulomb barrier, while the astrophysical factor $S(E)$ contains all the nuclear effects. The Sommerfeld parameter η is given by $2\pi\eta = 31.29 Z_1 Z_2 (\mu/E)^{1/2}$. Z_1 and Z_2 are the nuclear charges of the interacting nuclei. μ is their reduced mass (in units of a.m.u.), and E is the center of mass energy (in units of keV). Due to the low reaction yield, direct measurements at low energy are severely hampered by the background induced by cosmic rays. For this reason the LUNA collaboration carries out its measurements with the world's only underground accelerator facility, operating at the *"Laboratori Nazionali del Gran Sasso"* (LNGS). In fact, the ultra-low background at LNGS makes possible to study the nuclear reactions well below the Coulomb barrier.

The BBN production of ^{7}Li is dominated by the ^{3}He(α,γ)^{7}Be reaction, with subsequent decay of radioactive ^{7}Be to ^{7}Li. The ^{3}He(α,γ)^{7}Be reaction was studied at LUNA using two different experimental techniques: by detecting the promptly emitted γ-rays from the reaction and by measuring the ^{7}Be activity created in the experiment. The setup consisted of a windowless ^{3}He gas target, a $^4He^+$ beam and a high purity germanium detector (HPGe) to measure the prompt γ's yield. The ^{7}Be activity of the beam stopper were also measured with a HPGe detector installed underground. Figure 1 shows the LUNA result and literature data [1]. Note that only the LUNA data are well inside the BBN energy region. The LUNA result exacerbates the "lithium problem", excluding a nuclear solution to solve the tension between theory and observations.

Standard BBN production of ^{6}Li is dominated by just one nuclear reaction, ^{2}H(α,γ)^{6}Li. Before LUNA no direct measurements were performed inside the BBN energy region, making the BBN calculation concerning ^6Li very uncertain. The ^{2}H(α,γ)^{6}Li S-factor has been recently measured at low energy by LUNA, with an α beam and a windowless deuterium gas target. The main problems for this experiment were the very low cross section (about 60 pb at $E_{cm} = 133\,keV$) and the existence of a significant beam induced background. Figure 1 shows the LUNA result together with previous direct measurements [1, 2] and theoretical calculations [3]. The direct LUNA measurement excludes a nuclear solution to explain the debated overabundance of this isotope in metal poor stars.

The deuterium abundance is very important because it is determined from direct observations at the level of 1% [4] and it is very sensitive to Ω_b and to N_{eff}. The primordial abundance of deuterium depends on the four reactions shown in table 1. Its theoretical error is twice with respect to the uncertainty obtained by direct observations, and it is mainly due to the uncertainty of ^{2}H(p,γ)^{3}He cross section data [8]. The uncertainty due to this reaction is even larger if ab initio prediction is taken into account. Figure 1 shows the data of the $^2H(p,\gamma)^3He$ reaction in literature. Note that only a single dataset of S_{12} is currently available in the relevant BBN energy range, in which the authors state systematic uncertainty of 9% [6]. However the fit of experimental data is about 20% lower than the theoretical calculation [7]. In this plot the LUNA data obtained with the previous 50

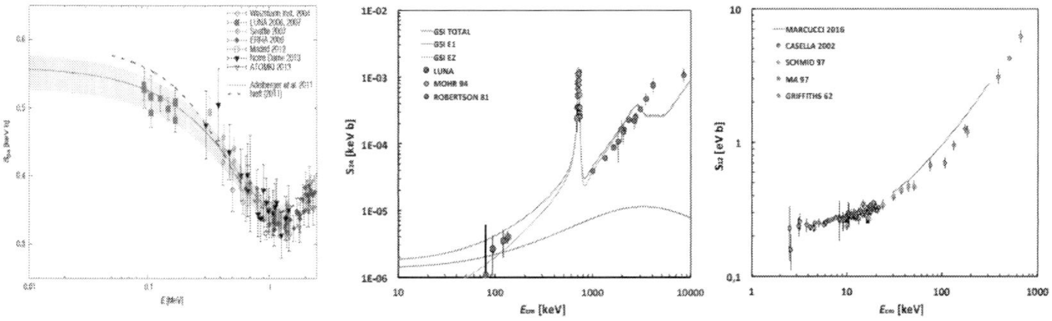

Figure 1. *Left: astrophysical S-factor data of the $^3He(\alpha,\gamma)^7Be$ reaction as a function of the center-of-mass energy. A theoretical curve and the ab initio prediction are also shown. Center: astrophysical S-factor data of the $^2H(\alpha,\gamma)^6Li$ reaction as a function of the center-of-mass energy. The LUNA data are shown with all the previous direct measurements. The continuous lines show the theoretical E1, E2, and total S_{24} factors describing recent Coulomb dissociation data. Right: S-factor data for the reaction $^2H(p,\gamma)^3He$. the red solid curve shows the prediction of recent ab initio theoretical calculation. See [1] and references therein.*

Table 1. *List of the leading reactions and corresponding rate symbols controlling the deuterium abundance after BBN. The last column shows the error on the ratio D/H coming from the uncertainty in the cross section of each reaction, for a fixed baryon density $\Omega_b h^2 = 0.02207$ [8].*

Reaction	Rate Symbol	$\sigma_{D/H} \cdot 10^5$
$p(n,\gamma)^2H$	R_1	± 0.002
$d(p,\gamma)^3He$	R_2	± 0.062
$d(d,n)^3He$	R_3	± 0.020
$d(d,p)^3H$	R_4	± 0.0013

kV pilot accelerator are also shown. The present LUNA 400 kV facility make possible to extend the measurements up to $E_{cm} = 266\ keV$, i.e. well inside the BBN energy range. Figure 2a) shows the scheme of the setup where a barrel BGO detector is implemented. The ongoing measurements are performed with the proton beam produced by the 400 kV accelerator, a windowless deuterium gas target and a 4π BGO crystal to detect prompt γs [5]. The study with the BGO detector will be accomplished by using a different layout in which a Germanium detector is faced to the gas target in a close geometry, as shown in figure 2b). With this configuration can be inferred the angular distribution of photons emitted by the $^2H(p,\gamma)^3He$ reaction, exploiting the doppler effect affecting the energy of γ's produced along the beam line. Figure 3 shows the result of a preliminary measurement measurement performed with the HPGe detector. The energy distribution of emitted photons is well in agreement with ab initio calculation (see figure caption).

3 Conclusions

The lack of data of the ^{2}H(p,γ)^{3}He reaction is presently the main obstacle to improve the accuracy of Ω_b and N_{eff} values. The study of the $^2H(p,\gamma)^3He$ reaction in the BBN energy range is in progress

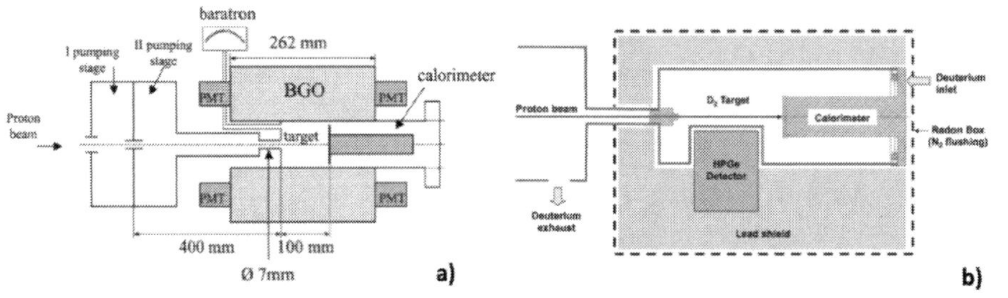

Figure 2. *a): Scheme of gas target setup and BGO detector. b): Scheme of gas target setup and HPGe detector.*

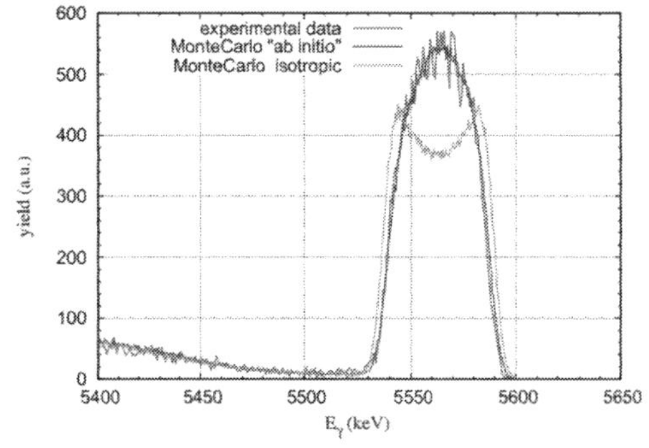

Figure 3. *Simulated spectra of the $^2H(p,\gamma)^3He$ reaction, assuming isotropic (green) and ab initio (blue) angular distribution at E_{cm} = 112.5 keV. The experimental data (red) are also shown. Data have been normalized to remark the close agreement with the ab initio angular distribution.*

with the LUNA400 facility at the underground Gran Sasso laboratory. The goal is to reach an accuracy at the 3% level, considerably better than the 9% uncertainty estimated in [6] and much lower than the 20% difference between data and nuclear calculations.

References

[1] C. Gustavino *et al*, Eur. Phys. J. A (2016) **52**, 74 (2016).
[2] M. Anders *et al*, PRL **113**, 042501 (2014).
[3] F. Hammache *et al*, Phys. Rev. C **82**, 065803 (2010).
[4] R. Cooke *et al*, ApJ. **781**, 31, (2014).
[5] C. Casella *et al*, Nuclear Physics A **706** 20316. (2002).
[6] L. Ma *et al*, Phys. Rev. C **55**, 588 (1997).
[7] L.E. Marcucci *et al*, PRL **116**, 102501 (2016).
[8] E. Di Valentino *et al*, Phys. Rev. D **90**, 023543 (2014).

DOI: 10.1051/epjconf/201713601010

Feasibility study of the $^{13}C(\alpha, n)^{16}O$ reaction at LUNA

I. Kochanek[1,a], A. Boeltzig[2], and G. F. Ciani[2] for the LUNA Collaboration

[1] *INFN, Laboratori Nazionali del Gran Sasso (LNGS), 67100 Assergi, Italy*
[2] *Gran Sasso Science Institute, INFN, Viale F. Crispi 7, 67100 L'Aquila, Italy*

Abstract. The $^{13}C(\alpha, n)^{16}O$ reaction determines the dominant neutron source of the s-process in thermally pulsing, low-mass, asymptotic giant branch (TP-AGB) stars. The temperature during the s-process in the ^{13}C pocket of 90×10^6 K corresponds to a Gamow window of 140-230 keV. Since this energy is far below the Coulomb barrier, the cross section of this reaction is extremely small and its rate can only be extrapolated from the measurements at higher energies. At present, the cross section at Gamow peak is uncertain by almost one order of magnitude.
An experimental campaign aimed at measuring low energy cross section in $^{13}C(\alpha, n)^{16}O$ is scheduled at the underground LUNA-400 accelerator in Gran Sasso Laboratory, Italy. The unique underground location of this facility offers significant improvement in sensitivity compared with previous investigations. It will allow to establish the interference pattern of the resonances and the absolute scale of this reaction.

1 Introduction

The $^{13}C(\alpha, n)^{16}O$ reaction is the major neutron source for the s-process in thermally pulsing, low mass, asymptotic giant branch (TP-AGB) stars [1]. The energy generation in such stars occurs in the H and He burning shells, separated by a thin He-rich intershell region, surrounding the inert C/O core. During H burning, the He concentration increases to the point of igniting He burning at the bottom of the intershell. Because of the heat generated during the He burning, the intershell becomes convective. The star expands and cools down thus inhibiting H burning in the outer layers. The He flash ends when most of the He has been consumed in the shell. The star contracts again and H burning is reactivated. These alternating H and He burning phases are repeated up to 40 times.

The main s-process is assumed to take place during the phase prior to the He flash when hydrogen is mixed from the convective envelope into the He intershell. The ^{13}C is produced by the reaction sequence $^{12}C(p, \gamma)^{13}N(\beta^- \nu)^{13}C$ that leads to the formation of the so-called ^{13}C pocket, a thin layer enriched in ^{13}C. The temperature during the s-process in the ^{13}C pocket reaches 90×10^6 K corresponding to a Gamow window of 140-230 keV for the reaction $^{13}C(\alpha, n)^{16}O$. Since this energy is far below the Coulomb barrier, the cross section is extremely small and not accessible to direct measurements. For this reason, its value has to be determined by extrapolation of measurements at higher energies, as shown in Figure 1.

The extrapolation is complicated by the unknown influence of a broad subthreshold state and by two subthreshold resonances. Our goal is to measure the S-factor at energies closer to the Gamow

[a] e-mail: iza.kochanek@lngs.infn.it

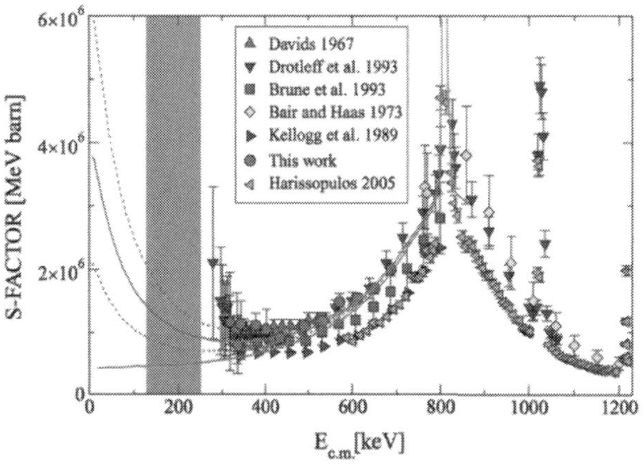

Figure 1. Astrophysical S-factor of the $^{13}C(\alpha, n)^{16}O$ reaction, the blue area defines the Gamow window [1], [2], [3], [4], [5], [6].

Table 1. $^{13}C(\alpha, n)^{16}O$ reaction rate for a current of 200 μA on a 100% ^{13}C enriched target.

E_α [keV]	cph
400	320
375	100
350	30
300	1.5
275	0.2
250	0.03
200	10^{-4}

window with respect to previous experiments and with high precision in order to facilitate the extrapolations. In Table 1 the rate in counts per hour (cph) for the $^{13}C(\alpha, n)^{16}O$ reaction is shown as a function of beam energy (for a current of 200 μA).

2 LUNA-400

The LUNA experiment is located in the Gran Sasso Underground Laboratory (LNGS) in Italy [7]. This is the largest underground laboratory in the world that hosts experiments in particle physics, particle astrophysics and nuclear astrophysics. The underground facilities have been built at the side of the ten kilometers long highway tunnel crossing the Gran Sasso Mountain, between L'Aquila and Teramo, about 120 km from Rome. The rock overburden of about 1400 m (3800 m water equivalent) reduces the muon component of the cosmic background by a factor of 10^6, the neutron component by a factor of 10^3, and the gamma component by a factor of 10 with respect to a laboratory on the Earth's surface. The reduced rate of low energy neutrons in LNGS, about 10^{-3} n/cm^2 per hour [8], can be compared to the reaction yield shown in Table 1.

The LUNA setup is composed of a 400 kV electrostatic accelerator providing high intensity proton or alpha beams. The beam can be delivered to a solid or gas target system. The low yield expected in this experiment requires targets with high resistance against ion bombardment.

3 Targets

The selection and the production of the ^{13}C target is of primary importance for this measurement, given the high beam intensity that they must withstand without deterioration. Two options are available within the LUNA-400 setup, gaseous or solid targets.

For the gas target, three molecules were considered CH_4, CO_2, CO all enrichment in ^{13}C up to 99%. The major problems for the gas target is so-called beam heating effect: the density of the target may decrease along the beam path because of heat transfer from the intense ion beam. The tests on

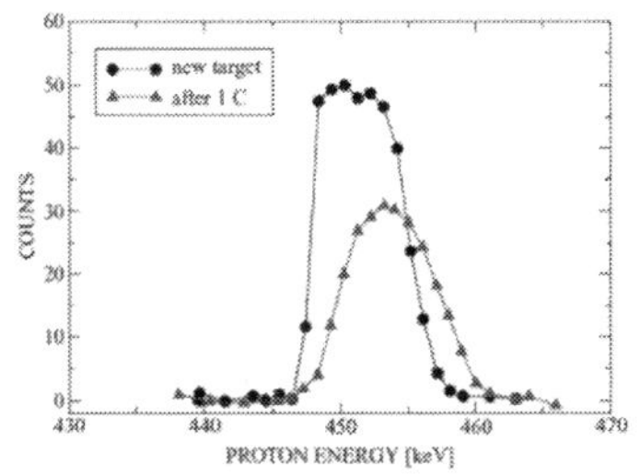

Figure 2. Reaction yield of the $^{13}C(p,\gamma)$ resonance at 448.5 keV for a fresh target and after an integrated charge of 1 C [1].

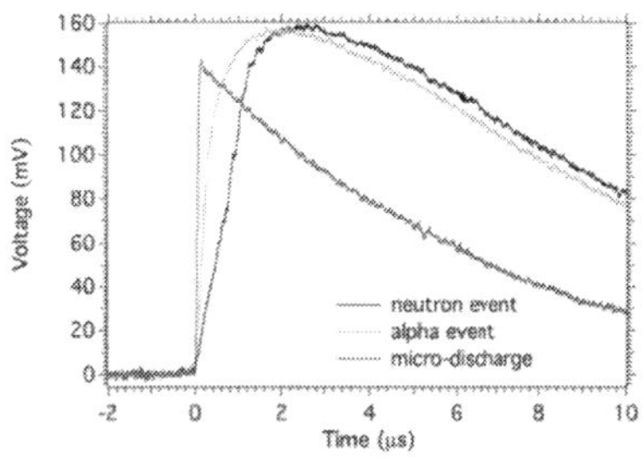

Figure 3. Example of the raw preamplifier traces showing the risetimes from three different sources within a ^{3}He tube: a microdischarge (red), an alpha-particle trace (gray) and a neutron trace (black).

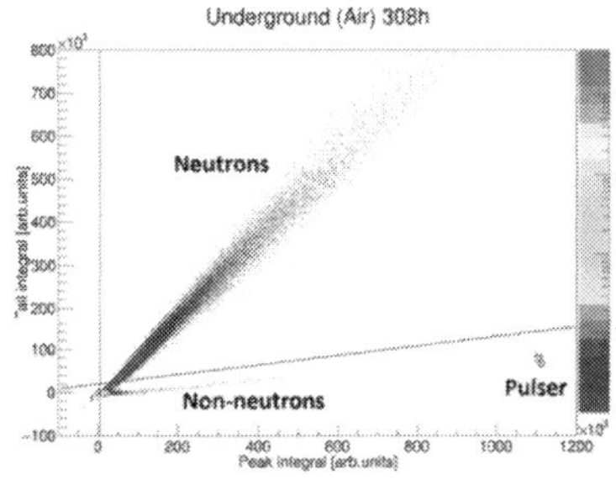

Figure 4. PSD method applied to data acquired underground in air with EJ-426 detector.

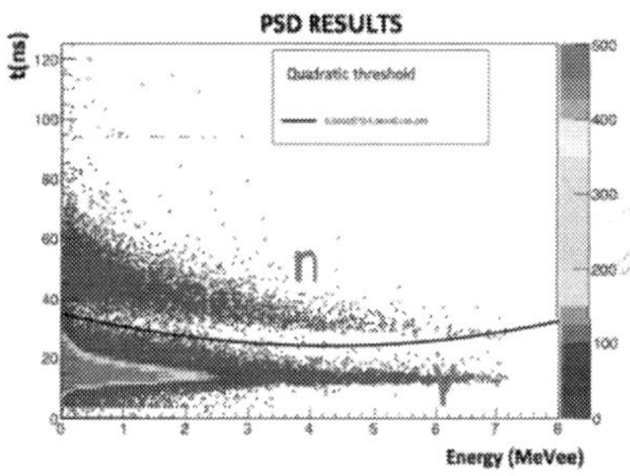

Figure 5. PSD results for BC-501A detector.

beam heating were carried out at LUNA using Nuclear Resonance Analysis (NRA) on a mixture of CH_4+N_2 (95%+5%). The analysis is still ongoing.

One of the most severe problems related to solid state targets exposed to α beam is the blistering effect: particles implanted into the target are trapped inside the material forming macroscopic bubbles. The bubbles increase in size during the implantation process and built up pressure. Eventually they explode and destroy the involved area of the target. Previous experiment [1] have shown that targets with a typical density of 22 $\mu g/cm^2$ can withstand a collected charge of 1 C equivalent to 1.5 h at an intensity of 200 μA (see Figure 2).

Several targets, with ^{13}C enrichment up to 99%, are under consideration with different bulk, substrate and deposition techniques. For the bulk materials we are evaluating Au, Ta, Ni, Cu, sapphire and diamond: the last two are known for their optimal mechanical strength and heat dissipation capabilities, while gold and tantalum do not contribute to beam-induced background in the region of interest. The substrates we are evaluating are Au and graphene or none (deposition directly on the bulk). The available deposition technique are implantation, polymerisation and electron gun evaporation.

A preliminary test on few solid targets will be performed at the Legnaro National Laboratory with the accelerator line *CN-7 MV*: this machine is powered by a Van de Graaff generator capable of providing ion beams with energies up to 7 MeV and currents up to 3 μA. For this tests the collaboration is preparing the following targets: Au + implantation, Ta + evaporation, Ta + polymer and sapphire + Au + evaporation.

3.1 Detectors

Given the low reaction yield, the efficient detection of neutrons is of paramount importance for the measurement of $^{13}C(\alpha, n)^{16}O$ reaction. A secondary requirement is the capability of rejection for non neutron signals. Finally a low-background detector would be desired. Several alternatives are under consideration: plastic scintillator, liquid scintillator and ^{3}He counters.

The most common reaction used for high efficiency thermal neutron detection is $^{3}He(n, p)^{3}H$, where both the proton and the ^{3}H are detected by a ^{3}He-filled proportional counters. They offer high detection efficiency with excellent gamma discrimination (Figure 3).

EJ-426-0 plastic scintillator-based detector is a homogeneous matrix of fine particles of Lithium-6-Fluoride (^{6}LiF) and Zinc Sulfide phosphor (ZnS:Ag) compactly dispersed in a colourless binder. Thermal neutrons are captured through the reaction $^{6}Li(n, {}^{3}H)\alpha$ (Q=4.78 MeV). The producer claims that the detector has low intrinsic background, is insensitive to gamma rays, has a 50% intrinsic efficiency for thermal (0.025 eV) neutrons, that falls sharply with increasing neutron energy and can be used with a Pulse Shape Discriminating (PSD) electronic chain to maximise its gamma rejection capability (Figure 4).

The BC501A is a liquid scintillator, made from a mixture of several organic molecules in liquid form with xylene as solvent. This scintillator is particularly effective in separating neutrons from γ-rays using PSD (Figure 5).

4 Conclusions

The proposed experiment will reduce the uncertainties in s-process in AGB stars by precisely measuring the cross-section of the reaction $^{13}C(\alpha, n)^{16}O$ close to the Gamow window. A feasibility campaign has started at LUNA by studying enriched ^{13}C targets, both solid and gaseous, and qualifying low background neutron detectors. A preliminary test on solid targets is already scheduled at Laboratori Nazionali di Legnaro. The $^{13}C(\alpha, n)^{16}O$ measurement in LUNA is scheduled for fall 2017.

References

[1] M. Heil, R. Detwiler, R.E. Azuma, A. Couture, J. Daly, J. Görres, F. Käppeler, R. Reifarth, P. Tischhauser, C. Ugalde et al., Phys. Rev. C **78**, 025803 (2008)
[2] C.N. Davids, Nuclear Physics A **110**, 619 (1968)
[3] H.W. Drotleff, A. Denker, K. H., S. M., W. G., H.J. W., G. U., R. C., T.H. P., The Astrophysical Journal **414**, 735 (1993)
[4] C.R. Brune, I. Licot, R.W. Kavanagh, Phys.Rev. **C48**, 3119 (1993)
[5] F.X. Haas, J.K. Bair, Phys.Rev. **C7**, 2432 (1973)
[6] S. Harissopulos, H.W. Becker, J.W. Hammer, A. Lagoyannis, C. Rolfs, F. Strieder, Phys. Rev. C **72**, 062801 (2005)
[7] C. Broggini, D. Bemmerer, A. Guglielmetti, R. Menegazzo, Ann. Rev. Nucl. Part. Sci. **60**, 53 (2010), 1010.4165
[8] A. Best, J. Görres, M. Junker, K.L. Kratz, M. Laubenstein, A. Long, S. Nisi, K. Smith, M. Wiescher, Nucl. Instrum. Meth. **A812**, 1 (2016), 1509.00770

EPJ Web of Conferences **136**, 01011 (2017) DOI: 10.1051/epjconf/201713601011
RICAP16

White Rabbit Facility

Roberto Ammendola[1], Aldo Morselli[1], Gonzalo Rodríguez[1,a], Gaetano Salina[1], and Valerio Verzi[1]

[1] *Sezione INFN di Tor Vergata, Roma, Italy*

Abstract. We describe a White Rabbit Facility for time synchronization we are preparing for the future Cherenkov Telescope Array observatory,

1 Introduction

Upcoming Gamma-Ray and Cosmic-Ray experiments require relative time calibration of all detector components with (sub-)nanosecond precision. White Rabbit (WR) [1] is a time-deterministic, low-latency Ethernet-based network that provides nanosecond accuracy and sub-nanosecond precision of clock synchronization for large distributed systems. Its aim is to combine the real-time performance of industrial networks and the flexibility of the Ethernet protocol with the accuracy of dedicated timing systems. White Rabbit is capable of synchronizing more than 1000 nodes, see figure 1, connected by fibers of up to 10km in length, with sub-nanosecond accuracy. It uses a robust, prioritized message delivery system. White Rabbit time synchronization packets are routinely flagged as high priority, but other critical Ethernet packets may be priority-flagged as well.

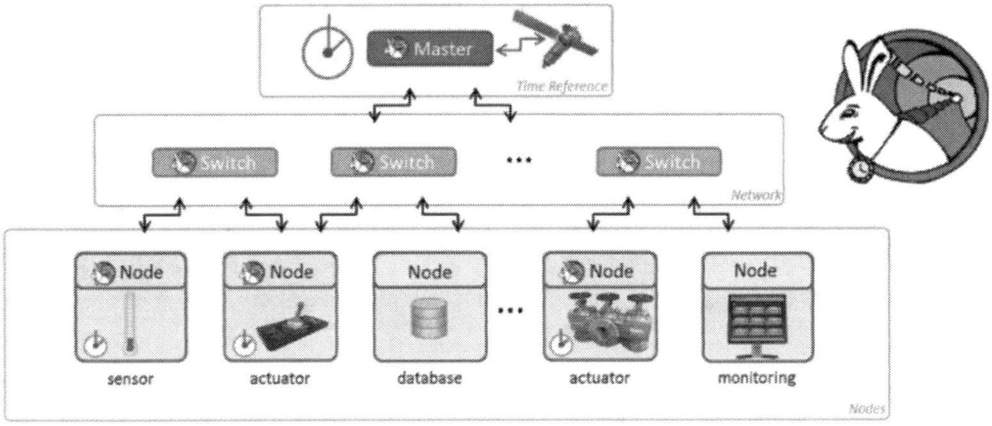

Figure 1: Scheme of White Rabbit network, from http://www.ohwr.org/projects/white-rabbit/wiki

[a] e-mail: grferna@roma2.infn.it

WR is currently being implemented as the next version of the IEEE 1588 standard, the Precision Time Protocol (PTP) and there is a large and active community of White Rabbit users and contributors (CERN, GSI, DESY). Several astroparticle experiments (HiSCORE, CTA, IceCube, LHAASO, km3Net, IceCube-Gen2) [2, 3] have also decided on the use of hardware based on the common and open WR solution. Non-White Rabbit devices may be connected to the WR network via a standard network interface card and operated within the WR network.

2 White Rabbit facility

The Roma Tor Vergata group is preparing a timing synchronization test facility based on the White Rabbit system. To fulfil the Cherenkov Telescope Array (CTA) requirements on event time stamping and dead time monitoring accuracy, the White Rabbit system has been chosen by the CTA Consortium. This facility will be used for the next gamma array telescope Cherenkov Telescope Array (CTA) [4, 5] but also for other future experiments. A common requirement for all the telescopes is the time stamping of the measured events. The required relative accuracy of these time stamps for CTA is < 10 ns, the goal accuracy is < 2 ns. This relative accuracy and precision must be achieved between all different types of telescopes and camera combinations.

Figure 2 a shows several PCs for mounting the SPECs 2 b. The baseline ingredients are (1) WR-Switches [6] (WRS) (Fig. 2 (c)) and (2) WR-Nodes (Fig. 2 (b)), connected by standard Gigabit Ethernet fibers. The WRS are arranged like in a normal ethernet-network; the central WRS (Grand Master Switch) acts as the time source (e.g. connected to a GPS antenna). White Rabbit is build on Gigabit Ethernet (1000base-BX10) and takes advantage of the Ethernet standards SyncE and Precision Time Protocol. It offers sub-ns precision, with excellent clock phase stability. It utilizes one fiber for each WR-node for both synchronization and user data, and compensates dynamically for clock drifts due to e.g. environmental influences (temperature).

(a) WR setup

(b) WR SPEC cards

(c) WR switch

Figure 2: WR facility

The WR-node allows to interface the user system (eg. the DAQ of a detector station or a telescope) to the WR-time system: by either time-stamping signals from the detector and/or by supplying clock-information (like PPS or periodic clock signals) to the detector. As the WR-node device we use for this work the Simple PCIe FMC carrier (SPEC) [7], shown in Fig. 2 (b) - a reliable workhorse of the

DOI: 10.1051/epjconf/201713601011

WR- community. It has a Spartan-6 FPGA (with the WR PTP Core, optional custom firmware and software) and can accommodate FMC-mezzanine cards. We use the FMC-DIO5Ch [8], a 5 channel digital I/O card, for analog/digital trigger input, control signals, and PPS/MHz clock output (e.g. for clock performance tests).

3 Results

The calibration of the WR setup (optical fibers, internal delays), and the monitoring their stability are key measurements to ensure a good performance . In figure 3 we study the delay between the PPS from master and the PPS from slave when different optical fiber's length are used. The measurements were done using an oscilloscope to visualize both PPS. We observed that the fluctuations are well below the nanosecond. Also, we study the stability of the system for a period of 21 hours. In the same figure we show the delay as a function of time, where we can check if the delay between the PPS is always below the nanosecond.

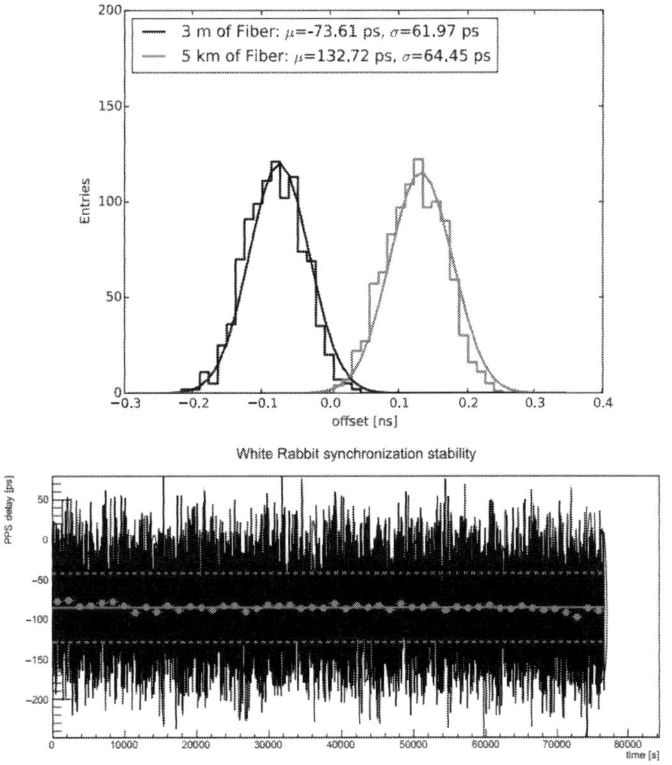

Figure 3: WR monitoring examples. Top figure shows the delay distribution between PPS from master and PPS from slave for two fibers of 3 m and 5 km. Bottom figure shows the stability of the synchronization for a 21 hours run, red dots are the averages over 25 minutes, dashed lines indicates 1 sigma deviation.

4 Conclusions

We have a multi purpose White Rabbit test facility at Tor Vergata Rome where the monitoring and the calibration of the system its done automatically using python scripts. In order to fulfill the CTA specification we have modify the White Rabbit firmware and implement the interface with the camera electronics. We also plan to implement the White Rabbit system providing the time synchronization for the camera demonstrator od ASTRI telescope [9].

References

[1] J. Serrano et al. The White Rabbit Project, ICALEPCS 2009, http://www.ohwr.org/projects/white-rabbit/wiki/

[2] M. Brückner et al., Time Synchronization with White Rabbit ? Experience from Tunka-HiSCORE, ICRC2015, Proceedings of the 34th ICRC 2015, The Hague, Netherlands, Proceedings of Science PoS (ICRC2015) 1041

[3] Oya, I. et al, arXiv:1509.01164, Proceedings of the 34th ICRC 2015, The Hague, Netherlands, Proceedings of Science PoS (ICRC2015) 975

[4] http://cta-observatory.org

[5] WR-provider, www.sevensols.com

[6] http://www.ohwr.org/projects/white-rabbit/wiki/Switch

[7] http://www.ohwr.org/projects/white-rabbit/spec/wiki

[8] http://www.ohwr.org/projects/white-rabbit/fmc-dio-5chttla/wiki

[9] Pareschi G. et al. (2015), proc. TAUP 2015

The mass composition of cosmic rays measured with LOFAR

Jörg R. Hörandel[1,2,a], A. Bonardi[1], S. Buitink[1,3], A. Corstanje[1], H. Falcke[1,2,4,5], P. Mitra[3], K. Mulrey[3], A. Nelles[1,2,8], J.P. Rachen[1], L. Rossetto[1], P. Schellart[1,9], O. Scholten[6,7], S. ter Veen[4], S. Thoudam[1,10], T.N.G. Trinh[6], and T. Winchen[3]

[1]*Department of Astrophysics, IMAPP, Radboud University Nijmegen P.O. Box 9010, 6500 GL Nijmegen, The Netherlands*
[2]*NIKHEF, Science Park Amsterdam, Amsterdam 1098 XG, The Netherlands*
[3]*Astrophysical Institute, Vrije Universiteit Brussel, Pleinlaan 2, Brussels 1050, Belgium*
[4]*ASTRON, Postbus 2, Dwingeloo 7990 AA, The Netherlands*
[5]*Max-Planck-Institut für Radio Astronomie, Bonn, Germany*
[6]*KVI-CART, Groningen University, P.O. Box 72, Groningen 9700 AB, The Netherlands*
[7]*Interuniversity Institute for High Energies, Vrije Universiteit Brussel, Pleinlaan 2, 1050 Brussels, Belgium*
[8]*now at: Department of Physics and Astronomy, University of California Irvine, Irvine, CA 92697-4575, USA*
[9]*now at: Department of Astrophysical Sciences, Princeton University, Princeton, NJ 08544, USA*
[10]*now at: Department of Physics and Electrical Engineering Linneuniversitetet, 35195 Växjö, Sweden*

Abstract. High-energy cosmic rays, impinging on the atmosphere of the Earth initiate cascades of secondary particles, the extensive air showers. The electrons and positrons in the air shower emit electromagnetic radiation. This emission is detected with the LOFAR radio telescope in the frequency range from 30 to 240 MHz. The data are used to determine the properties of the incoming cosmic rays. The radio technique is now routinely used to measure the arrival direction, the energy, and the particle type (atomic mass) of cosmic rays in the energy range from 10^{17} to 10^{18} eV. This energy region is of particular astrophysical interest, since in this regime a transition from a Galactic to an extra-galactic origin of cosmic rays is expected. For illustration, the LOFAR results are used to set constraints on models to describe the origin of high-energy cosmic rays.

1 Introduction

Cosmic rays (ionized atomic nuclei) impinge on the Earth with (kinetic) energies covering a wide range from MeV energies up to beyond 10^{20} eV. At energies below $\sim$ 100 MeV they are accelerated in energetic outbursts of the Sun. At higher energies, the are assumed to originate in our Milky Way, being accelerated in Supernova remnants (e.g. [1, 2]. At energies exceeding 10^{18} eV it becomes increasinlgy difficult to magnetically bind the particles to our Galaxy. Thus, particles with energies above $\sim 10^{18}$ eV are usually considered to be of extra-galactic origin. A transition from a Galactic to an extra-galactic origin of cosmic rays is expected at energies around 10^{17} to 10^{18} eV [3, 4].

In this paper we will shed new light on the understanding of the origin of cosmic rays in the transition region ($10^{17} - 10^{18}$ eV). This necessitates a precise measurement of the properties of cosmic

[a]e-mail: j.horandel@astro.ru.nl-http://particle.astro.ru.nl

rays, namely their arrival direction (on the sky), their (kinetic) energy, and their particle type (atomic mass A).

The flux of cosmic rays is steeply falling, approximately following a power law $\propto E^{-3}$. In our region of interest, cosmic rays are only measured indirectly, using large ground-based detector installations. High-energy cosmic rays impinging on the atmosphere, initiate cascades of secondary particles, the extensive air showers. The challenge of the indirect measurements is to derive the properties of the incoming cosmic rays from air-shower observations. Most challenging is the measurement of the particle type, since the sensitivity of air shower measurements is only proportional to $\ln A$. Intrinsic shower fluctuations allow to divide the measured cosmic rays in up to five mass groups for the best experiments [5].

The radio measurement of air showers to determine the properties of cosmic rays is briefly sketched in Sect. 2. The method is used to determine the properties of cosmic rays as outlined in Sect. 3. One of the key results obtained is the mass composition of cosmic rays in the transition region. Implications on our understanding of the origin of cosmic arys will be discussed.

2 Radio detection of air showers

Many secondary particles in extensive air showers are electrons and positrons. They emit radiation with frequencies of tens of MHz mainly due to interaction with the magnetic field of the Earth. Radio detection of air showers is suitable to measure the properties of cosmic rays with nearly 100% duty cycle [6, 7].

The LOFAR radio telescope [8, 9] is one of the leading installations for the LOFAR radio measurements of air showers. LOFAR is a digital radio telescope. Its antennas are spread over several European countries and are used together for interferometric radio observations in the frequency range of 30 – 240 MHz. The density of antennas increases towards the center of LOFAR, which is located in the Netherlands. Here, about 2400 antennas are clustered on an area of roughly 10 km^2 with increasing antenna density towards the center. This high density of antennas makes LOFAR the perfect tool to study features of the radio emission created by extensive air showers. The radio antennas have been calibrated with in-situ measurements, using a reference source and Galactic emission [10].

Air shower measurements are conducted based on a trigger received from an array of scintillators (LORA) [11, 12], which results in a read-out of the ring buffers that store the raw voltage traces per antenna for up to 5 s. LOFAR comprises two types of antennas. While air showers have also been measured in the high-band (110 – 240 MHz), most air showers are measured with the low-band antennas (LBA), which cover the frequency range from 10 – 90 MHz. The LBAs are arranged in compact clusters of 96 antennas, called stations. Of every station either the inner group or the outer ring of antennas (48 antennas each) can be used for cosmic-ray measurements at a given time.

In the last years the radio technique has been established as a precise method to measure the mass composition of cosmic rays. The LOFAR measurements together with the predictions of the CoREAS [13] simulation package result in a complete understanding of the emission mechanisms. With LOFAR the properties of the radio emission have been measured with high accuracy [14–16] in the frequency range 30 – 80 MHz, which allows us to establish key features, such as the lateral density distribution of the radio signals [17, 18], the shape of the shower front [19], and the polarization of the radio signal [20, 21]. These measurements help to understand the emission processes in the atmosphere and to quantify the contributions of the two mechanisms, being responsible for the radio emission of air showers – namely the geomagnetic effect (i.e. charge separation in the geomagnetic field) and the Askaryan effect (charge excess in the shower front). We obtained the first quantitative measurements in the frequency range 120 – 240 MHz [22]. We also recorded air showers during thunderstorm conditions [23, 24] and measured the structure of electric fields in the atmosphere.

 DOI: 10.1051/epjconf/201713602001

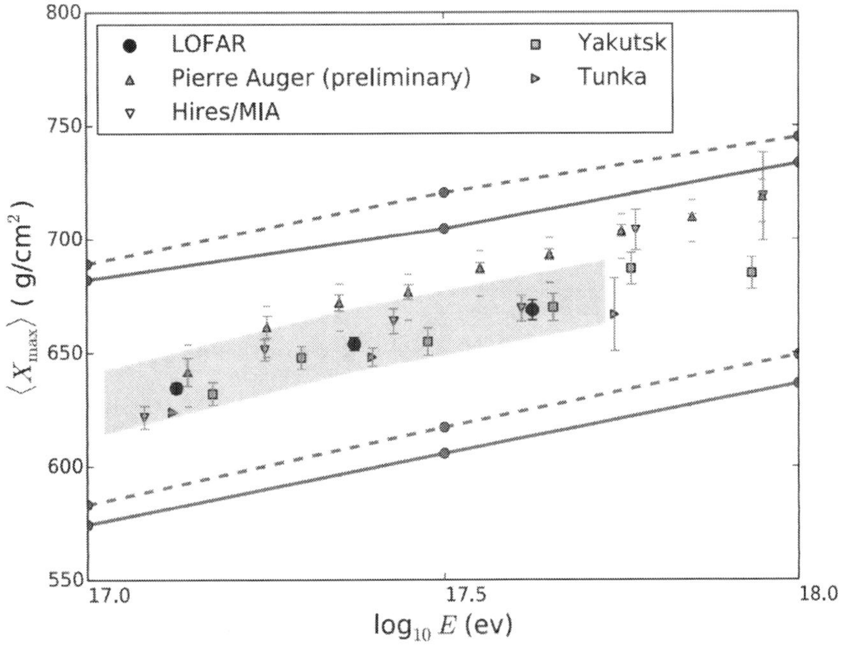

Figure 1. Average depth of the shwoer maximum $X_{\max}$ as a function of cosmic-ray energy [29]. The LOFAR radio results are compared to optical measurements. For details and references see [29]. The lines represent predictions for protons (green) and iron nuclei (red) for the hadronic interaction models QGSJETII.04 (solid) and EPOS-LHC (dashed).

3 Properties of cosmic rays

Direction

The excellent time resolution of LOFAR with ns accuracy allows to measure the shape of the shower front [19]. It is found that a hyperboloid describes the measurements best. Such a behavior has been found earlier by the LOPES experiment [25]. In order to estimate the accuracy of the measurement of the arrival direction of the shower, the same measured air showers have been reconstructed with different assumptions (plane, sphere, hyperboloid). These investigations indicate an uncertainty for the direction reconstruction of the order of 0.1°to 0.5°.

Energy

The recording of radio signals in the frequency range of interest (30 – 80 MHz) provides an excellent calorimetric measure of the energy contained in the electromagnetic component of the air shower and, thus, provides a good measure of the energy of the shower-inducing particle. The integral over the measured energy fluence distribution on the ground is proportional to the energy of the incoming cosmic ray. A resolution around 30% for the cosmic-ray energy is obtained with LOFAR [26]. Similar investigations at the Pierre Auger Observatory indicate that a resolution around 25% is possible for high-quality showers [27, 28].

DOI: 10.1051/epjconf/201713602001

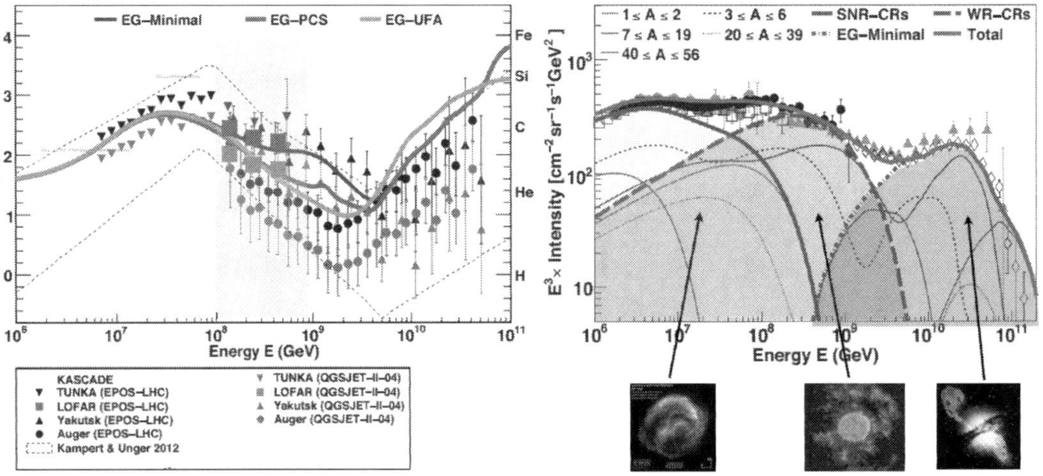

Figure 2. Left: Mean logarithmic mass of cosmic rays as a function of energy, for details, see [30]. Right: Three-component model of the origin of cosmic rays according to [30]: 'regular' supernovae, Wolf Rayet component, and an extra-galactic component.

Particle type

The good agreement between the measurements and the predictions of the CoREAS code is essential to identify the type of incoming cosmic ray. This is inferred from the (atmospheric) depth of the shower maximum X_{max}, one of the standard measures to estimate $\ln A$. To measure X_{max} [26, 29] we analyse simultaneously measurements of the radio emission and the particle detectors. The arrival direction and energy of each cosmic ray are determined first. Then, simulations for primary protons and iron nuclei are conducted for each measured shower with its corresponding direction and energy. Due to the intrinsic shower fluctuations is it sufficient to simulate only protons and iron nuclei to cover the parameter space in X_{max}. The predictions for the signals in the particle detectors and the radio antennas are compared on a statistical basis to the measured values. This method is used to determine X_{max} with an accuracy of better than ~ 20 g/cm^2 with the dense LOFAR core, thus, reaching the state of the art – the uncertainty of the Pierre Auger Observatory fluorescence detector. The X_{max} values obtained are depicted as a function of energy in Fig. 1 together with other measurements. The latter apply different techniques, namely measuring Cherenkov and fluorescence light from the air showers. The figure illustrates the good agreement between the radio measurements and the established optical methods.

The measured values for the depth of the shower maximum X_{max} are converted to the mean logarithmic mass of cosmic rays

$$\langle \ln A \rangle = \left(\frac{X_{max} - X_{max}^{p}}{X_{max}^{Fe} - X_{max}^{p}} \right) \times \ln A_{Fe}.$$

This necessiates predictions for the depth of the shower maximum for impinging protons and iron nuclei, X_{max}^{p} and X_{max}^{Fe}, respectively. These are illustrated as lines in Fig. 1: protons (green) and iron nuclei (red). The resulting mean mass is depicted in Fig. 2 (left) as a function of energy for the LOFAR results together with the world data set [30]. Two hadronic interaction models are used (EPOS-LHC and QGSJETII.04, dashed and solid lines in Fig. 1, respectively) to interpret the data. Both interaction models are tuned to LHC data but uncertainties remain, when extrapolating to the cosmic-ray parameter space.

Origin of cosmic rays

To understand the implications of the LOFAR measurements and the available world data set from direct and indirect measurements a model has been developed to consistently describe the observed energy spectrum and mass composition of cosmic rays with energies up to about 10^{18} eV [30]. We assume that the bulk of Galactic cosmic rays is accelerated by strong Supernova remnant shock waves [31]. Our study shows that a single Galactic component with rigidity-dependent energy cut-offs in the individual spectra of different elements cannot explain the observed all-particle spectrum at energies exceeding $\sim 2 \cdot 10^{16}$ eV. Similar findings have already been obtained earlier [32]. We discuss two approaches for a second component of Galactic cosmic rays: re-acceleration at a Galactic wind termination shock and Supernova explosions of Wolf-Rayet stars. The latter scenario can explain almost all observed features in the all-particle spectrum and the mass composition of cosmic rays up to $\sim 10^{18}$ eV, when combined with a canonical extra-galactic spectrum as expected from strong radio galaxies or a source population with similar cosmological evolution. The resulting spectrum is shown in Fig. 2 (right). In this two-component Galactic cosmic-ray model, the 'knee' at $\sim 4 \cdot 10^{15}$ eV and the 'second knee' at $\sim 4 \cdot 10^{17}$ eV in the all-particle spectrum are due to the fall-offs of the first and second Galactic cosmic-ray components, respectively.

4 Summary

The radio detection of extensive air showers enables us to measure the properties of cosmic rays above energies exceeding 10^{17} eV with high precision of $\sim 0.1° - 0.5°$for the arrival direction, $\sim 30\%$ for the energy, and to better than ~ 20 g/cm^2 for the depth of the shower maximum X_{max}.

To illustrate the potential of the LOFAR radio measurements we developed a model to consistently describe the observed energy spectrum and mass composition of cosmic rays from GeV energies up to 10^{20} eV. We adopt a three component model: 'regular' cosmic rays being accelerated in Supernova remnants up to $\sim 10^{17}$ eV, a second Galactic component, dominating the all-particle flux between $\sim 10^{17}$ and $\sim 10^{18}$ eV from cosmic rays being accelerated by exploding Wolf-Rayet stars, yielding a strong contribution of He and CNO elements, and, finally, an extra-galactic contribution at energies above $\sim 10^{18}$ eV.

Acknowledgements

JRH is grateful to the organizers of the RICAP conference for their kind invitation and the great hospitality.

References

[1] F. Aharonian et al., Astron. & Astroph. **449**, 223 (2006)
[2] H. Völk, E. Berezhko, Astron. & Astroph. **451**, 981 (2006)
[3] J. Blümer, R. Engel, J.R. Hörandel, Prog. Part. Nucl. Phys. **63**, 293 (2009)

[4] M. Nagano, A. Watson, Rev. Mod. Phys. **72**, 689 (2000)
[5] J.R. Hörandel, Nucl. Instr. & Meth. A **588**, 181 (2008)
[6] T. Huege, Phys. Rept. **620**, 1 (2016), arXiv:1601.07426
[7] F.G. Schröder, Prog. Part. Nucl. Phys. **93**, 1 (2017), arXiv:1607.08781
[8] M. van Haarlem et al., Astron. & Astrophys. **556**, A2 (2013), arXiv:1305.3550
[9] P. Schellart et al., Astron. & Astrophys. **560**, A98 (2013), arXiv:1311.1399
[10] A. Nelles et al., JINST **10**, P11005 (2015), arXiv:1507.08932
[11] S. Thoudam et al., Nucl.Instrum.Meth. **A767**, 339 (2014), arXiv:1408.4469
[12] S. Thoudam et al., Astropart. Phys. **73**, 34 (2016), arXiv:1506.09134
[13] T. Huege, M. Ludwig, C.W. James, AIP Conf. Proc. **1535**, 128 (2013), arXiv:1301.2132
[14] J.R. Hörandel et al., Proceedings of the 33rd International Cosmic Ray Conference, Rio de Janeiro # 865 (2013)
[15] J.R. Hörandel et al., Proceedings of the 34th International Cosmic Ray Conference, Den Haag PoS(ICRC2015)033 (2015)
[16] J.R. Hörandel, JPS Conf. Proc. **9**, 010004 (2016), arXiv:1509.04960
[17] A. Nelles et al., Astropart.Phys. **60**, 13 (2015), arXiv:1402.2872
[18] A. Nelles et al., JCAP **1505**, 018 (2015), arXiv:1411.7868
[19] A. Corstanje et al., Astropart.Phys. **61**, 22 (2015), arXiv:1404.3907
[20] P. Schellart et al., JCAP **1410**, 014 (2014), arXiv:1406.1355
[21] O. Scholten et al., Phys. Rev. **D94**, 103010 (2016), arXiv:1611.00758
[22] A. Nelles et al., Astropart.Phys. **65**, 11 (2014), arXiv:1411.6865
[23] P. Schellart et al., Phys.Rev.Lett. **114**, 165001 (2015), arXiv:1504.05742
[24] T.N.G. Trinh et al., Phys. Rev. **D93**, 023003 (2016), arXiv:1511.03045
[25] W. Apel et al., JCAP **1409**, 025 (2014), arXiv:1404.3283
[26] S. Buitinket al., Phys.Rev. **D90**, 082003 (2014), arXiv:1408.7001
[27] A. Aab et al., Phys. Rev. **D93**, 122005 (2016), arXiv:1508.04267
[28] A. Aab et al., Phys. Rev. Lett. **116**, 241101 (2016), arXiv:1605.02564
[29] S. Buitink et al., Nature **531**, 70 (2016), arXiv:1603.01594
[30] S. Thoudam, J.P. Rachen, A. van Vliet, A. Achterberg, S. Buitink, H. Falcke, J.R. Hörandel, Astron. & Astrophys. **595**, 33 (2016), arXiv:1605.03111
[31] S. Thoudam, J.R. Hörandel, Astron. & Astrophys. **567**, A33 (2014), arXiv:1404.3630
[32] A.M. Hillas, J. Phys. **G31**, R95 (2005)

Astrophysical interpretation of Pierre Auger Observatory measurements of the UHECR energy spectrum and mass composition

Armando di Matteo[1,ab] for the Pierre Auger Collaboration[2,cd]

[1] *INFN and Department of Physical and Chemical Sciences, University of L'Aquila, L'Aquila, Italy*
[2] *Observatorio Pierre Auger, Av. San Martín Norte 304, 5613 Malargüe, Argentina*

Abstract. We present a combined fit of a simple astrophysical model of UHECR sources to both the energy spectrum and mass composition data measured by the Pierre Auger Observatory. The fit has been performed for energies above 5 EeV, i.e. the region of the all-particle spectrum above the so-called "ankle"' feature. The astrophysical model we adopted consists of identical sources uniformly distributed in a comoving volume, where nuclei are accelerated with a rigidity-dependent mechanism. The fit results suggest sources characterized by relatively low maximum injection energies and hard spectral indices. The impact of various systematic uncertainties on the above result is discussed.

1 Introduction

Ultra-high-energy cosmic rays (UHECRs) are particles reaching the Earth from outer space with energies above 10^{18} eV. More than half a century after their discovery, their origin is still unknown, but there is a wide consensus that most of the highest-energy cosmic rays originate outside of our galaxy. If this is the case, their energy spectrum and mass composition is non-trivially affected by interactions with photon backgrounds during their propagation through intergalactic space, making it harder to infer properties of their sources from Earth-based observations. Also, whereas the energy of UHECRs can now be measured with resolution and systematic uncertainty less than 20%, determinations of their mass are still strongly model-dependent and only possible on a statistical basis.

The Pierre Auger Observatory [1] in western Argentina is the largest UHECR observatory in the world. It is operated by a collaboration of about 500 members from 86 institutions in 18 countries. The baseline array for the study of the highest-energy cosmic rays consists of 1 660 water-Cherenkov stations on a triangular grid with 1 500 m spacing covering a 3 000 km^2 area (the surface detector array, SD), overlooked by 24 telescopes in four locations at the periphery of the array (the fluorescence detector, FD). The Observatory also includes extra SD stations with closer spacing and three extra FD telescopes with higher elevation for the study of lower-energy cosmic rays, and various other facilities for atmospheric monitoring, R&D, and interdisciplinary studies.

[a] e-mail: armando.dimatteo@aquila.infn.it
[b] Now at Service de Physique Théorique, Université Libre de Bruxelles, Brussels, Belgium
[c] e-mail: auger_spokespersons@fnal.gov
[d] Full author list: http://www.auger.org/archive/authors_2016_06.html

 DOI: 10.1051/epjconf/201713602002

Table 1. The propagation models used in this work (see ref. [7] and references therein for details) and the resulting best-fit parameter values and fit deviances (assuming EPOS LHC showers and no systematic errors)

model	MC code	photodisint.	EBL	γ	$\log_{10}\left(\frac{R_{cut}}{V}\right)$	$D_{\min}{}_{D(X_{\max})}^{D(J)}$
SPG	*SimProp*	PSB	Gilmore	$+0.94^{+0.09}_{-0.10}$	$18.67{\pm0.03}$	$178.5^{18.8}_{159.8}$
SPD	*SimProp*	PSB	Domínguez	$-0.45{\pm0.41}$	$18.27^{+0.07}_{-0.06}$	$193.4^{21.1}_{172.3}$
STG	*SimProp*	TALYS	Gilmore	$+0.69^{+0.07}_{-0.06}$	$18.60{\pm0.01}$	$176.9^{19.5}_{157.4}$
CTG	CRPropa	TALYS	Gilmore	$+0.73^{+0.07}_{-0.06}$	$18.58{\pm0.01}$	$195.3^{33.6}_{161.7}$
CTD	CRPropa	TALYS	Domínguez	$-1.06^{+0.29}_{-0.22}$	$18.19^{+0.04}_{-0.02}$	$192.3^{21.2}_{171.1}$
CGD	CRPropa	GEANT4	Domínguez	$-1.29^{+0.38}_{-\infty?}$	$18.18^{+0.06}_{-0.04}$	$192.5^{19.2}_{173.3}$

The FD can only operate during clear moonless night (duty cycle $\approx$ 15%), but it provides us with near-calorimetric measurements of shower energies. These are used to calibrate the energy scale of the SD, which has duty cycle $\approx$ 100%. The FD also provides us with measurements of the shower maximum depth $X_{\max}$, the most important observable sensitive to the mass composition of UHECRs.

We present the result of a simple phenomenological model of UHECR sources to Pierre Auger Observatory measurements of the energy spectrum and $X_{\max}$ distributions for energies above $10^{18.7}$ eV, as a demonstration of the constraining power of Auger data. The source model is not necessarily intended to be astrophysically plausible. The data above $10^{18.7}$ eV consist of 15 bins for the energy spectrum [2] and 110 non-empty bins for the $X_{\max}$ distributions [3]. Most of these results were already presented in refs. [4, 5]. An updated version of this work will be published in ref. [6].

2 The models we used

2.1 The astrophysical sources

In this work, we assume that all UHECR sources are identical, with constant comoving density, and they emit hydrogen-1, helium-4, nitrogen-14 and iron-56 with a broken exponential rigidity cutoff, $Q_i(E_{inj}) = Q_0 p_i (E_{inj}/\text{EeV})^{-\gamma}$ for $E_{inj} \leq Z_i R_{cut}$ and $Q_0 p_i (E_{inj}/\text{EeV})^{-\gamma} \exp(1 - E_{inj}/Z_i R_{cut})$ for $E_{inj} \geq Z_i R_{cut}$. The free parameters of the fit are the normalization constant Q_0, the spectral parameters γ and R_{cut}, and three of the mass fractions p_i (the fourth being bound by $\sum_i p_i = 1$). The choice of cutoff shape is motivated by numerical convenience rather than astrophysical plausibility, but we will also show the effects of using a different cutoff shape.

2.2 The propagation through intergalactic space

We simulate the propagation of UHECRs using two publicly available Monte Carlo codes (*SimProp* v2r3 and CRPropa 3), along with two models for the extragalactic background light (EBL) spectrum and evolution (Gilmore et al. 2012 and Domínguez et al. 2011) and three models of photodisintegration cross sections (PSB, TALYS and GEANT4), in the combinations listed in table 1. An overview of the differences between the two simulation codes and the effects of different EBL and photodisintegration models can be found in ref. [7].

2.3 Interactions in the atmosphere

We model the $X_{\max}$ distribution for each primary energy and mass number as a Gumbel distribution [8] with parameter values found by fitting it to the results of CONEX simulations of air showers assuming EPOS LHC [9], SIBYLL 2.1 [10] or QGSJET II-04 [11] as the hadronic interaction model. We then multiply these distributions by the detector acceptance and convolve them by the detector resolution [3].

3 Our results

3.1 The reference fit

Using the SPG model of UHECR propagation, the EPOS LHC model of air interactions, and neglecting the systematic uncertainties in the measurements, the best fit to the measured energy spectrum and $X_{\max}$ distributions is found with a relatively hard source spectral index $\gamma \approx 1$, low cutoff rigidity $R_{\text{cut}} \approx 5$ EV (see table 1), and heavy composition (62.0% helium, 37.2% nitrogen, and 0.8% iron). Similar results have already been found by other authors, e.g. [12, 13]. The deviance (generalized χ^2) per degree of freedom of our fit is $D/n = 178.5/119$, corresponding to a p-value of 2.6%. The best-fit region extends to very low γ, R_{cut}, because, in the energy range of interest, changes in either spectral parameter can be nearly compensated by changes in the other spectral parameter and the mass composition.

In this scenario, the high-energy cut-off in the all-particle spectrum at Earth is mostly given by the photodisintegration of medium-heavy elements, whereas the injection cut-off does limit the flux of secondary protons with $E > Z_{\text{inj}}R_{\text{cut}}/A_{\text{inj}} \approx 2.4$ EeV. Since the cutoff rigidity corresponds to an energy per nucleon way below the threshold for pion production on the CMB, the resulting flux of cosmogenic neutrinos at EeV energies is negligible. Also, particles with magnetic rigidity $E/Z \lesssim 5$ EV can be deflected by intergalactic and galactic magnetic fields by several tens of degrees even when originating from relatively nearby sources [14], making it very hard to infer source positions.

There also is a second local minimum at $\gamma \approx 2$, $R_{\text{cut}} \approx 70$ EV, but due to the absence of a low rigidity cutoff this model predicts a higher admixture of protons at high energies than indicated by the narrowness of the observed $X_{\max}$ distributions.

3.2 Effects of systematic uncertainties

Most of the physical quantities relevant to the propagation of UHECRs in intergalactic space are well known, but some are still very uncertain. For example, recent models of the EBL still differ by a factor of 2 in the far infrared, and photodisintegration branching ratios have only been measured for a few channels [7]. To assess the sensitivity of our fit to these uncertainties, we repeated it using various combinations of simulation codes and EBL and photodisintegration models. The results are shown in table 1. The best-fit parameter values in the various models differ by much more than their statistical uncertainties, but they are all aligned in a hyperbola-shaped region of the (γ, R_{cut}) plane where the injection spectra in the energy range we are interested in are similar.

Details of hadronic interactions in kinematic regions relevant to air shower development are not accessible to accelerator-based measurements and extrapolations are necessary. In our reference fit we used the EPOS LHC model; using SIBYLL 2.1 or QGSJET II-04 instead, which predict shallower $X_{\max}$ values, would result in unacceptable fits even at very low γ. Note that the differences between these models may understate the actual uncertainties in hadronic interactions [15].

We also repeated the fit shifting all energy or $X_{\max}$ measurements within their measurement systematic uncertainty. The resulting best-fit deviance (as a function of γ, all other parameters being re-fitted to minimize the deviance) is shown in fig 1, left panels.

Finally, using a different shape (simple exponential) for the injection cutoff function results in different numerical values for the parameters ($\gamma = 0.53$, $R_{\text{cut}} = 10^{18.63}$ V) but they correspond to very similar injection spectra (see fig 1, right panel) with little difference in the fit deviance ($D = 177.2$).

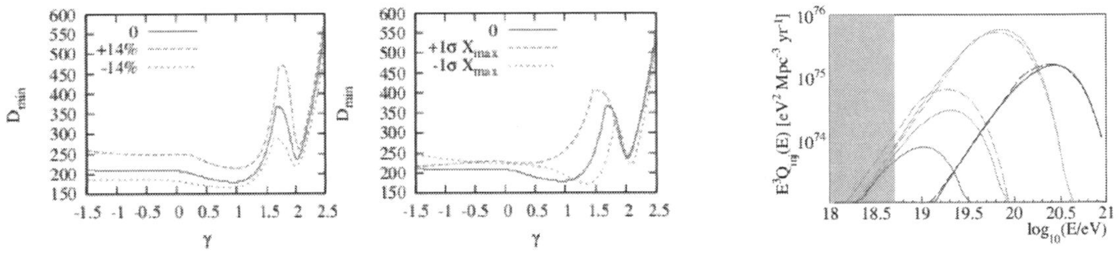

Figure 1. Left: best-fit deviance as a function of the source spectral index γ when the energy or $X_{\max}$ data are shifted by their systematic uncertainty. Right: comparison of best-fit injection spectra assuming two different cutoff shapes, showing that the differences resulting from the two models are slight (solid: simple exponential, dashed: broken exponential; red: ^{1}H, grey: ^{4}He, green: ^{14}N, blue: ^{56}Fe).

4 Discussion and conclusions

We found that our fit results are very strongly sensitive to systematic uncertainties in $X_{\max}$ predictions and measurements: shallower predictions or deeper measurements require a lower injection spectral index and cutoff rigidity and result in a worse fit. The planned upgrade AugerPrime will also measure another independent mass-sensitive observable, the muon number, hopefully helping us alleviate the uncertainties in primary mass determinations. To a lesser extent, our fit is sensitive to the interaction rates in UHECR propagation, which depend on the EBL intensity and photodisintegration cross sections: lower interaction rates tend to require higher γ and R_{cut} and result in better fits. The systematic uncertainty on the energy scale and the shape of the injection cutoff have comparatively minor impacts on the fit.

In a forthcoming work [6], we will publish an update of this fit, in which we will use the latest SD data, correctly take into account the SD energy resolution and Poisson statistics, include silicon-28 among the possible injected elements, study the effects of possible redshift evolutions of source emissivity (e.g. $\propto (1+z)^m$), and qualitatively discuss the effects of possible extra sub-ankle components.

References

[1] A. Aab et al. (Pierre Auger), Nucl. Instrum. Meth. **A798**, 172 (2015), 1502.01323
[2] A. Schulz (Pierre Auger), proc. 33rd ICRC p. 0769 (2013), 1307.5059
[3] A. Aab et al. (Pierre Auger), Phys. Rev. **D90**, 122005 (2014), 1409.4809
[4] A. di Matteo (Pierre Auger), PoS **ICRC2015**, 249 (2016), 1509.03732
[5] D. Boncioli et al. (Pierre Auger), submitted to Nucl. Phys. B Proc. Suppl. (2015), 1512.02314
[6] A. Aab et al. (Pierre Auger), in preparation (2016)
[7] R. Alves Batista et al., JCAP **1510**, 063 (2015), 1508.01824
[8] M. De Domenico, M. Settimo, S. Riggi, E. Bertin, JCAP **1307**, 050 (2013), 1305.2331
[9] T. Pierog et al., Phys. Rev. **C92**, 034906 (2015), 1306.0121
[10] E.J. Ahn et al., Phys. Rev. **D80**, 094003 (2009), 0906.4113
[11] S. Ostapchenko, Phys. Rev. **D83**, 014018 (2011), 1010.1869
[12] R. Aloisio, V. Berezinsky, P. Blasi, JCAP **1410**, 020 (2014), 1312.7459
[13] A.M. Taylor, M. Ahlers, D. Hooper, Phys. Rev. **D92**, 063011 (2015), 1505.06090
[14] R. Šmída, R. Engel, PoS **ICRC2015**, 470 (2016), 1509.09033
[15] R.U. Abbasi, G.B. Thomson (2016), 1605.05241

 DOI: 10.1051/epjconf/201713602003

The Pierre Auger Observatory Upgrade

Giovanni Marsella[1,2,a] for the Pierre Auger Collaboration[b]

[1] *Universitá degli Studi del Salento*
[2] *INFN sez. Lecce*

Abstract. It is planned to operate the Pierre Auger Observatory until at least the end of 2024. An upgrade of the experiment has been proposed in order to provide additional measurements to allow one to elucidate the mass composition and the origin of the flux suppression at the highest energies, to search for a flux contribution of protons up to the highest energies and to reach a sensitivity to a contribution as small as 10% in the flux suppression region, to study extensive air showers and hadronic multi-particle production. With operation planned until 2024, event statistics will more than double compared with the existing Auger data set, with the critical added advantage that every event will now have mass information. Obtaining additional composition-sensitive information will not only help to better reconstruct the properties of the primary particles at the highest energies, but also improve the measurements in the energy range just above the ankle. Furthermore, measurements with the new detectors will help to reduce systematic uncertainties related to the modelling hadronic showers and to limitations in the reconstruction algorithms.

A description of the principal proposed Auger upgrade will be presented. The Auger upgrade promises high-quality future data, and real scope for new physics.

1 Introduction

Since its starting time, the Pierre Auger Observatory [1] gave a strong contribution to the understanding of ultra high-energy cosmic rays (UHECRs). The principal motivations of the proposed AUGER upgrade [2] is to provide additional measurements to allow the collaboration to address the following issues:

- The origin of the flux suppression at the highest energies and the measurement of the mass composition beyond the reach of the Fluorescence Detector (FD).
- Search for a proton contribution in the flux suppression region ($E > 5\times10^{19}$eV), aiming at reaching a sensitivity to a proton contribution as small as 10% in the flux suppression region, search of point sources and estimate the physics potential of existing and future cosmic ray, neutrino, and gamma-ray detectors.
- Fundamental particle physics at energies beyond man-made accelerators and studies of extensive air showers and hadronic multiparticle production.

[a]e-mail: giovanni.marsella@le.infn.it
[b]Full author list: http://auger.org/archive/authors_2016_06.html

The proposed upgrade will consist in: (i) the addition of a plastic scintillator plane (SSD) above the existing Water-Cherenkov Detectors (WCD); (ii) the installation in the Surface Detector (SD) stations of new electronics that will process both WCD and SSD signals; (iii) the installation of an Underground Muon Detector (UMD) in the existing SD infill area; (iv) a change in the operation mode of the Fluorescence Detector (FD) to extend measurements into periods with higher night sky background.

2 Surface Scintillator Detector

The key element of the upgrade will be the installation of a new detector consisting of a plastic scintillator plane above each of the existing water-Cherenkov detectors. This scintillation detector will provide a complementary measurement of the shower particles: they will be sampled with two detectors having different responses to muons and electromagnetic particles, allowing for the reconstruction of the different shower components. The design of the surface scintillator detectors (SSDs) is simple and reliable. The SSD unit will consist of a box of 3.8 m × 1.3 m, housing two scintillator modules, each covering an area of 1.9 m^2, see Fig. 1 (left). They will be easily deployed over the full 3000 km^2 area of the overall Auger Surface Detector (SD). The 10 mm thick scintillators are read out by wavelength-shifting fibers guiding the light of the two modules to a PMT. The response over the area of the scintillator is uniform within 5%. In Fig. 1 (right) the charge distribution of signals in the SSD triggered by coincidences in the larger WCD is shown. A clean separation of Minimum Ionizing Particle (MIP) signals from the background is evidenced. An engineering array of 10 detectors has been installed at the Auger site since September 2016.

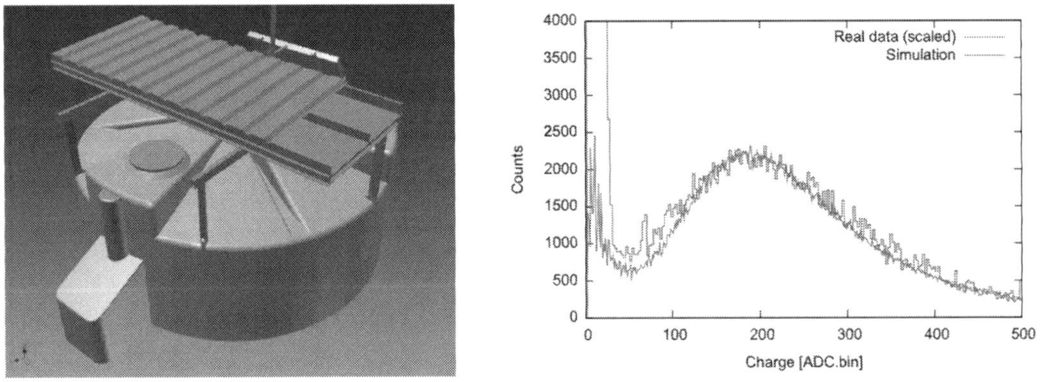

Figure 1. Left panel: 3D view of a SSD mounted on a WCD. A double roof, with the upper layer being corrugated aluminum is used to reduce the temperature variations. Right panel: Charge distribution for Minimum Ionizing Particles taken with a 2 m^2 prototype installed in the Auger array. The data correspond to one minute of data taking and are well reproduced by the detector simulation based on Geant 4.

3 Surface Detector Upgrade

The surface detector stations will be upgraded with new electronics that will process both WCD and SSD signals. It will increase the data quality with faster sampling of ADC traces (120 MHz), giving a better timing accuracy, and increased dynamic range (12 bits). To enhance the local trigger and processing capabilities a more powerful local station processor and FPGA have been implemented. This also improves calibration and monitoring capabilities of the surface detector stations. The surface detector electronics upgrade (SDEU) can be easily deployed, and will have only minimal impact on the continuous data taking of the Surface Detector. To increase the dynamic range also a 4th PMT of 1" of diameter, corresponding to 1/9 of of the WCD PMTs, will be added. With this solution the total dynamic range of the SD will correspond to 22 effective ADC bits, ranging from fractions of VEM (energy from Vertical Equivalent Muon) to 20000 VEM. This range is due to the 3 different ADC channels, each 12 bits wide and shifted by 5 bits corresponding to the relative gain, of the 30 dB amplified SD standard PMT, the 0dB SD standard PMT and the small PMT respectively. This leads to less than 2% saturated events at the highest energies and unambiguous determination of the particle density down to less than 300 m from the shower core.

4 Underground Muon Detector

An Underground Muon Detector (UMD) is being installed in the existing SD infill area of 23.5 km^2. The UMD will provide important direct measurements of the shower muon content and its time structure. It will also serve as a verification and fine-tuning of the methods used to extract the muon information with the SSD and WCD measurements. The performance and characteristics of the AMIGA project [3] match these requirements, and thus the completed AMIGA array will serve as the UMD. 61 muon detectors (30 m^2) are planned and will be deployed on a 750 m spaced grid. The muon detectors are shielded by 1.3 m of soil.

5 Fluorescence Detector Upgrade

The Fluorescence Detector [4] provides exceptional information such as model-independent energy reconstruction and mass composition measurement. The main limitation of the FD is its duty cycle that nowadays reaches 15%. The current criteria for FD measurement are:

- The sun more than 18° below the horizon
- The moon remains below horizon for longer than 3 hours
- The illuminated fraction of the moon must be below 70%

The operation mode of the Fluorescence Detector (FD) will be changed to extend measurements into periods with higher night sky background by lowering the PMT HV supply thus reducing the PMT gain by a factor of ten. A successful test has already been done over a period of seventeen nights. This will allow to increase the current duty cycle of the FD by about 50%.

The combination of SD and FD upgrades are also expected to increase the sensitivity to the flux of photons and neutrinos (see Fig. 2).

6 Conclusions

The AUGER upgrade will allow to study the mass composition above 5×10^{19} eV. The different proposed models, GZK energy loss model versus a picture defining the maximum energy of sources, will be analyzed with a higher and more precise statistics. The sensitivity to the proton flux above 5×10^{19} eV will be enhanced to reach a sensitivity to a contribution as small as 10% . Finally the upgrade will enhance the possibility of studying new particle physics beyond the reach of LHC, due to the combination of detectors allowing for a better particle identification. Actually 12 Engineering

DOI: 10.1051/epjconf/201713602003

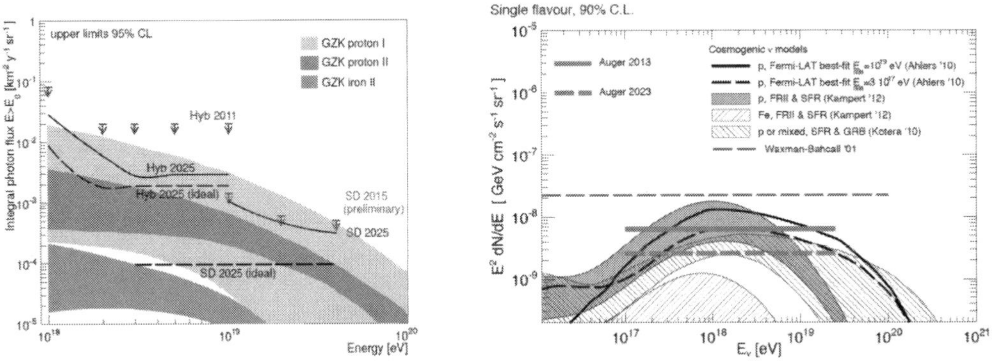

Figure 2. Expected sensitivity on the flux of photons and neutrinos. In addition to the conservative estimates based on the increase of statistics, also the projected photon sensitivity for the ideal case of being able to reject any hadronic background due to the upgraded surface detector array is shown [2] .

Array stations have been installed and are starting taking data. In early 2017 the detector evaluation will be completed and at the beginning of 2018 the deployment of the AUGER upgrade will start. The detector is expected to take data at least until 2024 with an exposure up to 40000 km^2 sr yr, allowing to collect a statistics comparable with the one reached so far by the AUGER experiment.

References

[1] **Pierre Auger** Collaboration, A. Aab et al., *The Pierre Auger Cosmic Ray Observatory, Nucl. Instrum. Meth.* **A 798**, 172-213 (2015) [http://arxiv.org/abs/1502.013231502.01323].

[2] **Pierre Auger** Collaboration, A. Aab et al., *The Pierre Auger Observatory Upgrade*, [http://arxiv.org/abs/1604.036371604.03637].

[3] **Pierre Auger** Collaboration, B. Wundheiler et al., *The AMIGA Muon Counters of the Pierre Auger Observatory: Performance and Studies of the Lateral Distribution Function, Proc. of 34th Int. Cosmic Ray Conf., The Hague, PoS(ICRC2015)324* (2015).

[4] **Pierre Auger** Collaboration, J. Abraham et al., *The Fluorescence Detector of the Pierre Auger Observatory, Nucl. Instrum. Meth.* **A620** (2010) 227–251, [http://arxiv.org/abs/0907.42820907.4282].

DOI: 10.1051/epjconf/201713602004

The AMS-02 detector on the International Space Station - The status after the first 5 years on orbit

Matteo Duranti[1,2,a] *on behalf of the* AMS Collaboration

[1] *INFN Sezione Perugia*
[2] *Università degli Studi di Perugia*

Abstract. The Alpha Magnetic Spectrometer, AMS-02, detector is operating on the International Space Station (ISS) since May the 19th, 2011. More than 80 billion events have been collected by the instrument in the first 5 years of data taking. This unprecedented amount of data is being used to perform accurate measurements of the different Cosmic Rays (CR) components. In this contribution a review of the published results will be presented.

1 The AMS experiment and its scientific objectives

The Alpha Magnetic Spectrometer, AMS-02, is a general purpose high energy particle physics detector. It was launched to space with the Space Shuttle STS-134 mission and installed onboard the ISS in 2011, on May $19th$, to conduct a unique long duration mission (up to the lifetime of the ISS) of fundamental physics research in space. More than 80 billion events have been collected by the instrument in the first 5 years of data taking.
The experimental challenge of the experiment is the accurate measurement of the Charged Cosmic Rays (CCR) composition and energy spectra, up to the TeV scale, that could reveal the presence of primordial anti-matter or give the signature of 'exotic' sources. An accurate and high statistics measurement of all the various components is needed to have a better and coherent picture describing the observed CR phenomenology.
A good example to show the need of the full set of measurements is the indirect search of Dark Matter (DM). The annihilation of DM particles can produce matter-antimatter pairs (electron-positron, $e^+ - e^-$, proton-antiproton, $p - \bar{p}$, etc...). These particles propagate through the Galaxy mixing up with the *standard* particles and can be revealed as excesses with respect to the expected fluxes. Rarer is the standard CR component (for example e^+ and $\bar{p}$) and more statistical significant is any observed excess. The 'natural' candidate measurements where to search for excesses are the positron fraction, $e^+/(e^- + e^+)$, and the antiproton to proton ratio, $\bar{p}/p$.
To search for an excess with respect to an expected spectrum is obviously crucial to constrain as much as possible the models used to predict the purely primary or secondary spectra. The precise knowledge of the electron flux and its spectral features, for example, is mandatory to understand the features in the positron fraction, while the measurement of the proton flux and of the nuclear components (for

[a] e-mail: matteo.duranti@pg.infn.it

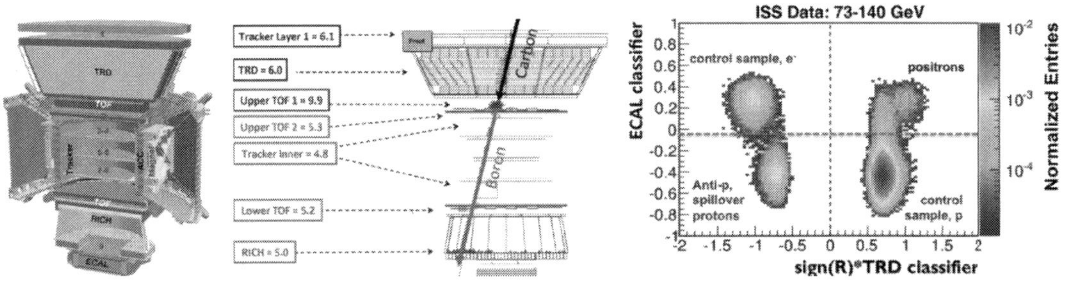

Figure 1. Left - Schematic view of the AMS-02 spectrometer. Center - A Carbon nucleus, fragmenting into a Boron one, as identified by the AMS detector on the ISS in the bending (y-z) plane. Tracker planes measure the particle charge and momentum. The TRD measure the charge of the particle. The TOF measures the charge and ensures that the particle is downward-going. The RICH independently measures the charge and velocity. The discrepancy between the charge in the first layer of the Tracker and in the TRD with the one in the upper part of the TOF and in the remaining detectors, clearly identify the event as a Carbon nucleus fragmenting into a Boron one. Right - Distribution of the events, in the energy range 73-140 GeV, in the *ECAL classifier - TRD classifier* plane. Negative from positive events are distinguished by means of the sign of the Rigidity in the spectrometer, *sign(R)*. The separation of the various components of e^-, e^+, p and $\bar{p}$ is clearly visible.

example the Boron/Carbon) is needed to predict the spectrum of the secondary positrons (from the collision of the CRs with the ISM) and their propagation through the Galaxy.
Unfortunately, at low energy, the spectra are also deformed by the effect of the solar wind. The solar modulation models are still quite rough or under development ([1–3]) and so a large campaign of measurement, for example of the fluxes as a function of time, can widely improve the models and the overall picture.
To reach its scientific goals, the AMS-02 instrument, whose schematic layout is reported in Fig.1 (left), has been conceived with a large redundancy in the measurement of particle properties with complementary techniques by different sub detectors. The apparatus is built around a magnetic spectrometer made of nine planes of precision silicon tracker and a permanent magnet. A Transition Radiation Detector (TRD), four planes of Time of Flight counters (TOF), an array of Anti-Coincidence Counters (ACC) surrounding the inner tracker, a Ring Imaging CHerenkov detector (RICH), and an Electromagnetic CALorimeter (ECAL) add completeness and redundancy to the whole detector.
Details about the performances of the different sub-detectors can be found in the first published result [4] and in the reference therein.
The identification of the various nuclear components is granted by the measurement of the charge of the impinging particles, all along the detector. Essentially all the sub-detectors have some charge measurement capability: the most accurate is the inner tracker, providing a $\sim$ 0.1 c.u. charge resolution, with its 7 layers (each layer, alone, has a resolution of about 0.3 c.u.); the RICH and the TRD have a resolution of $\sim$ 0.3 c.u., while each pair of planes of the TOF has a resolution of $\sim$ 0.16 c.u.. This permits to follow the particle traversing the detector and to search for fragmentation that could spoil the charge reconstruction and constitute a source of background in the less abundant nuclear components. This is sketched in Fig.1 (center): a Carbon nucleus fragments into a Boron one interacting with the material of the upper pair of planes of the TOF (Upper TOF). The discrepancy between the charge measured by the first layer of the Tracker, L1, and the TRD, $\sim$ 6, the one in the Upper TOF, $\sim$ 10 (due to the large energy deposit of the interaction), and the one in all the remaining detectors, $\sim$ 5, is a clear signature of fragmentation. The dynamic range of the various detectors permits, with a resolution slightly degrading with Z, the charge measurement up to the higher charges: Iron and

above (the Zinc peak is clearly visible) for Tracker, TOF and RICH, and up to $Z \sim 6$ for the TRD. The discrimination power to the rare electromagnetic components, e^+ and e^-, from the overwhelming background of protons, and from the even rarer antiprotons component is sketched in Fig.1 (right). The two key detectors to distinguish e^+/e^- from $\bar{p}/p$ are the ECAL (via a *classifier* exploiting the 3D imaging capability of the detector and built using a BDT approach [5]), and the TRD (via a *log-likelihood* to the energy deposit in its straw tubes). The discrimination between matter and anti-matter (i.e. e^+ from e^- or $\bar{p}$ from p), is provided by the spectrometer. The redundancy is also a key capability that permits to build, for example using the TRD classifier and the sign of the rigidity, almost pure control samples of e^- and p, to study the performances of the ECAL.
The redundancy and the complementarity of the various sub-detectors of the apparatus, give to the experiment a full coverage of the CCR, both in terms of composition and spectral features.

2 Experimental results

The analysis of the enormous amount of data being collected by the experiment, permitted the publication of high statistics and very accurate measurement of the more important CCR components:

- $e^+/(e^+ + e^-)$ in the range 0.5-500 GeV, [4, 6], based on 30 months and ~ 10.9 million positron and electron events;
- positron flux in the range 0.5-500 GeV and electron flux in the range 0.5-700 GeV, [7], based on 30 months and ~ 0.6 million positrons and ~ 9.2 million electrons ;
- *all-electron*, $e^+ + e^-$, flux, in the range 0.5-1000 GeV, [8], based on 30 months and ~ 10.6 million positron and electron events;
- p flux in the range 1-1800 GV, [9], based on 30 months and ~ 300 million protons;
- He flux in the range 1.9-3000 GV, [10], based on 30 months and ~ 50 million helium events;
- antiproton flux and $\bar{p}/p$ ratio in the range 1-450 GV, [11], based on the first 4 years of the collected data, corresponding to ~ 350000 antiproton and ~ 2.4 billion proton events;

The high statistics, low systematics, measurement of the e^+ and e^- components, see Fig.2, allowed the determination, with very high accuracy, of the spectral features of the $e^+/(e^+ + e^-)$ and of the fluxes of the single components:

- the accuracy of the measurement of the high energy rise in the positron fraction (already seen by PAMELA, [12], and FERMI, [13]), allowed the systematic study of the slope, that was found to be consistent with a zero crossing, i.e. with a maximum in the positron fraction, at ~ 275 GeV.
- both the electron and positron fluxes have a spectral index increasing with energy after few GeV's. The *hardening* of both spectra gives the hint that the rise in the positron fraction is not due to a lack of high energy electrons, since their spectrum is hardening, but to a fresh source of positrons;
- the absence of structures in the all-electron flux. The flux is smooth and well described by a single power law, after ~ 30 GeV and doesn't show the peak claimed by ATIC, [14].

The measurement of the proton and helium fluxes, instead, not only permitted the confirmation of the spectral features already seen by PAMELA, [15], but, given the unprecedented accuracy, permitted a systematic study of the change in the spectral indexes, allowing the determination of the $\Delta\gamma$'s and of the transition rigidity values, R_0, reported in Fig.3.
The measurement of the $\bar{p}/p$, finally, widely extended the experimental energy reach and the accuracy of the previous measurements, [16, 17] and resulted, at high energies, in a ratio almost independent from energy.

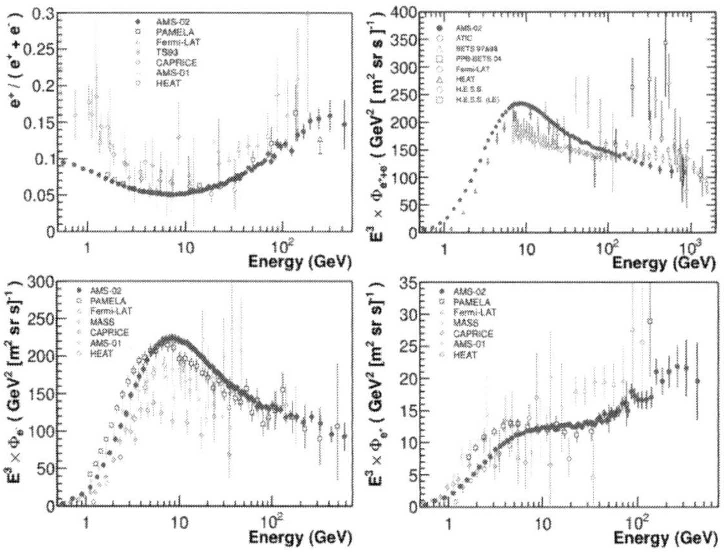

Figure 2. Positron fraction in the energy range 0.5 - 500 GeV (top-left), all-electron flux in the range 0.5 - 1000 GeV (top-right), electron flux in the range 0.5 - 700 GeV (bottom-left) and positron flux in the range 0.5 - 700 GeV (bottom-right), as measured by the AMS02 experiment. The results from previous experiments are reported for comparison, [12, 14, 18–28].

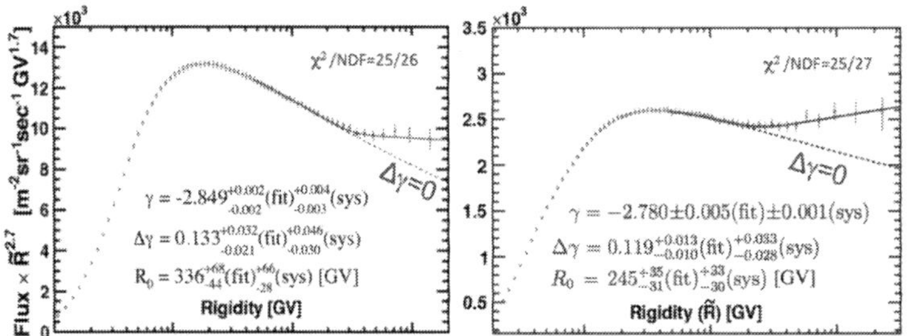

Figure 3. Left - Protons flux as measured by AMS-02 up to 1.8 TeV. Right - Helium flux up to 3 TeV. The main parameters coming from the study of the spectral features are reported, together with the fit to the experimental points.

The big amount of collected data is being analyzed for the measurements of the other nuclear components (Li, B, C, O fluxes, B/C, C/O, etc..) and for the study of the other characteristics of the already published spectra (for example the variability of the electron, positron, proton, helium and antiproton fluxes and ratios as a function of time).

3 Conclusions

Data recorded in the first $\sim$ 5 years of mission are being analyzed to cover the physics program of the AMS-02 experiment: all the measurements performed so far show interesting features, not expected

 DOI: 10.1051/epjconf/201713602004

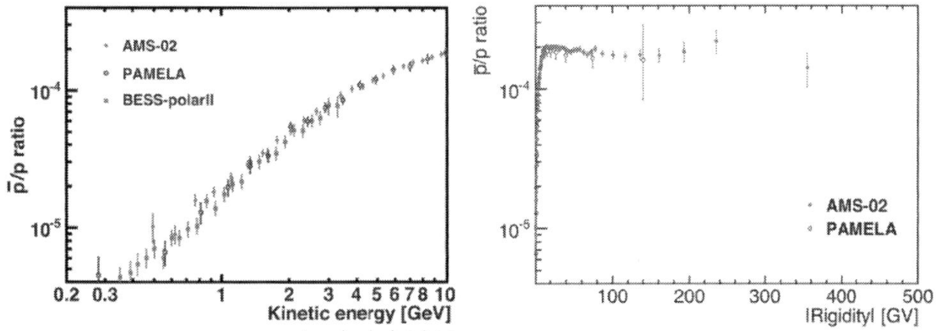

Figure 4. $\bar{p}/p$ ratio as a function of the Rigidity (on the right) and of the Kinetic Energy (for low energies, on the left). At high rigidities the ratio is almost independent from the energy. Results from previous experiments reported for comparison, [16, 17, 29].

by the standard model of CRs, extending previous observations and adding new findings whose experimental accuracy challenges the actual theoretical predictions. AMS-02 represents the first instrument with the capability to simultaneously study all the different nuclear CR species, including the less abundant ones and up to Iron, electrons and anti-matter with unprecedented accuracy and in an extended energy range. These unique features will allow AMS-02 to shed light on new phenomena and improve the current understanding of CR origin and propagation.

Acknowledgments

This work has been supported by acknowledged persons and institutions in the AMS-02 publications [4, 6–11], as well as by the Italian Space Agency under ASI-INFN Agreement No. 2013-002-R.0.

References

[1] L.J. Gleeson, W.I. Axford, Astrophysical Journal **154**, 1011 (1968)

[2] M.S. Potgieter, Living Reviews in Solar Physics **10**, 3 (2013)

[3] M. Gervasi, P. Rancoita, I. Usoskin, G. Kovaltsov, Nuclear Physics B - Proceedings Supplements **78**, 26 (1999)

[4] M. Aguilar, G. Alberti, B. Alpat, A. Alvino, G. Ambrosi, K. Andeen, H. Anderhub, L. Arruda, P. Azzarello, A. Bachlechner et al. (AMS Collaboration), Phys. Rev. Lett. **110**, 141102 (2013)

[5] B.P. Roe, H.J. Yang, J. Zhu, Y. Liu, I. Stancu, G. McGregor, Nuclear Instruments and Methods in Physics Research Section A: Accelerators, Spectrometers, Detectors and Associated Equipment **543**, 577 (2005)

[6] L. Accardo, M. Aguilar, D. Aisa, B. Alpat, A. Alvino, G. Ambrosi, K. Andeen, L. Arruda, N. Attig, P. Azzarello et al. ((AMS Collaboration)), Phys. Rev. Lett. **113**, 121101 (2014)

[7] M. Aguilar, D. Aisa, A. Alvino, G. Ambrosi, K. Andeen, L. Arruda, N. Attig, P. Azzarello, A. Bachlechner, F. Barao et al. ((AMS Collaboration)), Phys. Rev. Lett. **113**, 121102 (2014)

[8] M. Aguilar, D. Aisa, B. Alpat, A. Alvino, G. Ambrosi, K. Andeen, L. Arruda, N. Attig, P. Azzarello, A. Bachlechner et al. ((AMS Collaboration)), Phys. Rev. Lett. **113**, 221102 (2014)

[9] M. Aguilar, D. Aisa, B. Alpat, A. Alvino, G. Ambrosi, K. Andeen, L. Arruda, N. Attig, P. Azzarello, A. Bachlechner et al. (AMS Collaboration), Phys. Rev. Lett. **114**, 171103 (2015)

[10] M. Aguilar, D. Aisa, B. Alpat, A. Alvino, G. Ambrosi, K. Andeen, L. Arruda, N. Attig, P. Azzarello, A. Bachlechner et al. (AMS Collaboration), Phys. Rev. Lett. **115**, 211101 (2015)
[11] M. Aguilar, L. Ali Cavasonza, B. Alpat, G. Ambrosi, L. Arruda, N. Attig, S. Aupetit, P. Azzarello, A. Bachlechner, F. Barao et al. (AMS Collaboration), Phys. Rev. Lett. **117**, 091103 (2016)
[12] O. Adriani, G.C. Barbarino, G.A. Bazilevskaya, R. Bellotti, M. Boezio, E.A. Bogomolov, L. Bonechi, M. Bongi, V. Bonvicini, S. Bottai et al., Nature **458**, 607 (2009)
[13] M. Ackermann, M. Ajello, A. Allafort, W.B. Atwood, L. Baldini, G. Barbiellini, D. Bastieri, K. Bechtol, R. Bellazzini, B. Berenji et al. (Fermi LAT Collaboration), Phys. Rev. Lett. **108**, 011103 (2012)
[14] J. Chang, J.H. Adams, H.S. Ahn, G.L. Bashindzhagyan, M. Christl, O. Ganel, T.G. Guzik, J. Isbert, K.C. Kim, E.N. Kuznetsov et al., Nature **456**, 362 (2008)
[15] O. Adriani, G.C. Barbarino, G.A. Bazilevskaya, R. Bellotti, M. Boezio, E.A. Bogomolov, L. Bonechi, M. Bongi, V. Bonvicini, S. Borisov et al., Science **332**, 69 (2011)
[16] O. Adriani, G.A. Bazilevskaya, G.C. Barbarino, R. Bellotti, M. Boezio, E.A. Bogomolov, V. Bonvicini, M. Bongi, L. Bonechi, S.V. Borisov et al., Soviet Journal of Experimental and Theoretical Physics Letters **96**, 621 (2013)
[17] O. Adriani, G.C. Barbarino, G.A. Bazilevskaya, R. Bellotti, M. Boezio, E.A. Bogomolov, L. Bonechi, M. Bongi, V. Bonvicini, S. Bottai et al., Physical Review Letters **102**, 051101 (2009), 0810.4994
[18] O. Adriani, G.C. Barbarino, G.A. Bazilevskaya, R. Bellotti, M. Boezio, E.A. Bogomolov, M. Bongi, V. Bonvicini, S. Borisov, S. Bottai et al., Phys. Rev. Lett. **106**, 201101 (2011)
[19] M. Ackermann, M. Ajello, W.B. Atwood, L. Baldini, J. Ballet, G. Barbiellini, D. Bastieri, B.M. Baughman, K. Bechtol, F. Bellardi et al. (Fermi LAT Collaboration), Phys. Rev. D **82**, 092004 (2010)
[20] J.J. Beatty, A. Bhattacharyya, C. Bower, S. Coutu, M.A. DuVernois, S. McKee, S.A. Minnick, D. Müller, J. Musser, S. Nutter et al., Phys. Rev. Lett. **93**, 241102 (2004)
[21] F. Aharonian, A.G. Akhperjanian, U. Barres de Almeida, A.R. Bazer-Bachi, Y. Becherini, B. Behera, W. Benbow, K. Bernlöhr, C. Boisson, A. Bochow et al. (H.E.S.S. Collaboration), Phys. Rev. Lett. **101**, 261104 (2008)
[22] Aharonian, F., Akhperjanian, A. G., Anton, G., Barres de Almeida, U., Bazer-Bachi, A. R., Becherini, Y., Behera, B., Bernlöhr, K., Bochow, A., Boisson, C. et al., A&A **508**, 561 (2009)
[23] R.L. Golden, S.J. Stochaj, S.A. Stephens, F. Aversa, G. Barbiellini, M. Boezio, U. Bravar, A. Colavita, F. Fratnik, P. Schiavon et al., Astrophysical Journal Letters **457**, L103 (1996)
[24] M. Boezio, P. Carlson, T. Francke, N. Weber, M. Suffert, M. Hof, W. Menn, M. Simon, S.A. Stephens, R. Bellotti et al., The Astrophysical Journal **532**, 653 (2000)
[25] M. Aguilar, J. Alcaraz, J. Allaby, B. Alpat, G. Ambrosi, H. Anderhub, L. Ao, A. Arefiev, P. Azzarello, L. Baldini et al., Physics Letters B **646**, 145 (2007)
[26] S. Torii, T. Tamura, N. Tateyama, K. Yoshida, J. Nishimura, T. Yamagami, H. Murakami, T. Kobayashi, Y. Komori, K. Kasahara et al., The Astrophysical Journal **559**, 973 (2001)
[27] K. Yoshida, S. Torii, T. Yamagami, T. Tamura, H. Kitamura, J. Chang, I. Iijima, A. Kadokura, K. Kasahara, Y. Katayose et al., Advances in Space Research **42**, 1670 (2008)
[28] C. Grimani, S.A. Stephens, F.S. Cafagna, G. Basini, R. Bellotti, M.T. Brunetti, M. Circella, A. Codino, C. De Marzo, M.P. De Pascale et al., Astronomy and Astrophysics **392**, 287 (2002)
[29] K. Abe, H. Fuke, S. Haino, T. Hams, M. Hasegawa, A. Horikoshi, K.C. Kim, A. Kusumoto, M.H. Lee, Y. Makida et al., Physical Review Letters **108**, 051102 (2012), 1107.6000

DOI: 10.1051/epjconf/201713602005

Nuclei Measurements with the Alpha Magnetic Spectrometer on the International Space Station

Melanie Heil[1a]

[1]*Massachusetts Institute of Technology, 77 Massachusetts Avenue, Cambridge, MA 02139-4307*

Abstract. The exact behavior of nuclei fluxes in cosmic rays and how they relate to each other is important for understanding the production, acceleration and propagation mechanisms of charged cosmic rays. Precise measurements with the Alpha Magnetic Spectrometer on the International Space Station of light nuclei fluxes and their ratios in primary cosmic rays with rigidities from GV to TV are presented. The high statistics of the measurements require detailed studies and in depth understanding of associated systematic uncertainties.

1 Introduction

The Alpha Magnetic Spectrometer (AMS) is a multi-purpose magnetic spectrometer measuring cosmic rays up to TeV energies on the International Space Station (ISS). Its precision, large acceptance and ability to identify particle types over a large energy scale during its long duration mission make it unique in astro-particle physics. The AMS detector [1] consists of a transition radiation detector (TRD) [2], a silicon tracker [3], a permanent magnet, a Time-of-Flight detector (TOF) [4], a Ring Imaging Cherenkov detector (RICH) [5] and an electromagnetic calorimeter (ECAL) [6]. With these detectors AMS identifies particles and nuclei by redundant measurements of their charge and energy or momentum. In its first 5 years onboard the ISS AMS has recorded over 80 billion events. To match these statistics, detailed systematic error studies are important.

2 Particle Identification

Particle types are identified by charge measurements along the particle trajectory inside AMS. Redundant measurements by tracker, TOF, TRD, RICH and ECAL allow for precise particle identification. The best charge resolution is achieved by the combination of the seven inner tracker planes, which give a charge resolution of $\Delta Z = 0.05$ for protons and $\Delta Z = 0.07$ for Helium.

The TOF measures the particle velocities with a resolution of $\Delta\beta/\beta^2$ of 4% for protons and 2% for Helium. It thereby discriminates between upward and downward going particles. For our flux measurements only downward going particles are selected.

[a] Corresponding author: melanie.heil@cern.ch

The tracker together with the magnet determines the rigidity (momentum/charge) of the particles. The coordinate resolution in the bending direction is 10 μm and 7.5 μm for protons and helium, respectively. With this resolution we obtain a maximum detectable rigidity of 2 TV for protons and 3.2 TV for Helium using the full 3m lever arm of the tracker. In order to increase the statistics of the measurement for Z>2 nuclei, for rigidities below 1 TV we also analysed particles crossing only tracker planes L1 to L8 by which we gain a significant increase in acceptance.

3 Flux Measurements and Results

The isotropic particle or nuclei flux Φ_i for the ith rigidity bin (R_i , $R_i + \Delta R_i$) is

$$\Phi_i = N_i / (A_i \, \varepsilon_i \, T_i \, \Delta R_i) \tag{1}$$

where N_i is the number of events corrected for the bin-to-bin migration with the rigidity resolution function, A_i is the effective acceptance, ε_i is the trigger efficiency, and T_i is the collection time. The AMS fluxes [7,8] and flux ratios [9] were measured in bins chosen according to the tracker resolution function and available statistics. While the proton and Helium flux where determined with data collected in the first 30 month of data taking of AMS on the ISS, the Boron-to-Carbon ratio (B/C) was measured based on events collected in the first 5 years.

The effective acceptance A_i was calculated using Monte Carlo samples and then corrected for small differences found between the data and Monte Carlo event selection efficiencies, such as the efficiencies of track and beta measurement quality cuts. The trigger efficiency ε is measured from data with the unbiased trigger events. The trigger efficiency of protons ranges from 90 to 95% [7], while the trigger efficiency for helium and higher charge nuclei is well above 95% [8,9]. The bin-to-bin migration of events was corrected using the rigidity resolution functions obtained from Monte Carlo simulations [7-9].

Extensive studies were made of the systematic errors. The errors include the uncertainties in the trigger efficiency, the acceptance, the background contamination also accounting for interactions in the detector, the geomagnetic cutoff factor [10], the event selection, the unfolding, the rigidity resolution function, and the absolute rigidity scale.

The trigger efficiency error is dominated by the statistics available from the 1% prescaled unbiased event sample [7-9]. It is negligible (less than 0.2%) below 500 GV and reaches 1.5% at 1.8 TV for protons and 1% at 3 TV for helium and nuclei. The geomagnetic cutoff factor was varied from 1.0 to 1.4 and the resulting fluxes showed a systematic uncertainty of 2% at 1 GV and negligible above 2 GV.

As discussed above, the effective acceptance was corrected for small differences between the data and the Monte Carlo samples related to the event reconstruction and selection. The total corrections are less than 5% over the whole rigidity range, while the corresponding systematic uncertainties are less than 1% above 2 GV for protons and less than 2 % for helium and nuclei.

To accurately determine the acceptance of protons, helium and nuclei, their interactions with the detector materials have to be studied. The detector is mostly made of carbon and aluminum. For protons the corresponding inelastic cross-sections of p+Cand p+Al are known within better than 10%. To estimate the systematic error of the proton flux due to the uncertainty in the inelastic cross sections, dedicated samples of protons were simulated with the p + C and p+Al cross sections varied by ±10%. From the analysis of these samples together with the current knowledge of the cross sections, a systematic error of less than 1% was obtained. The inelastic cross sections of nuclei (Z>1) with Carbon and Aluminum have, if at all, only been measured below 10 GV. To accurately determine the effect on the acceptance of nuclei interactions in the detector, we have developed a method to determine the magnitude and rigidity dependence of the survival probability of the nuclei when traversing the detector materials. For this we used a sample of primary cosmic rays collected with AMS horizontal, that is, when the ISS was oriented such that AMS was pointing within 90° ± 10° of the local zenith [8]. Using the measured interaction probabilities and, in addition, above 100 GV the rigidity dependence of the cross sections from the Glauber-Gribov model, the systematic error on the flux due to uncertainties of nuclei inelastic cross sections was evaluated to be below 2% over the entire rigidity range.

The rigidity resolution function for protons was verified with data from both the ISS and the test beam and compared to the simulated events [7]. The rigidity resolution function for Z>1 nuclei is similar to that of protons. It was obtained from the simulations and extensively verified with the data [8, 9]. First, the differences of the coordinates measured in the inner tracker layers to those obtained from the track fit excluding the tested layer were compared between data and simulation. Second, the differences between the coordinates measured in L1 and L9 and those obtained from the track fit using the information from only the inner tracker were compared between events in data and simulation. The systematic errors on the fluxes due to uncertainties in the rigidity resolution function were obtained by varying the width of the Gaussian core of the resolution function and the amplitude of the non-Gaussian tails by their corresponding uncertainty over the entire rigidity range in the unfolding procedures. They were found to be small below 400 GV and reached 3.5% to 5% at the highest measured rigidities.

There are two contributions to the systematic uncertainty on the rigidity scale. The first is due to residual tracker misalignment. For the ISS data this error was estimated by comparing the E/p ratio for electron and positron events, where E is the energy measured with the ECAL and p is the momentum measured with the tracker [7]. It was found to be 1/26 TV^{-1}, limited by the current high-energy positron statistics. The second systematic error on the rigidity scale arises from the magnetic field map measurement and its temperature corrections and amounts to less than 0.5% for rigidities above 2 GV.

The contributions of individual sources to the systematic errors are added in quadrature to arrive at the total systematic uncertainty of the proton and helium flux. The Monte Carlo event samples have sufficient statistics such that they do not contribute to the errors.

To ensure that the treatment of systematic errors described above is correct, we performed several additional independent verifications. We checked that there is no dependence of the integral of the fluxes above 30 GV, *i.e.*, above the maximum geomagnetic cutoff, on the angle θ between the incoming particle direction and the AMS z-axis; this verifies the systematic errors assigned to the acceptance. We verified that the monthly integral fluxes above 45 GV are within the systematic errors; this verifies that the detector performance is stable over time. We checked that the ratios of fluxes obtained using events which pass through different sections of L1 to the average flux is in good agreement and within the assigned systematic errors; this verifies the errors assigned to the tracker alignment. Lastly, we verified that the fluxes obtained using the rigidity measured by only the inner tracker are in good agreement with the fluxes measured using the full lever arm. The flux ratios use the two different event samples corresponding to the inner tracker acceptance and to the L1 to L9 acceptance used for the results presented here. This verifies the systematic errors from the acceptance, the unfolding procedure, and the rigidity resolution function.

Figure 1(a) shows the proton flux as a function of rigidity with the total errors [7]. In this and the subsequent figures, the points are placed along the abscissa at the bins mean rigidity $\tilde{R}$ calculated for a flux $\propto R^{-2.7}$. Figure 1(b) shows the helium flux as a function of rigidity with the total errors [8].

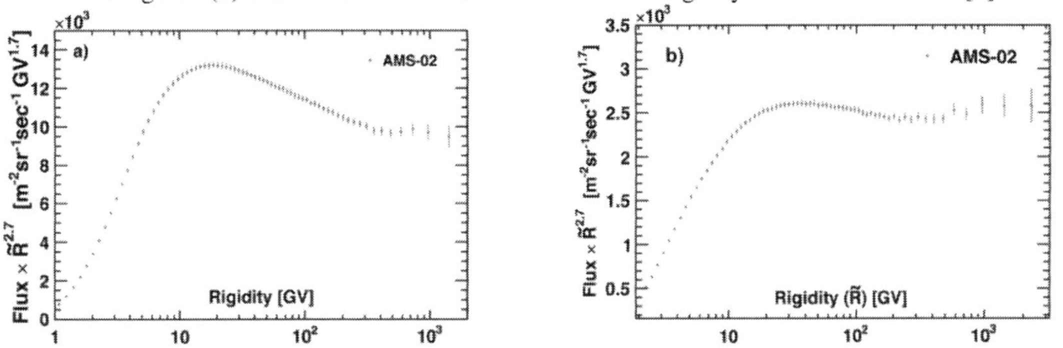

Figure 1. a) AMS proton flux and b) AMS Helium Flux, both multiplied by $R^{-2.7}$ for illustration purposes.

A power law with a constant spectral index does neither fit the proton flux nor the helium flux at the 99.9% C.L. for $R > 45$ GV. Instead a description with a double power law and a smooth transition between the two power laws describes the data well with the transition rigidity between 250 GV and 350 GV. Figure 2 a) shows the spectral indexes of the proton and the helium flux as a function of independent rigidity intervals and Figure 2 b) show the proton/helium ratio as a function of rigidity together with a single power law fit above 45 GV.

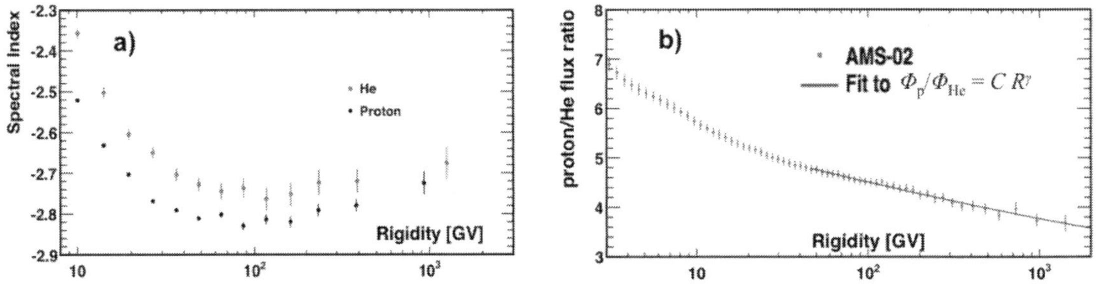

Figure 2. a) The spectral indexes of the AMS proton and helium fluxes. b) The AMS proton/helium ratio with a single power law fit above 45 GV.

Figure 3 shows the AMS result on B/C [9] as a function of rigidity with total errors. The B/C ratio is well described by a single power law above 65 GV with an index of -0.333.

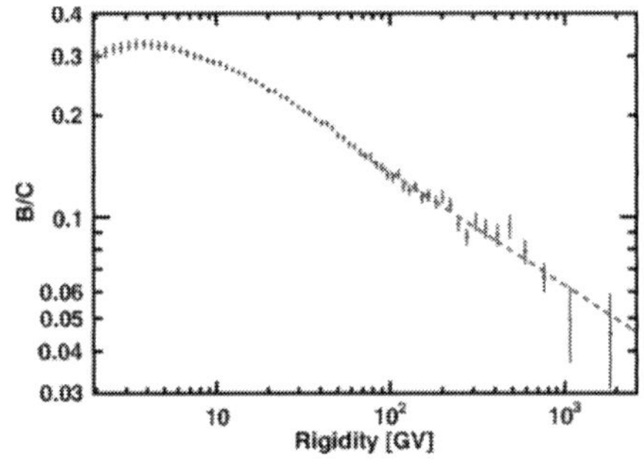

Figure 3. The AMS B/C ratio as a function of rigidity with a single power law fit above 65 GV.

4 Conclusions

The AMS results for protons, helium and B/C are reaching an unprecedented accuracy of ~1% for cosmic ray measurements. These precise measurements of different types of cosmic rays require a comprehensive model to describe all of their behaviour at the same time.

References

1. M. Aguilar *et al.*, Phys.Rev.Lett. **110,** 141102 (2013), and references therein.
2. Th. Kirn *et al.*, Nucl. Instrum. Methods Phys. Res., Sect. A **706**, 43 (2013).
3. B. Alpat *et al.*, Nucl. Instrum. Methods Phys. Res., Sect. A **613**, 207 (2010).
4. V. Bindi *et al.*, Nucl. Instrum. Methods Phys., Sect. A **743**, 22 (2014).
5. M. Aguilar-Benitez *et al.*, Nucl. Instrum. Methods Phys. Res., Sect. A **614**, 237 (2010).
6. C. Adloff *et al.*, Nucl. Instrum. Methods Phys. Res., Sect. A **714**, 147 (2013).
7. M. Aguilar *et al.*, Phys.Rev.Lett. **114,** 171103 (2015), and references therein.
8. M. Aguilar *et al.*, Phys.Rev.Lett. **115,** 211101 (2015), and references therein.
9. M. Aguilar *et al.*, Phys.Rev.Lett. **117,** 231102 (2016), and references therein.
10. D.F. Smart, M. A. Shea, Advances in Space Research **36,** 2012–2020 (2005).

The proton and helium anomalies in the light of the Myriad model

Pierre Salati [a,1,b], Yoann Génolini [1,c], Pasquale Serpico[1], and Richard Taillet[1]

[1] *LAPTh, Université de Savoie Mont Blanc, CNRS, 9 Chemin de Bellevue, B.P.110, Annecy-le-Vieux F-74941, France*

Abstract. A hardening of the proton and helium fluxes is observed above a few hundreds of GeV/nuc. The distribution of local sources of primary cosmic rays has been suggested as a potential solution to this puzzling behavior. Some authors even claim that a single source is responsible for the observed anomalies. But how probable these explanations are? To answer that question, our current description of cosmic ray Galactic propagation needs to be replaced by the Myriad model. In the former approach, sources of protons and helium nuclei are treated as a jelly continuously spread over space and time. A more accurate description is provided by the Myriad model where sources are considered as point-like events. This leads to a probabilistic derivation of the fluxes of primary species, and opens the possibility that larger-than-average values may be observed at the Earth. For a long time though, a major obstacle has been the infinite variance associated to the probability distribution function which the fluxes follow. Several suggestions have been made to cure this problem but none is entirely satisfactory. We go a step further here and solve the infinite variance problem of the Myriad model by making use of the generalized central limit theorem. We find that primary fluxes are distributed according to a stable law with heavy tail, well-known to financial analysts. The probability that the proton and helium anomalies are sourced by local SNR can then be calculated. The p-values associated to the CREAM measurements turn out to be small, unless somewhat unrealistic propagation parameters are assumed.

1 Gambling with the discreteness of cosmic ray sources

The proton and helium spectra are well described by a power-law distribution up to an energy of ~ 350 GeV/nuc, above which a hardening is observed. This putative anomaly was reported by the PAMELA collaboration [1] and has been recently confirmed by the precision AMS-02 measurements [2, 3]. The proton and helium fluxes measured by the CREAM balloon borne detector [4] are clearly in excess of a simple power-law behavior.

Various explanations have been proposed to account for the observed hardening, such as the existence of distinct populations of primary cosmic ray (CR) accelerators with different injection spectra [5], sources characterized by a double injection spectrum like the magnetized winds of exploding Wolf-Rayet and red supergiant stars [6], or the possibility of a retro-action of cosmic rays themselves

[a]Presentation given by P. Salati at the 6th RICAP Conference on June 23rd, 2016.

[b]e-mail: salati@lapth.cnrs.fr
[c]e-mail: genolini@lapth.cnrs.fr

on the properties of the plasma, with the consequence of a softer diffusion coefficient K at high energies [7].

Another possibility lies in the presence of local accelerators which, if considered as point-like objects, may yield a contribution somewhat different from what is expected from sources continuously spread in space and time inside our neighborhood. According to [8], a few nearby remnants can be responsible for the observed spectral changes. In a more consistent analysis [9] where all sources, either remote or local, have the same power-law injection spectrum, the proton and helium data are well fitted with local injectors borrowed from the Green catalog and the ATNF pulsar database. There is no need to modify CR transport and even the MED CR propagation model [10] yields a χ^2 of 1.3 per dof. The fit improves considerably if the somewhat unrealistic CR model A is assumed [9].

The question in which we are interested here is to know whether such an explanation is natural or not. In the conventional approach, sources are described by a continuous jelly in space and time. Here, the local sources are treated as point-like objects and their distribution is such that they yield a larger flux. But is this probable? To address this question and the more general problem of the stochasticity of CR fluxes yielded by injectors localized in space and time, we need the so-called Myriad model.

2 The Myriad model: a framework for statistics

Once injected in the interstellar medium, primary cosmic rays propagate inside the turbulent Galactic magnetic fields. CR transport is basically understood as a diffusion process taking place within a magnetic halo which is generally pictured as a circular slab, matching the shape of the Milky Way, inside which a thin Galactic disk of gas and stars is sandwiched. The CR density $\psi = dn/dE$ is related to the CR flux $\Phi \equiv (v_{CR}/4\pi)\,\psi$ and fullfills the diffusion equation

$$\frac{\partial \psi}{\partial t} - K\,\Delta\psi = q_{\rm acc}(\mathbf{x}, t)\,, \tag{1}$$

where $q_{\rm acc}(\mathbf{x}, t)$ accounts for the *continuous* space-time distribution of accelerators. The solution to the master equation (1) is given by the convolution, over the magnetic halo (MH) and the past, of the CR propagator $G(\mathbf{x}, t \leftarrow \mathbf{x}_S, t_S)$ with the source distribution $q_{\rm acc}(\mathbf{x}_S, t_S)$

$$\psi(\mathbf{x}, t) = \int_{-\infty}^{t} dt_S \int_{\rm MH} d^3\mathbf{x}_S \; G(\mathbf{x}, t \leftarrow \mathbf{x}_S, t_S)\, q_{\rm acc}(\mathbf{x}_S, t_S)\,. \tag{2}$$

The propagator describes the probability that a CR species injected at point $\mathbf{x}_S$ and time t_S diffuses at location $\mathbf{x}$ at time t. Assuming that $q_{\rm acc}$ does not depend in time leads to the steady state canonical solution produced by most CR numerical codes, where the flux at the Earth is constant in time.

In the Myriad model, the jelly of sources $q_{\rm acc}$ of the conventional approach is replaced by a constellation of point-like objects located each at $\mathbf{x}_i$ and t_i. The CR flux at the Earth yielded by a population $\mathcal{P}$ of such a myriad of injectors becomes

$$\Phi_{\mathcal{P}}(\odot, 0) = \frac{v_{CR}}{4\pi} \sum_{i \in \mathcal{P}} G(\odot, 0 \leftarrow \mathbf{x}_i, t_i)\, q_{\rm acc}(\mathbf{x}_i, t_i) \equiv \sum_{i \in \mathcal{P}} \varphi_i\,. \tag{3}$$

Although we can have some information on the closest and youngest sources, we have little knowledge of the actual population $\mathcal{P}$ in which we live. This is a problem insofar as the flux $\Phi_{\mathcal{P}}$ depends precisely on how the sources are distributed in space and time around us. To tackle the calculation of the flux in these conditions, we can adopt a statistical point of view and consider the ensemble of all possible populations $\mathcal{P}$ of sources. Without loss of generality, we may assume that each source accelerates

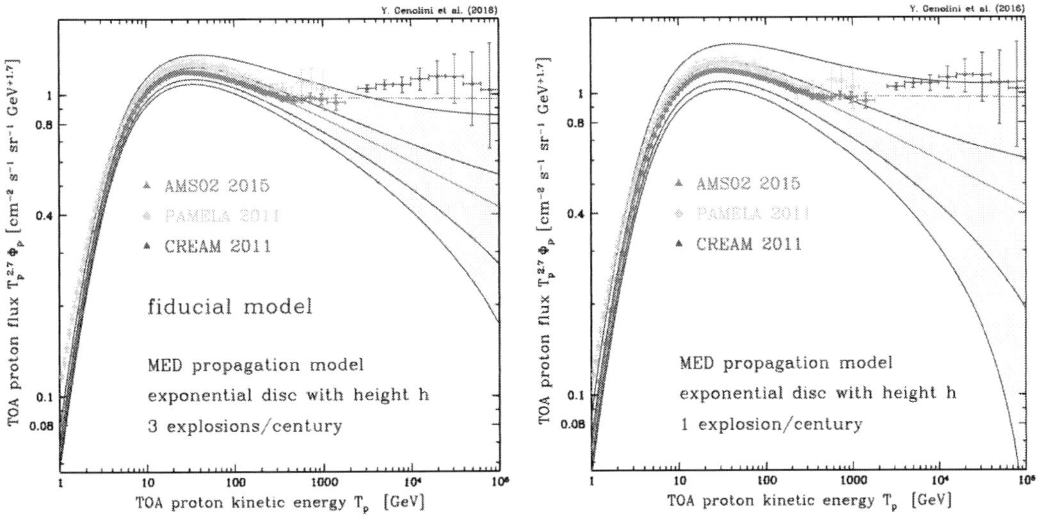

Figure 1. In both panels, the TOA proton flux is plotted as a function of TOA proton kinetic energy. The 1-σ (pale yellow) and 2-σ (dark yellow) bands predictions of the Myriad model have been derived assuming a stable law with index 5/3. The average flux is featured by the solid red line and is assumed to follow a simple power-law. Actually, we have used here a fit to the AMS-02 data from which the hardening has been removed. Measurements from AMS-02 [3], PAMELA [1] and CREAM [4] are also shown for comparison. The dotted long-dashed red curve is a fit to the AMS-02 data with hardening taking place at a rigidity of 355 GV. The MED CR propagation model [10] has been assumed here, with sources exponentially distributed along the vertical direction with scale height 100 pc. The fiducial case of 3 explosions per century is presented in the left panel. In the right panel, 1 explosion per century is assumed, hence wider uncertainty bands.

the same CR yield q_{SN}, lies within the magnetic halo and is younger than some critical value $\mathcal{T}$ that sets the size of the phase space volume over which the analysis is carried out. Sources older than $\mathcal{T}$ have no effect on the flux at the Earth if that age is taken to be a few times the CR confinement time inside the magnetic halo. We also assume a constant explosion rate $\nu \sim 1$ to 3 per century. Each population contains a number $\mathcal{N} = \nu \times \mathcal{T}$ of sources yielding each a flux φ_i at the Earth whose sum is the flux $\Phi_{\mathcal{P}}$. We may finally assume that each source is *independently* and *randomly* distributed in phase space according to the probability distribution function (pdf) $\mathcal{D}(\mathbf{x}_S, t_S)$. In the Myriad model, the individual flux φ yielded by a single source is a random variable whose pdf $p(\varphi)$ is the key of the statistical analysis. It is related to the phase space pdf $\mathcal{D}(\mathbf{x}_S, t_S)$ through

$$dP = p(\varphi)\, d\varphi = \int_{\mathcal{V}_\varphi} \mathcal{D}(\mathbf{x}_S, t_S)\, d^3\mathbf{x}_S\, dt_S \,, \tag{4}$$

where $\mathcal{V}_\varphi$ denotes the space-time region inside which a source contributes a flux in the range between φ and $\varphi + d\varphi$.

All the conditions are now met to use the central limit theorem in order to derive the pdf of the total flux Φ. Notice that $p(\varphi)$ scales as $\varphi^{-8/3}$ in the large flux limit so that the generalized version of the theorem must be applied, with the consequence that the variable $(\Phi - \langle\Phi\rangle)/\Sigma_\Phi$ is distributed according to the stable law S(5/3, 1, 1, 0; 1). Here, $\langle\Phi\rangle$ denotes the *statistical ensemble* average flux. It is equal to the flux yielded by the smooth distribution of sources $q_{\rm acc}(\mathbf{x}_S, t_S) = \mathcal{N}\, q_{\rm SN}\, \mathcal{D}(\mathbf{x}_S, t_S)$ of the conventional approach. The typical flux variance Σ_Φ scales like $q_{\rm SN}\, K^{-3/5}\, \nu^{3/5}$. Notice that a similar result was found by [11] in a pioneering analysis of the CR electron and positron fluxes produced by discrete stochastic sources.

3 Computing the odds of the Galactic lottery

We can apply the Myriad model to calculate the pdf of the proton flux at the Earth and estimate the p-value associated to the AMS-02 and CREAM measurements. The 1-σ (pale yellow) and 2-σ (dark yellow) theoretical uncertainty bands are featured in figure 1 for two values of the explosion rate. The corresponding p-values are listed in table 1. We can conclude from these results that the observed CR proton and helium anomalies have little chance to originate from a statistical fluctuation in the positions of the local sources. The p-values are larger in the last row of table 1 where the CR propagation model A of [9] is assumed with a half-height of the magnetic halo of 1.5 kpc and a diffusion spectral index of 0.85. Both values are now in tension with recent observations. Notice that the statistical analysis sketched above needs to be refined in two respects: (i) the behavior of the pdf $p(\varphi)$ can be dominated for intermediate values of the flux φ by sources distributed along the Galactic plane, hence a stable law with index 4/3 instead of 5/3 and (ii) a light cone cut-off needs to be implemented in relation (4) since CR diffusion cannot be faster than light. These refinements have been studied in detail in a comprehensive analysis [12] where stable laws are shown to provide a robust description of the statistical behavior of the CR flux.

Table 1. The p-values in % of the three last AMS-02 (red) and four first CREAM (blue) proton flux measurements have been calculated in the framework of the Myriad model. They are obtained from the convolution of the experimental uncertainty with a stable law with index 5/3. The three first rows correspond to the MED CR propagation model [10] where the SN explosion rate has been decreased from 3 to 1 per century. The last row refers to model A found in [9] to explain the CREAM data as resulting from known local sources.

Kinetic energy [TeV]	0.724	0.96	1.41	3.16	5.02	7.94	12.6
MED model with $\nu = 3$ [%]	10.2	8.68	7.67	1.6	1.23	1.18	0.98
MED model with $\nu = 2$ [%]	12.3	10.6	9.34	2.14	1.64	1.57	1.31
MED model with $\nu = 1$ [%]	16.2	14.2	12.6	3.52	2.7	2.59	2.16
model A [9] with $\nu = 0.8$ [%]	27.2	25.8	24.5	14.3	12.8	13.3	12.9

Acknowledgements

P.S. would like to thank the organizers of the 6th RICAP Conference for the friendly and inspiring atmosphere of the meeting. The authors wish to thank Philippe Briand for his kind help on the generalized central limit theorem. This work was supported by Institut Universitaire de France (IUF).

References

[1] O. Adriani et al. (PAMELA), Science **332**, 69 (2011), 1103.4055
[2] M. Aguilar et al. (AMS Collaboration), Physical Review Letters **115**, 211101 (2015)
[3] M. Aguilar et al., Physical Review Letters **114**, 171103 (2015)
[4] Y.S. Yoon et al., Astrophys. J.**728**, 122 (2011), 1102.2575
[5] V.I. Zatsepin, N.V. Sokolskaya, Astron. Astrophys. **458**, 1 (2006), astro-ph/0601475
[6] P.L. Biermann et al., Astrophys. J.**725**, 184 (2010), 1009.5592
[7] P. Blasi, E. Amato, P.D. Serpico, Phys. Rev. Lett. **109**, 061101 (2012), 1207.3706
[8] S. Thoudam, J.R. Hörandel, M.N.R.A.S.**421**, 1209 (2012), 1112.3020
[9] G. Bernard et al., Astron. Astrophys. **555**, A48 (2013), 1207.4670
[10] F. Donato et al., Phys. Rev. **D69**, 063501 (2004), astro-ph/0306207
[11] P. Mertsch, JCAP **1102**, 031 (2011), 1012.0805
[12] Y. Genolini, P. Salati, P. Serpico, R. Taillet (2016), 1610.02010

The Cosmic Ray spectrum in the energy region between 10^{12} and 10^{16} eV measured by ARGO–YBJ

Paolo Montini[1,a] for the ARGO–YBJ collaboration

[1] *DIpartimento di Matematica e Fisica - Università degli Studi Roma TRE and INFN Sezione di Roma TRE*

Abstract. The ARGO-YBJ experiment has been in full and stable data taking at the Yangbajing cosmic ray observatory (Tibet, P.R. China, 4300 m a.s.l.) for more than five years. The detector has been designed in order to explore the Cosmic Ray (CR) spectrum in an energy range from few TeV up to several PeV. The high segmentation of the detector allows a detailed measurement of the lateral particle distribution which can be exploited on order to identify showers produced by primaries of different mass. The results of the measurement of the all-particle and proton plus helium energy spectra in the energy region between 10^{12} and 10^{16} eV are discussed.

The measurement of Cosmic Ray (CR) energy spectrum and composition gives important information concerning the production, acceleration and the propagation of high energy particles in our Galaxy. The CR all–particle energy spectrum is roughly described by a power–law with a *knee* at energies around 3 PeV. It is commonly believed that the origin of the knee is related to a change of the elemental composition of CRs, in particular to a decrease of the flux of light elements (H and He nuclei). The determination of the individual abundances of elements at energies above 100 TeV must be inferred from the measurements of extensive air showers (EAS). The development of EASs is subject to large fluctuations. Owing to the high altitude location (atmospheric depth $606\,\mathrm{g/cm^2}$), the ARGO-YBJ experiment is able to sample the EAS induced by high energy CRs not far from the maximum of its longitudinal development where the fluctuations are reduced.

1 The detector

The ARGO–YBJ experiment was a full–coverage EAS detector operated at the Yangbajing cosmic ray observatory (Tibet, P.R. China, 4300 m a.s.l.) and it was in full and stable data taking from November 2007 up to February 2013. The detector was made of a single layer of 1836 Resistive Plate Chambers (RPCs) with $\sim 93\%$ active area surrounded by a partially instrumented ($\sim 23\%$) guard ring in order to improve the event reconstruction. The detector was equipped with two independent readout systems: each RPC is simultaneously read–out by 80 copper strips ($6.75 \times 61.80\,\mathrm{cm^2}$) logically arranged in 10 independent pads ($55.6 \times 61.8\,\mathrm{cm^2}$) and by two large electrodes called Big Pads ($139 \times 123\,\mathrm{cm^2}$). Each Big Pad collects the total charge developed by the particles impinging on the detector surface (analog readout) [1]. The analog readout system can be operated at different gain scales in order to measure showers induced by primaries in a wide energy region. Data coming from the most sensitive

[a] Now at Dipartimento di Fisica - Sapienza Università di Roma and INFN sezione di Roma e-mail: paolo.montini@roma1.infn.it

scales perfectly overlap with the digital readout, thus providing a powerful inter–calibration [1] . At the highest scale the analog readout samples the shower front up to a particle density of $2 \cdot 10^4\,m^{-2}$, thus extending the dynamic range of the detector up to PeV energies. A dedicated calibration procedure has been implemented for each gain scale [1, 2]. The full-coverage technique enables a detailed imaging of the shower front which is a fundamental tool that allows a deep investigation of the shower properties even in the core region. The high segmentation of the Big Pad system allows the measurement of the shower size and of the lateral distribution of particles in the shower front that can be exploited in order to estimate the primary energy and mass.

2 Data Analysis

The analysis has been carried out on events collected during 2010 by using the analog readout system. For each event the core position, arrival direction, shower size (N_8), particle density on the carpet and lateral distribution are reconstructed. The shower size N_8 has been defined as the number of particles within a radius of 8 m from the shower core. It is well correlated with energy for a given mass and not affected by bias effects due to the finite detector size [3]. The determination of the energy and of the primary mass from the measured quantities can be faced out by using the Bayesian unfolding. The Monte Carlo simulations are therefore used in order to evaluate a probabilistic response matrix which can be inverted by means of an iterative algorithm based on the Bayes's theorem. A detailed description of this procedure can be found in [4–6]. Showers produced by H, He, CNO, NeMgSi, and Fe have been simulated in the energy range $1 - 31.6 \times 10^4$ TeV with an E^{-1} differential spectrum by using the CORSIKA (v. 7.3) code [7] including the QGSJET-II.04 and FLUKA interaction models. A smaller data set have been simulated using SYBILL 2.1 for systematic studies. Showers have been sampled at the Yangbajing altitude and randomly distributed over an area of $250 \times 250\,m^2$ centered on the ARGO–YBJ detector. The detector response has been simulated by using a GEANT3 based code. The present analysis is based on the data collected with two analog scales (low gain and high gain) covering the energy range from about 20 TeV to a few PeV. A sample of quasi–vertical showers ($\vartheta_{rec} \leqslant 35°$) has been selected within an area of $40 \times 40\,m^2$ around the detector center ensuring that a large fraction of the shower is fully contained in the full–coverage area. Additional selection criteria based on the shower size improve the correlation between shower size and primary energy and avoid any contribution due to the electronic noise. In figure 1a the shower size N_8 of data and MC events is reported, showing a good agreement between data and simulations. In figure 1b the selection efficiency is shown for proton, helium nuclei, CNO and NeMgSi mass groups and iron nuclei. The plot shows that in the energy region 300 TeV – 10 PeV the selection efficiency is almost the same for all the species, demonstrating the selection criteria do not affect the spectrum measurement.
In a shower produced by heavy nuclei a substantial amount of secondary particles is spread further away from the core region. On the contrary, in a shower produced by light elements, the largest amount of particles is concentrated in a small region around the shower core.The ratio between the particle density measured at several distances from the core and the one measured very close to the core can be exploited in order to identify showers produced by light elements. Several studies performed on simulated events have shown that the quantities $\beta_5 = \rho_5/\rho_0$, and $\beta_{10} = \rho_{10}/\rho_0$, where ρ_0, ρ_5 and ρ_{10} are respectively the particle density measured in the core region, at 5 m from the core and at 10 m from the core, are sensitive to primary mass. In a probabilistic approach the probability $P(N_8, \beta_5, \beta_{10}|E, A)$ of measuring a shower size N_8 and a certain value of β_5 and β_{10} giving a primary energy E and mass A, relates the characteristics of the primary particle to the experimental observables. The bayesian unfolding algorithm has been therefore tuned in order to take into account also the information coming from the two quantities β_5 and β_{10}. The fraction of selected showers induced by light elements (p and

DOI: 10.1051/epjconf/201713602007

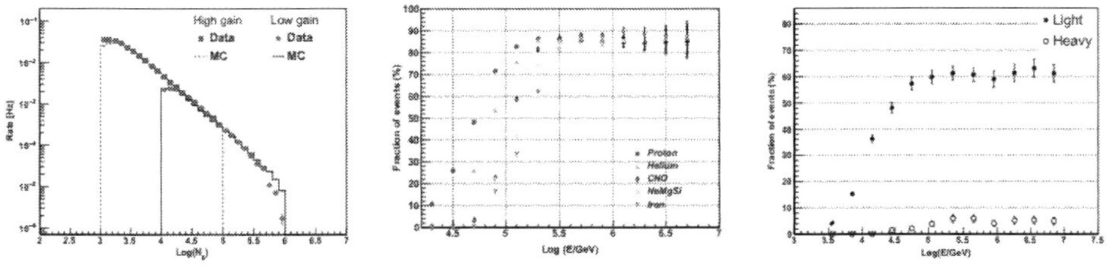

(a) (b) (c)

Figure 1: Shower size distribution for data and simulations (a), Fraction of selected MC showers produced by P, He, CNO, NeMgSi and Fe(b). Fraction of selected MC showers produced by light elements and the corresponding contamination

He) and the corresponding contamination by heavier nuclei has been evaluated in order to check the discrimination power. In figure 1c the values obtained are reported as a function of the energy. The fraction of selected light elements increases with energy and is around 60% at energies above 50 TeV, while contamination is well below 10% over the whole energy range.

3 All–particle and P+He energy spectra

In the figure 2a the all–particle spectrum measured in this work is reported. The measurements are affected by a statistical uncertainty of the order of 1% at the lowest energies, gradually increasing up to ~ 8% at energies higher than 1 PeV. The systematic uncertainty is of the order of 15% mainly due to the limited Monte Carlo statistics (10%) and to variations of the bin edges (10%) used in the determination of the probability response matrix. The systematic uncertainty related to the hadronic interaction model used in simulations has been derived by comparing the results obtained by QGSJET and SIBYLL. In particular, simulations with SIBYLL systematically yield to a flux ~ 7% higher. Systematic effects introduced by variation of the fiducial cuts and by the unfolding procedure have also been studied and give a minor contribution ($\leqslant$ 1%) to the total uncertainty.
The proton plus helium spectrum, including both statistical and systematic errors is also reported in figure 2a, spanning the energy range between 20 TeV and 5 PeV. Statistical errors are of the order of 1% at the lowest energies and increase with energy up to 18% at PeV energies. The contributions to the total systematic uncertainty come from event selection, estimation of the conditional probabilities, hadronic interaction model, composition model, unfolding procedure. As for the all–particle spectrum the major contribution to the systematic uncertainty comes from the determination of the probability response matrix and is about 10% for energy below 300TeV, 8% in the region 300–500 TeV and it turns to about 21% at the PeV energies. A minor contribution comes from the selection criteria (2.5%) and the unfolding procedure ($\leqslant$ 1%). Simulations with SIBYLL yield to a flux ~ 4% and ~ 10% higher in the energy region below and above 500 TeV respectively.

4 Conclusions

The ARGO–YBJ experiment allows a deep investigation of the properties of EASs providing a detailed measurement of the distribution of the charged particles in the shower front. The detector is able to investigate the CR energy spectrum in a wide energy range. The measurements of the

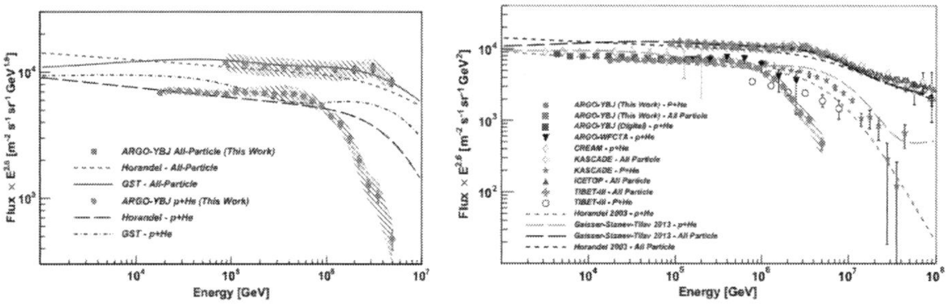

(a) (b)

Figure 2: CR all–particle and p+He energy spectra measured by ARGO–YBJ (a) compared with previous results of ARGO–YBJ [6, 8], other experimental results [9] and theoretical models [10] (b).

all–particle and p+He energy spectra are presented. As shown in figure 2b the all–particle spectrum is in good agreement with other experimental results [9]. The measurement is also in agreement with an independent analysis of ARGO–YBJ data [11]. The accurate reconstruction of the lateral distribution has been exploited in order to discriminate showers produced by primaries of different mass groups. The ARGO–YBJ experiment measured the proton plus helium flux over two energy decades, from 3 TeV to 5 PeV. There is a strong evidence of a deviation from a single power law at energies around 1 PeV, suggesting that the *knee* of the all–particle spectrum is due to heavier elements. Similar conclusion has been suggested also by the results of the hybrid experiment ARGO–WFCTA based on a Wide FoV Cherenkov telescope. These results open new scenarios about the evolution of the p+He energy spectrum towards the highest energies and the origin of the knee.

References

[1] B. Bartoli et al., Astropart. Phys. **67**, 47 (2015)
[2] B. Bartoli et al., Nucl. Instrum. Methods Phys. Res, Sect. A **783**, 68 (2015)
[3] P. Bernardini et al., PoS(ICRC2015)388 (2015)
[4] G. D'Agostini, Nucl. Instrum. Meth. **A362**, 487 (1995)
[5] B. Bartoli et al., Phys. Rev. D **85**, 092005 (2012)
[6] B. Bartoli et al., Phys. Rev. D **91**, 112017 (2015)
[7] D. Heck, J. Knapp, J. Capdevielle, G. Schatz, T. Thouw, Report **FZKA** 6019 (1998)
[8] B. Bartoli et al., Phys. Rev. D **92**, 092005 (2015)
[9] Y. S. Yoon et al., ApJ **728**, 122 (2011); T. Antoni et al., Astroparticle Physics **24**, 1 (2005); W. Apel et al., Astropart. Phys. **47**, 54 (2013); M. Amenomori et al., Astrophys. J. **678**, 1165 (2008); M.G. Aartsen et al., Phys. Rev. D **88**, 042004 (2013)
[10] J.R. Hörandel, Astropart. Phys. **19**, 193 (2003), T.K. Gaisser et al. Front. Phys. **8(6)**, 748 (2013)
[11] A. D'Amone et al., PoS(ICRC2015)366

Cosmic-ray energy densities in star-forming galaxies

Massimo Persic[1,2,a] and Yoel Rephaeli[3,b]

[1] *INAF/Osservatorio Astronomico di Trieste and INFN, via G.B. Tiepolo 11, I-34143 Trieste, Italy*
[2] *Physics & Astronomy Department, University of Bologna, via C. Ranzani 1, I-40127 Bologna, Italy*
[3] *School of Physics & Astronomy, Tel-Aviv University, Tel-Aviv 69978, Israel; and UC San Diego, USA*

Abstract. The energy density of cosmic ray protons in star forming galaxies can be estimated from π^0-decay γ-ray emission, synchrotron radio emission, and supernova rates. To galaxies for which these methods can be applied, the three methods yield consistent energy densities ranging from $U_p \sim 0.1 - 1\,\mathrm{eV\,cm^{-3}}$ to $U_p \sim 10^2 - 10^3\,\mathrm{eV\,cm^{-3}}$ in galaxies with low to high star-formation rates, respectively.

1 Introduction

Active star formation (SF) in galaxies leads to acceleration of protons and electrons (CRp, CRe) via the Fermi-I diffusive shock acceleration mechanism in supernova (SN) remnants (Fermi 1954; Ginzburg & Syrovatskii 1964; Bell 1978; Protheroe & Clay 2004). Timescales of starburst (SB) activity in galaxies are comparable to galactic dynamical timescales, $\tau_{SB} \sim \tau_{dyn} \sim 10^8$ yr. On the other hand, in a SB region the characteristic timescales for protons to gain energy (by acceleration) and to lose it by collisions with interstellar gas particles (leading to heating and pion production) are typically much shorter than τ_{SB} (e.g., Persic & Rephaeli 2010).

The respective lengths of these timescales suggest that in galaxies, during typical episodes of SF, particle distributions are likely to attain asymptotic steady-state form. Additionally, under virial equilibrium, CR and magnetic fields are in an approximate minimum-energy state; this is equivalent to having CR and magnetic fields in (approximate) energy equilibrium (e.g., Longair 1994).

The equipartition assumption enables deduction of the CRp energy density in star forming galaxies (SFGs), U_p (the main contribution to the particle energy density), indirectly from the electron energy density which can be estimated from radio synchrotron measurements if the proton-to-electron (p/e) energy density ratio is known or can be reliably estimated.

Another way to derive U_p is based on measuring the GeV-TeV γ-ray emission, which is largely from CRp interactions with ambient gas that yield neutral (π^0) and charged pions; π^0 decays into γ-rays. Thus, γ-ray measurements allow a direct estimate of U_p; this is now possible given the recent detection of γ-ray emission from SFGs.

[a] e-mail: persic@oats.inaf.it
[b] e-mail: yoelr@wise.tau.ac.il

Also, since energy-loss timescales are shorter than SF timescales, U_p can be estimated from the observed rate of core-collapse SN and the deduced CRp characteristic residence time in the galactic disk, given a realistic estimate of the fraction of SN kinetic energy that is channeled into particle acceleration.

This paper describes measures of galactic CR energy densities based on the methods outlined above. Although not strictly independent, these methods are based on very different observationally deduced quantities – radio emission, γ-ray emission, and the rate of core-collapse SN. We find that the three methods yield consistent U_p for a sample of 14 galaxies (and galactic environments) with widely varying levels of SF activity, from quiescent to intense starbursts. These are the only galaxies for which γ-ray data (and/or radio data and SN rates) are available. After reviewing the γ-ray, radio, and SN methods (sect. 2,3,4), the respective results are discussed in section 5 and summarized in section 6.

2 Estimating U_p from γ-ray emission

Detection of GeV-TeV emission from several SFGs provided the basis for observationally-based estimates of U_p in the central SB region and across the galactic disk. In most SB galaxies, such as the two nearby ones M 82 and NGC 253, the central SB region with a radius of ~ 200 pc, is identified as the main site of particle acceleration. Here, the injection particle spectrum is assumed to have the non-relativistic strong-shock index $q = 2$. A theoretical N_p/N_e ratio, predicted from charge neutrality of the injected CRs, is likely to hold in this source region, as is the assumption of energy equipartition of particles and magnetic fields.

Adopting the convection-diffusion model for energetic electron and proton propagation and accounting for all the relevant hadronic and leptonic processes, the steady-state energy distributions of these particles, in both the nuclear SB region and across the full disk, can be determined in the context of a detailed numerical treatment (e.g., Paglione e al. 1996; Torres 2004; Persic, Rephaeli & Arieli 2008; de Cea et al. 209; Rephaeli, Arieli & Persic 2010). The procedure is similar when SF is not undergoing a burst confined to the nuclear region but occurs throughout the whole disk.

The relevant electron energy loss processes are bremsstrahlung, Compton, and synchrotron, whereas for protons the main losses are γ-ray emission from π^0 decay following p-p collisions. Bremsstrahlung losses dominate at lower energies, whereas π^0-decay losses dominate at higher energies; in the GeV-TeV region, emission is mainly from pp–induced π^0 decay (e.g., Rephaeli et al. 2010).

For a source with gas number density n_g, proton energy density U_p, and volume V, the integrated hadronic emission from *pp*-induced π^0 decay is $L^{[q]}_{\geq E} = \int_V g^{[q]}_{\geq E}\, n_g\, U_p\, dV\ \ \mathrm{s}^{-1}$, with the integral emissivity $g^{[\eta]}_{\geq\epsilon}$ in units of photon $\mathrm{s^{-1}(H\text{-}atom)^{-1}(eV/cm^3)^{-1}}$ (Drury et al. 1994). Therefore, U_p can be determined, when $L_{\geq\epsilon}$ and $n_{gas}(r)$ are observationally known, and the particle steady-state energy distributions are numerically determined by solving the convection-diffusion model for CRe and CRp propagation.

For the two local SB galaxies M 82 and NGC 253, GeV-TeV fluxes and spectra agree with numerical modeling with $U_p \sim 200\,\mathrm{eV\,cm^{-3}}$ in the central SB region (Acciari et al. 2009; Acero et al. 2009; Abdo et al. 2010a). For the highest-SFR galaxy in the nearby universe, Arp 220, recent GeV data (Peng et al. 2016) match theoretical predictions (Torres 2004) with $U_p \sim 2000\,\mathrm{eV\,cm^{-3}}$. Several low-SFR galaxies were also detected in the GeV band: *(i)* M 31, with $U_p \simeq 0.35\,\mathrm{eV\,cm^{-3}}$ (Abdo et al. 2010b); *(ii)* the Large Magellanic Cloud (LMC), with $U_p \simeq 0.2 - 0.3\,\mathrm{eV\,cm^{-3}}$ (Abdo et al. 2010c); *(iii)* the Small Magellanic Cloud (SMC), with $U_p \simeq 0.15\,\mathrm{eV\,cm^{-3}}$ (Abdo et al. 2010d); *(iv)* the composite SB/Seyfert-2 galaxies NGC 1068 and NGC 4945 (Lenain et al. 2010). *(v)* For the Milky Way, modeling of the diffuse HE emission leads to $U_p \sim 1\,\mathrm{eV\,cm^{-3}}$ (Strong et al. 2010; Ackermann et al. 2011): this value agrees well with $U_p \sim 1\,\mathrm{eV\,cm^{-3}}$ measured at Earth (e.g., Webber 1987), and is consistent with the somewhat higher value $(6 \pm 3)\,\mathrm{eV\,cm^{-3}}$ inferred for the ~200 pc region of the Galactic center (Aharonian et al. 2006, based on HESS data).

Published U_p values based on GeV-TeV data modeling are reported in Table 2.

3 Estimating U_p from radio emission

Determining U_p from the measured radio flux clearly requires knowledge of the mean magnetic field in the emitting region, B, and a way to couple CRp and CRe. To overcome this (implied) indeterminacy, the assumption of field and particle energy equipartition is commonly made. In addition, the contribution of secondary electrons (from π^-decay) [1] to the (steady state) electron density has to be included. (Throughout this section we follow Persic & Rephaeli 2014.)

While the exact form of the particle steady-state spectral density does not generally have a single power-law form (e.g., Rephaeli 1979, Rephaeli & Persic 2015), the radiative yields are largely by protons and electrons with energies higher than a few Gev, for which Coulomb losses (which flatten the spectral density) are subdominant. In this limit, the total electron spectral density can be approximated by

$$N_e(\gamma) = N_{e,0}\,(1+\chi)\,\gamma^{-q_e}, \qquad (1)$$

where the electron Lorentz factor γ is in the range $\gamma_1 \leq \gamma \leq \gamma_2$, $N_{e,0}$ is a normalization factor of the primary electrons, and χ is the secondary-to-primary electron ratio. The electron spectral index is $q_e \geq 2$, with the minimal value of 2 corresponding to the strong-shock limit of the Fermi-I acceleration mechanism. Outside the acceleration region $q_e > 2$. Ignoring the contribution of low-energy electrons with $\gamma < \gamma_1$, the electron energy density is $U_e \simeq N_{e,0}\,(1+\chi)\,m_e c^2 \int_{\gamma_1}^{\gamma_2} \gamma^{1-q_e} d\gamma = N_{e,0}\,(1+\chi)\,m_e c^2\,\gamma_1^{2-q_e}[1-(\gamma_2/\gamma_1)^{2-q_e}]/(q_e-2)$.

For a population of electrons described by Eq. (1), traversing a homogeneous magnetic field of strength B that permeates a region of (spherically equivalent) radius r_s located at a distance d from the observer, and emitting a 5 GHz synchrotron radiation flux of f_5 Jy, the standard synchrotron formula (e.g., Blumenthal & Gould 1970) yields

$$N_{e,0}(1+\chi) = 1.6\times10^{-16}\,a_{q_e}^{-1}\,\psi_5\,1250^{\frac{q_e}{2}}\,B^{-\frac{q_e+1}{2}}, \qquad (2)$$

where quantities are expressed in c.g.s. units, the factor a_{q_e} is defined in the latter paper, and $\psi_5 \equiv (\frac{r_s}{0.1\,\mathrm{kpc}})^{-3}(\frac{d}{\mathrm{Mpc}})^2(\frac{f_5}{\mathrm{Jy}})$. From the above expressions we derive

$$U_e = 1.3\times10^{-22}\,1250^{\frac{q_e}{2}}\,\psi_5\,B^{-\frac{q_e+1}{2}} \times \times \frac{\gamma_1^{2-q_e}[1-(\gamma_2/\gamma_1)^{2-q_e}]}{(q_e-2)\,a_{q_e}}\ \ \mathrm{erg\,cm^{-3}}. \qquad (3)$$

Since U_e includes both primary and secondary electrons, the rough assumption that both populations can be characterized with nearly the same power-law index (see Persic & Rephaeli 2014) means that the primary electron energy density is $U_e/(1+\chi)$. Denoting the primary p/e energy density ratio by $\kappa(q_p, q_e)$, the proton energy density is $U_p \simeq \kappa(q_p, q_e)\,U_e/(1+\chi)$. Since tight coupling is expected in the very dense environment of SB nuclei, particle and magnetic field energy densities can be assumed to be close to equipartition. If so, we can express the field in terms of the total particle energy density; since $\gamma_2 >> \gamma_1$,

[1] Secondary positrons (from π^+decay) almost immediately annihilate with thermal electrons.

 DOI: 10.1051/epjconf/201713602008

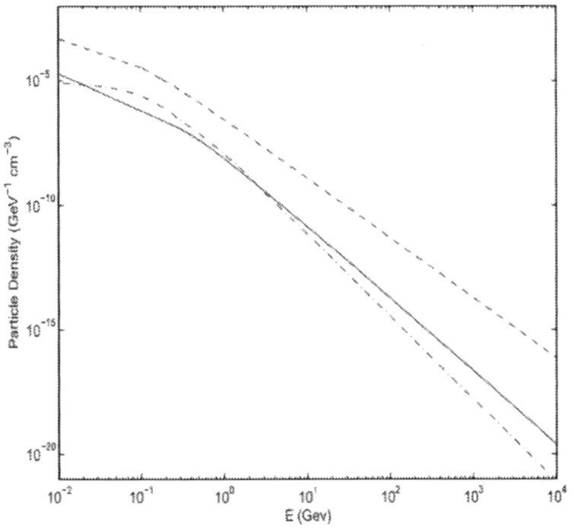

Figure 1. Steady-state spectra of primary protons (dashed line), primary electrons (solid line), and secondary electrons (dash-dotted line) in the central source region of NGC 253 (from Rephaeli et al. 2010).

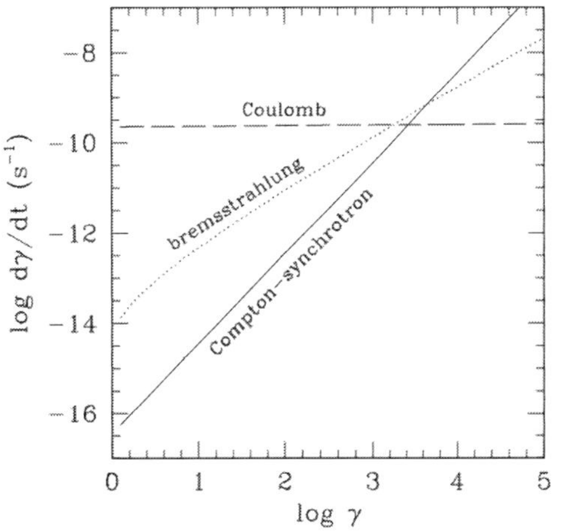

Figure 2. Electron energy loss rates by Coulomb, bremsstrahlung, and Compton-synchrotron processes in a central SB region of a (M 82-like) galaxy for B=100 μG, $L_{\rm IR}$ = 10^{44} erg s^{-1}, r_s = 0.2 kpc, n = 100 cm^{-3}, n_e = 200 cm^{-3} (from Persic & Rephaeli 2014).

[2] we obtain

$$U_{\rm p} = \frac{2.5 \times 10^{10}}{1 + (1+\chi)/\kappa} \left[3.3 \times 10^{-21} \left(1 + \frac{\kappa}{1+\chi}\right) \times \right.$$
$$\left. \times\, \gamma_1^{2-q_e} \frac{1250^{q_e/2}\, \psi_5}{(q_e - 2)\, a_{q_e}}\right]^{4/(5+q_e)} \mathrm{eV\,cm^{-3}}. \qquad (4)$$

In general, q_e, q_p, χ, γ_1, and κ need to be specified in order to compute U_p.

- The value of q_e is deduced from the (nonthermal) radio spectral index, α, through $q_e = 2\alpha + 1$.
- The proton spectral index has been measured to be q_p = 2.2 in the nearby SBGs NGC 253, NGC 3034, and NGC 4945 (Ackermann et al. 2012). Suprathermal particles injected into a supernova shock have a power-law spectrum with index $q = (R+2)/(R-1)$, where R is the shock compression ratio. Taking $R \simeq 3.6$ in the nuclear SB regions yields $q_p \simeq 2.2$.
- The secondary-to-primary electron ratio, χ, depends on the effectiveness of creating electron-positron pairs from *pp* interactions of primary protons with ambient protons. Therefore, χ will depend on the injection p/e number ratio, ζ, that sets the number of primary protons per primary electron, on the *pp* cross section at $\gtrsim$ 1 TeV, and on the gas density. The inverse of the product of the latter two quantities is the characteristic mean free path of energetic protons, whose *pp* interactions yield $\pi^{\pm}$ which decay into $e^{\pm}$ with a branching ratio of 2/3. The probability for a proton to undergo *pp* interactions during its 3D random walk through a region of radius r_s and density n, is $\chi = \frac{2}{3}\zeta(q)\sqrt{3}\, r_s\, n\, \sigma_{\rm pp}(n)$. In a typical SB nucleus with r_s = 0.2 kpc, n = 200 cm^{-3}, and q = 2.2, we get of $\chi \sim 1$, in agreement with results from numerical models (see Fig. 1, from Rephaeli et al. 2010).

[2]The exact value of γ_2 is of little significance for the range of values of q_e of interest here; in our calculations we take $\gamma_2 = 10^5$.

- For consistency with the assumed power-law form of the electron spectrum, we take the low-energy limit γ_1 to be the value of the Lorentz factor at which the sum of the Coulomb (or electronic excitation, in ionized gas) and bremsstrahlung loss rates equals the synchrotron-Compton loss rate. This is also based on the fact that even for the relatively high values of the magnetic field in SB nuclei, the measured radio emission (upon which our normalization of the electron density is based) samples electrons with $\gamma > 10^3$. Equating the sum of the first two loss rates with the latter yields an estimate of γ_1. In Fig. 1 we display the energy-loss rates expressed in Eqs. (18)-(20) of Persic & Rephaeli (2014) for typical SB nuclei parameters.
- An approximate expression for κ at injection is $\kappa(q) \simeq (\frac{m_p}{m_e})^{(3-q)/2}$ in the relativistic limit.

Thus, for given values of q_e, q_p, χ, γ_1, and κ, the proton energy density can be estimated if the radio synchrotron flux, source size, and distance are known (see col. 3 in Table 1).

4 Estimating U_p from SN rates

With SN shocks the likely sites of particle acceleration, the combination of core-collapse SN rate and CR proton residence time in the galactic disk, $\tau_{\rm res}$, provide a basis for another independent estimate of the proton energy density. The value of $\tau_{\rm res}$ is determined by two timescales: *(i)* energy-loss timescale for p-p interactions, $\tau_{\rm pp} = (\sigma_{\rm pp} c n_p)^{-1}$ that, for protons with kinetic energy $E \gtrsim 10$ TeV for which $\sigma_{\rm pp} \simeq$ 50 mb, can be written as $\tau_{\rm pp} \sim 2 \times 10^5 (\frac{n_p}{100\,{\rm cm}^{-3}})^{-1}$ yr; and *(ii)* CRp advection timescale, $\tau_{\rm out}$, that characterizes proton escape out of the disk mid-plane region in a fast ($v_{\rm out} \sim$ 2500 km s^{-1} for M 82: Chevalier & Clegg 1985; Seaquist & Odegard 1991; Strickland & Heckman 2009) SB-driven wind.

Table 1. Star-forming galaxies: CRp energy densities and SN rates.

Object	γ-ray meth.	radio meth.	ν_{SN}
Arp 220 E		390	0.49
Arp 220 W		730	0.72
Arp 220	1970		2.8
Arp 299-A		365	0.46
NGC 224 (M 31)	0.36		0.01
NGC 253	200	235	0.12
NGC 3034 (M 82)	200	250	0.25
NGC 3628		100	0.035
NGC 4945	200	270	0.3
NGC 5236 (M 83)		260	0.1
NGC 6946		110	0.09
LMC	0.25		0.002
SMC	0.15		0.001
Milky Way	1		0.02

Notes. Energy densities and SN rates are given in, respectively, eV cm^{-3} and yr^{-1}. Data on core-collapse SN rates are from Torres (2004: Arp 220), Persic & Rephaeli (2010: M 82, NGC 253), Diehl et al. (2006: Milky Way), van den Bergh & Tammann (1991: M 31, SMC, LMC; see also Pavlidou & Fields 2001), Lenain et al. (2010: NGC 4945), Strickland et al. (2004: NGC 3628), Blair et al. (2014: NGC 5236, MGC 6946). Arp 220E+W SN rates are computed using $\nu_{SN} = (2.4 \pm 0.1) \times 10^{-2}$ $[L_{FIR}/(10^{10}L_{\odot})]$ yr^{-1} (Mannucci et al. 2003), taking L_{FIR} from Torres (2004).

For a homogeneous distribution of SNe within a central SB region of radius r_s the latter timescale is $\tau_{out} = 3 \times 10^4 (\frac{r_s}{0.3\,\mathrm{kpc}})(\frac{v_{out}}{2500\,\mathrm{km\,s^{-1}}})^{-1}$ yr. The overall residence time is then $\tau_{res}^{-1} = \tau_{pp}^{-1}(n_{HI}) + \tau_{out}^{-1}(r_s, v_{out})$. During τ_{res}, a number $\nu_{SN}\tau_{res}$ of SN explode and deposit the kinetic energy of their ejecta, $E_{ej} = 10^{51}$ erg per SN (Woosley & Weaver 1995), into the interstellar medium. Arguments based on the CR energy budget in the Galaxy and SN statistics yield an estimate of an efficiency factor of $\eta \sim 0.05$ in converting kinetic energy to particle acceleration (e.g., Peng et al. 2016). Accordingly,

$$U_p \sim 100 \frac{\nu_{SN}}{0.2\,\mathrm{yr}^{-1}} \frac{\tau_{res}}{3 \times 10^4\,\mathrm{yr}} \frac{\eta}{0.05} \frac{E_{ej}}{10^{51}\,\mathrm{erg}} \times \left(\frac{r_s}{0.25\,\mathrm{kpc}}\right)^{-3} \mathrm{eV\,cm^{-3}}. \quad (5)$$

In Fig. 3 we show values of U_p vs. SN rates for the above galaxy sample; the scaling relation in Eq.(5) is plotted as a thin dotted line which represents the explicit scaling $U_p = 100 (\frac{\nu_{SN}}{0.2\,\mathrm{yr}^{-1}})^{\delta}$ eV cm^{-3} with $\delta \sim 1.5$. that incorporates the reciprocal dependences $\tau_{res} \propto \nu_{SN}^{-\alpha}$ with $\alpha \sim 2.5$, and $r_s \propto \nu_{SN}^{-\beta}$ with $\beta \sim 1$).

5 Uncertainties

The values of U_p listed in Table 1 do not include substantial observational and modeling uncertainties. In this section we attempt to estimate the level of precision with which U_p was determined based on only limited information on the errors in the various observational parameters.

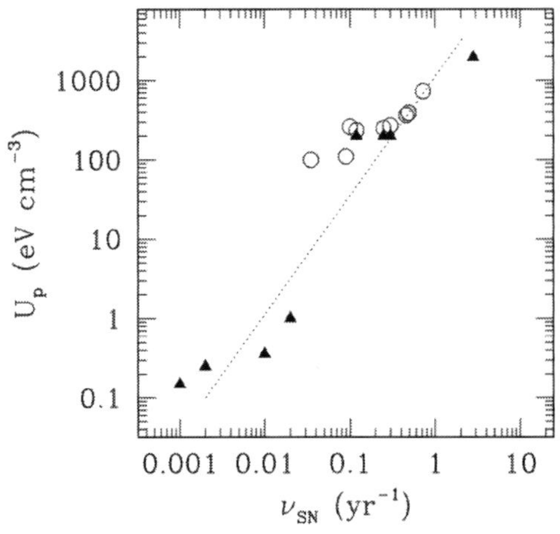

Figure 3. Correlation between CRp energy density and SN rate for the sources in Table 1. Results deduced from γ-ray (radio) measurements are denoted by triangles (circles). The thin dotted line represents the scaling relation $U_p = 100 (\frac{\nu_{SN}}{0.2\,\mathrm{yr}^{-1}})^{1.5}$ eV cm^{-3}, that corresponds to Eq.(5) (as specified in the text).

The use of γ-ray measurements to estimate U_p, the only direct method to measure U_p, entails a substantial uncertainty due to the poorly known spectral slope and, especially for fainter and/or more distant galaxies, the density distribution of ambient gas.

Uncertainties in estimating U_p from radio measurements are more tractable due to the fact that the quantities in Eq.(4) are usually well determined for our sample galaxies, except for the p/e energy density ratio κ, for which a CRp spectral index, q_p, need to be assumed. The uncertainty on the latter parameter, $\delta q_p \simeq 0.1$, translates to a $\sim$50% uncertainty in κ, i.e. typically an uncertainty of $\sim$10% in U_p as deduced from Eq.(4). The effective source radii, r_s, are deduced from high-resolution optical and radio data, so their values are relatively well determined, with typical uncertainties of a few 10%. Thus, all the galaxy quantities relevant to computing the CRp density are presumed to be measured at a precision level of $\sim$ 30%; accordingly, U_p can be estimated to within an overall uncertainty of 50%.

A precise measurement of the actual rate of core-collapse SNe is obviously crucial essential for reliable estimates of U_p. This is quite difficult especially in central SB regions where optical extinction is very strong. Moreover, radio counts of SN remnants require information on their ages in order to determine actual SN rates. For galaxies in our sample galaxies, available observational results indicate that ν_{SN} are known to within a factor $\sim$2. Also, uncertainties in U_p can also be due to uncertainties in τ_{res}; the latter mostly arise from the uncertainty in the fast wind velocity, which in turn is probably known to within a factor of $\sim$2. Published estimates of the SN energy that is channeled to CR, ~5% – 10% of the total kinetic energy of the SN ejecta, agree within a factor $\lesssim 2$ (Woosley & Weaver 1995; Lingenfelter et al. 1998; Higdon et al. 1998).

6 Discussion

The three methods discussed here are not independent. The γ-ray and radio methods are related through the common dependence on the p/e ratio at injection, on the secondary-to-primary electron ratio, and on the assumption of particle-field equipartition. The SN method is not independent of the γ-ray method either, since both depend on the CRp residence time in the emission region - although, unlike the γ-ray and radio methods, it does not depend on the particle radiative yield but on the statistics of core-collapse SN. Also, the γ-ray, radio, and SN methods are not on equal footing. By these methods the value of $U_{\rm p}$ is either *measured*, *inferred*, or *estimated*, respectively. This is due to *(i)* π^0-decay γ-ray emission is the most robust *measure* of $U_{\rm p}$ when the distribution of target gas is known and the particle propagation mode and energy losses are well known, whereas *(ii)* radio emission enables deduction of $U_{\rm p}$ from the radio spectral distribution, and assumed p/e energy density ratio and particle-field energy equipartition; and *(iii)* assuming a SN origin for CR, SN statistics for a given (region of a) galaxy leads to an estimate of $U_{\rm p}$ there.

Nonetheless, it seems that the three methods yield roughly consistent results. Specifically, as expected, there is a very significant difference between the very high CRp energy densities ($U_{\rm p} \sim O(10^2)\,{\rm eV\,cm^{-3}}$), deduced for a central SB region as compared with the low values ($U_{\rm p} \sim O(10^{-1})\,{\rm eV\,cm^{-3}}$) for the rest of the galactic disk.

Finally, note that the distribution of data points in Fig. 3, with scaling $U_{\rm p} \propto \nu_{\rm SN}^{1.5}$, is consistent (within uncertainties) with the observational relation $L_\gamma \propto {\rm SFR}^{1.4}$ (Abdo et al. 2010b, for essentially this same sample of galaxies). This clearly supports a mostly pionic, SN-powered origin of the γ-ray emission in star forming galaxies (e.g., Pavlidou & Fields 2001).

7 Conclusion

For a sample of galaxies for which pointed GeV-TeV data are available (from *Fermi*/LAT) and from Cherenkov telescopes), we compared values of $U_{\rm p}$ derived from the radio, γ-ray, and SN methods. Reasonable agreement among estimates based on the three methods is reached, showing clearly that there is a very substantial gap between the very high values ($U_{\rm p} \sim O(10^2)$ eV cm^{-3}) in central SB regions such as Arp 220, M 82, NGC 253 and NGC 4945, and the low values ($U_{\rm p} \sim O(10^{-1})\,{\rm eV\,cm^{-3}}$) in the less active SF regions in the disks of, e.g., Local Group galaxies.

The results of this study extend the validity of our previous findings:

(i) The Fermi-I acceleration mechanism, assumed to be at play in the environment of SN remnants, leads to acceleration timescales for CRp in galaxies such that particles and fields attain equilibrium over typical SF timescales, in agreement with observational evidence.

(ii) CR energy densities and magnetic fields, inferred from radio data under the assumption of energy equipartition, can be used as proxies of the actual quantities that are measured directly only from γ-rays. This could be particularly useful in the case of distant ($z \gtrsim 1$) galaxies, whose (unbeamed) γ-ray fluxes are too faint to be measured but whose radio fluxes are within reach of sub-mJy surveys (e.g., Tozzi et al. 2009).

References

[1] Abdo A.A., et al. (LAT Collab.) 2010a, ApJL, 709, L152
[2] Abdo A.A., et al. (LAT Collab.) 2010b, A&A, 523, L2
[3] Abdo A.A., et al. (LAT Collab.) 2010c, A&A, 512, A7
[4] Abdo A.A., et al. (LAT Collab.) 2010d, A&A, 523, A46
[5] Acciari V.A., et al. (VERITAS Collab.) 2009, Nature, 462, 770
[6] Acero F. et al. (HESS Collab.) 2009 Science, 326, 1080
[7] Ackermann M et al. (LAT Collab.) 2011, ApJ, 726:81
[8] Ackermann M et al. (LAT Collab.) 2012, ApJ, 755:A164
[9] Peng, F.K., et al. 2016, ApJL, 821:L20
[10] Aharonian F., et al. (HESS Collab.) 2006, Nature, 439, 695
[11] Bell A.R. 1978, MNRAS 182, 443
[12] Blair, W.P. 2014, ApJ, 788:55
[13] Blumenthal, G.R., Gould, R.J. 1970, Rev. Mod. Phys., 42, 237
[14] Chevalier R.A., Clegg A.W., 1985, Nature, 317, 44
[15] de Cea Del Pozo E., Torres D.F., Rodriguez Marrero A.Y., 2009, ApJ, 698, 1054
[16] Diehl R., Halloin H., Kretschmer K., et al., 2006, Nature, 439, 45
[17] Domingo-Santamaría E., Torres D.F., 2005, A&A, 444, 403
[18] Drury L.O'C., Aharonian F.A., Völk H.J. 1994, A&A, 287, 959
[19] Fermi E. 1954, ApJ, 119, 1
[20] Ginzburg V.L., Syrovatskii S.I., 1964, The Origin of Cosmic Rays (New York: Macmillan)
[21] Higdon J.C., Lingenfelter R.E., Ramaty R., 1998, ApJ, 509, L33
[22] Lenain J.-P., Ricci C., Türler M., Dorner D., Walter R. 2010, A&A, 524, 72L
[23] Lingenfelter R.E., Ramaty R., Kozlovsky B. 1998, ApJ, 500, L153
[24] Longair M.S., 1994, High Energy Astrophysics (2nd ed.; Cambridge: Cambridge Univ. Press)
[25] Mannucci F., Della Valle M., Panagia N., et al. 2005, A&A, 433, 807
[26] Paglione T.A.D., Marscher A.P., Jackson J.M., Bertsch D.L., 1996, ApJ, 460, 295

[27] Pavlidou V., Fields B.D. 2001, ApJ, 558, 63
[28] Persic M., Rephaeli Y., 2014, A&A, 567, A101
[29] Persic M., Rephaeli Y. 2010, MNRAS, 403, 1569
[30] Persic M., Rephaeli Y., Arieli Y., 2008, A&A, 486, 143
[31] Protheroe R.J., Clay R.W. 2004, PASA, 21, 1
[32] Rephaeli Y., 1979, ApJ, 227, 364
[33] Rephaeli Y., Arieli Y., Persic M. 2010, MNRAS, 401, 473
[34] Rephaeli Y., Persic M. 2015, ApJ, 805:111
[35] Seaquist, E.R., Odegard, N. 1991, ApJ, 369, 320
[36] Strickland D.K., Heckman T.M., 2009, ApJ, 697, 2030
[37] Strickland D.K., et al. 2004, ApJ, 606, 829
[38] Strong A.W., Porter T.A., Digel S.W., et al. 2010, ApJ, 722, L58
[39] Torres D.F., 2004, ApJ, 617, 966
[40] Tozzi, P., et al. 2009, ApJ, 698, 740
[41] van den Bergh S., Tammann G.A., 1991, ARAA, 29, 363
[42] Webber W.R. 1987, A&A, 179, 277
[43] Woosley S.E., Weaver T.A., 1995, ApJS, 101, 181

Two-zone model for the broadband Crab nebula spectrum: microscopic interpretation

F. Fraschetti[1,a] and M. Pohl[2,3]

[1] *Depts. of Planetary Sciences and Astronomy, University of Arizona, Tucson, AZ, 85721, USA*
[2] *DESY, 15738 Zeuthen, Germany*
[3] *Institute of Physics and Astronomy, University of Potsdam, 14476 Potsdam, Germany*

Abstract. We develop a simple two-zone interpretation of the broadband baseline Crab nebula spectrum between 10^{-5} eV and $\sim$ 100 TeV by using two distinct log-parabola energetic electrons distributions. We determine analytically the very-high energy photon spectrum as originated by inverse-Compton scattering of the far-infrared soft ambient photons within the nebula off a first population of electrons energized at the nebula termination shock. The broad and flat 200 GeV peak jointly observed by Fermi/LAT and MAGIC is naturally reproduced. The synchrotron radiation from a second energetic electron population explains the spectrum from the radio range up to $\sim$ 10 keV. We infer from observations the energy dependence of the microscopic probability of remaining in proximity of the shock of the accelerating electrons.

1 Introduction

The steady-state gamma-ray emission of the Crab Nebula has been observed in the past decade by a number of ground-based observatories, i.e., HEGRA [1], MAGIC [2], HESS II [3] and VERITAS [4]. The photon differential spectrum as determined by such experiments is consistent with a log-parabola distribution from $\sim$ 50 GeV up to 100 TeV. We propose a two-zone model with two distinct populations of energetic electrons having two distinct log-parabola distributions injected at the same energy to explain the broadband spectrum from 10^{-5} eV up to 100 TeV; however, the MeV region exceeds our prediction. In the literature, the broad and flat VHE peak at 200 GeV joining the TeV-band spectrum observed with MAGIC (0.05 – 30 TeV) with the GeV-band spectrum obtained with Fermi/LAT (1 – 200 GeV) has been empirically modelled with a modified log-parabola with an additional free-parameter that is not physically justified [2]. We show that such VHE peak is naturally explained by inverse-Compton scattering off monochromatic Infra-Red ambient photons within the nebula of a log-parabola distribution of electrons. Synchrotron radiation explains the radio/X-ray spectrum.

2 Outline of the model for steady injection of energetic particles

We assume that the number of electrons accelerated at the nebula termination shock (TS) with Lorentz factor $> \gamma$ is given by a log-parabola

$$N(>\gamma) = N_0(\gamma/\gamma_0)^{-[s-1+r\log(\gamma/\gamma_0)]}, \qquad (1)$$

[a] e-mail: ffrasche@lpl.arizona.edu

where N_0 is a normalization constant, γ_0 is the electron injection Lorentz factor and the spectral parameters s and r are related to the microscopic statistical parameters.

2.1 Inverse Compton flux

The target photon population within the nebula is assumed to be isotropic in the frame of the plasma flowing downstream of the shock at the scale of the VHE electrons mean-free path. The cold photons are IC scattered by the isotropic energetic electrons.

For a photon ambient population with a phase space photon density (number of photons per unit of volume and energy) given by $n(\varepsilon)$ in the plasma frame, the spectrum of the scattered photons with initial energy between ε and $\varepsilon + d\varepsilon$ and final energy ε_1 per single electron is [5]

$$\frac{dN_{\gamma,e}}{dtd\varepsilon_1} = \frac{3\sigma_T c}{4\gamma^2}\frac{n(\varepsilon)}{\varepsilon}d\varepsilon G(q,\Gamma), \tag{2}$$

where σ_T is the Thomson cross-section, c the speed of light in vacuum and

$$G(q,\Gamma) = 2q\ln q + (1+2q)(1-q) + \frac{\Gamma^2 q^2(1-q)}{2(1+\Gamma q)}, \quad \Gamma = \frac{4\varepsilon\gamma}{m_e c^2}, \quad q = \frac{\varepsilon_1}{\Gamma(m_e\gamma c^2 - \varepsilon_1)}. \tag{3}$$

and where Eqs (2, 3) are derived under the only restriction $\gamma \gg 1$. The calculation of the IC power emitted by a single electron in the plasma frame over the target photon distribution $n(\varepsilon)$ requires integration over the kinematically permitted photon final energies:

$$P_{IC} = \int_0^\infty d\varepsilon \int_0^{\varepsilon_1^M} d\varepsilon_1 \frac{dN_{\gamma,e}}{dtd\varepsilon_1}\varepsilon_1 = 12\sigma_T c\gamma^2 \int_0^\infty d\varepsilon\, \varepsilon n(\varepsilon) J(\Gamma), \tag{4}$$

where $\varepsilon_1^M = \Gamma m_e \gamma c^2/(1+\Gamma)$ and $J(\Gamma) = \int_0^1 dq\, qG(q,\Gamma)/(1+q\Gamma)^3$. The factor $J(\Gamma)$ can be approximated as follows: $J(\Gamma) \simeq 1/9$ in the Thomson limit ($\Gamma \ll 1$) and $J(\Gamma) \simeq [\ln\Gamma - 11/6]/2\Gamma^2$ in the extreme Klein-Nishima limit ($\Gamma \gg 1$) [6].

We use the simplifying assumption of a monochromatic target photon field, $n(\varepsilon) = n_0\delta(\varepsilon - \varepsilon_0)$, where ε_0 is determined by the joint fit of Fermi/LAT and MAGIC data. Thus, the power P_{IC} is given in the extreme Klein-Nishima limit by

$$P_{IC} = \frac{3}{8}\sigma_T c(mc^2)^2 \int_0^\infty d\varepsilon \frac{n(\varepsilon)}{\varepsilon}\left[\ln\frac{4\varepsilon\gamma}{m_e c^2} - \frac{11}{6}\right] \simeq \frac{3}{8}\sigma_T c(mc^2)^2\frac{n_0}{\varepsilon_0}\left[\ln\frac{4\varepsilon_0\gamma}{m_e c^2} - \frac{11}{6}\right]. \tag{5}$$

Given the differential electron distribution from Eq. 1,

$$\frac{dN(\gamma)}{d\gamma} = \frac{N_0}{\gamma_0}|-s+1-2r\log(\gamma/\gamma_0)|\left(\frac{\gamma}{\gamma_0}\right)^{-[s+r\log(\gamma/\gamma_0)]}, \tag{6}$$

the total IC power is found by folding $dN(\gamma)/d\gamma$ with P_{IC} in Eq. 5, $P_{IC}^{tot}(\nu) = \int d\gamma P_{IC} dN(\gamma)/d\gamma$. Finally, defining the total IC flux at Earth from a source at distance d as $\nu F_\nu^{IC} = P_{IC}^{tot}(\nu)/4\pi d^2$ leads to

$$\nu F_{IC,\nu}(\varepsilon) = \frac{\mathcal{A}'(\varepsilon_0)}{4\pi d^2}\left[\ln\frac{4\varepsilon_0\varepsilon}{(m_e c^2)^2} - \frac{11}{6}\right]\frac{N_0}{\gamma_0}\left[-s+1-2r\log(\varepsilon/\bar{\varepsilon}_0)\right](\varepsilon/\bar{\varepsilon}_0)^{-[s-1+r\cdot\log(\varepsilon/\bar{\varepsilon}_0)]} \tag{7}$$

where we have defined $\bar{\varepsilon}_0 = h\bar{\nu}_0 = m_e c^2\gamma_0$ and $\mathcal{A}'(\varepsilon_0) = (3/8)(h/m_e c)\sigma_T(m_e c^2)^3 n_0\gamma_0/h\varepsilon_0$ and we have used the monochromatic approximation $h\nu \simeq m_e c^2\gamma$ for the extreme Klein-Nishima limit. Equation 7 is a log-parabola distribution multiplied by two logarithmically dependent pre-factors in the squared brackets of the right hand side.

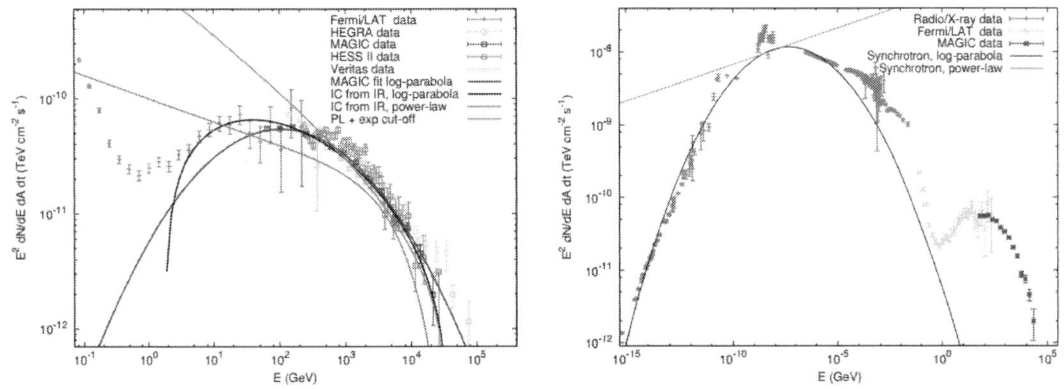

Figure 1. Left Theoretical photon differential spectrum (black) in Eq. 7 from inverse-Compton off a $\varepsilon_0 = 10^{-2}$ eV monochromatic target photon field. The theoretical curve with best-fit parameters (see Table 1) is compared with the data from Fermi/LAT [7], HEGRA [1], MAGIC [2], HESS II [3] and VERITAS [4] and the log-parabola fit (blue) of the MAGIC data with indexes $\alpha = 2.47$ and $\beta = -0.24$ (see Eq. 9). The red-curve shows the same spectrum as the black curve with $r = 0$. A power-law with exponential cut-off (grey) with slope 2.2 and energy cut-off $E_c = 6$ TeV is reported for illustrative purpose. **Right** Theoretical photon differential spectrum (in black) from Eq. 8 with best-fit parameters (see Table 1). The red-curve shows the same spectrum as the black curve with $r = 0$. The radio and soft X-rays data from ISO-SCUBA, Spitzer, XMM-Newton, SPI, IBIS/ISGRI (same set used for the broadband spectrum in [2]) are shown in red. Fermi/LAT and MAGIC data are overlaid for comparison.

2.2 Synchrotron flux

The synchrotron power emitted by a single electron, averaged over an isotropic electron distribution, is given in the plasma frame by $P_{sync} = \frac{\sigma_T c}{6\pi}\gamma^2 B^2$, where B is the external magnetic field assumed to be uniform. The total synchrotron power is found by folding the single electron power with the differential electron spectrum in Eq. 6: $P^{tot}_{sync} = \int d\gamma P_{sync} dN(\gamma)/d\gamma$. We use the monochromatic approximation with a characteristic synchrotron frequency $\nu_s = 0.29(3e\gamma^2 B)/(4\pi m_e c) = \gamma^2\nu_s^0$, so we recast $P^{tot}_{sync} = \int d\gamma P_{sync} dN(\gamma)/d\gamma|_{\nu=\nu_s}$. Thus, for $\nu \simeq \nu_s$, the total flux at Earth νF^{sync}_ν, related to synchrotron power $\nu F^{sync}_\nu = P^{tot}_{sync}(\nu)/4\pi d^2$ can be approximated as

$$\nu F^{sync}_\nu(\varepsilon) = \frac{\sigma_T c}{12\pi 4\pi d^2} N_0 B^2 \gamma_0^2 \left(\frac{\varepsilon}{\varepsilon_s^0\gamma_0^2}\right)^{-\frac{1}{2}[s-3+\frac{r}{2}\log(\varepsilon/\varepsilon_s^0\gamma_0^2)]} \left|-s+1-r\log(\varepsilon/\varepsilon_s^0\gamma_0^2)\right| \quad (8)$$

where we have defined $\varepsilon_s^0 = h\nu_s^0$.

3 Comparison with broadband observations

3.1 GeV and multi-TeV spectrum

We assume that energetic electrons inverse-Compton scatter off a gas of monochromatic IR photons within the nebula with energy $\varepsilon_0 = 10^{-2}$ eV and number density $n_0 = 5\times 10^2$ cm^{-3} (the distance from the source is $d = 1.9$ kpc). Figure 1, Left panel, compares the resulting analytic photon differential spectrum in the VHE range given by Eq. 7 with the spectra from various ground-based VHE observatories above 10 GeV: HEGRA (300 GeV- 100 TeV) [1], MAGIC (50 GeV - 30 TeV) [2], HESS II

(230 GeV - 25 TeV) [3] and VERITAS (115 GeV - 42 TeV) [4]. The best-fit parameters in Fig. 1 of the joint Fermi-LAT and MAGIC data are reported in Table 1. In Fig. 1, Left panel, the overlaid log-parabola empirical fit of the solely MAGIC data provided by [2] has the form, modulo an ε^2 factor,

$$\frac{dN}{d\varepsilon dA dt}(\varepsilon) = f_0(\varepsilon/\varepsilon_0^M)^{-\alpha+\beta\cdot\log(\varepsilon/\varepsilon_0^M)} \tag{9}$$

with power-law index $\alpha = 2.47 \pm 0.01$, curvature $\beta = -0.24 \pm 0.01$, scale $\varepsilon_0^M = 1$ TeV (note that $\varepsilon_0^M = \bar{\varepsilon}_0$) and normalization factor $f_0 = 3.23 \times 10^{-11}$ TeV^{-1} cm^{-2} s^{-1}. We note that, not suprisingly, the best-fit value of $2 - \alpha$ and β of MAGIC data are inconsistent with our value of joint Fermi/LAT MAGIC fit $-(s-1)$ and r. The grey curve in Fig. 1 shows that a power-law with exponential cut-off is unable to reproduce the VHE spectrum from the peak up to the highest energies (here the cut-off is set to $E_c = 6$ TeV from [2]). It turns out that there is no parameter combination for a simple power-law distribution of the electrons ($r = 0$) able to reproduce the broadband observations in the assumption of monochromatic photon target gas, but a different distribution of energetic particles is needed.

Figure 1, Left panel, shows that the MAGIC empirical log-parabola in Eq. 9 and our solution in Eq. 7 equally well reproduce the MAGIC data down to the IC peak at $\simeq$ 200 GeV. Our theoretical curve is also able to reproduce the broadness of the IC peak and the high-energy Fermi/LAT data; the logarithmic pre-factor in the second squared parentheses in Eq. 7 governs the width of the GeV peak that a simple log-parabola curve fails to reproduce.

The simple theoretical curve derived here is in good agreement with both the Fermi/LAT rising data up to the IC peak and the VHE data. The declining part of the Fermi-LAT spectrum below $\simeq$ 1 GeV is likely to arise from a distinct particle population or distinct acceleration process. The best-fit value of the VHE band in a magnetic field with strength $B = 140\,\mu$G gives a characteristic synchrotron photon energy $\varepsilon_s^0\gamma_0^2 = 2.7$ eV. The corresponding $\nu F_\nu^{sync} = P_{sync}^{tot}(\nu)/4\pi d^2$ (see Eq.8) with parameters fixed by fit of IC spectrum is inconsistent with radio and X-ray data.

3.2 Radio and X-ray spectrum

By using a constant and uniform magnetic field with strength $B = 140\,\mu$G, the radio and soft X-ray data (10^{-5} eV up to $\sim$ 10 keV) can be reproduced with the steady synchrotron emission from a log-parabola given by Eq. 8, as shown in Fig.1, right panel. However, the best-fit parameters differ from those in the VHE range (see Table 1) with the same injection $\gamma_0 \simeq 2 \times 10^6$. Such a parameter set leads to a predicted IC flux inconsistent with VHE observations. In addition, Fig. 1 shows that the range $0.1 - 10^3$ MeV requires a third energetic particles population or emission process that is not discussed in this work. The best-fit values of the indexes s, r are about twice for synchrotron than for IC (see Table 1). The reason is in the monochromatic approximation leading to the difference of a factor $1/2$ in the exponential of the log-parabola in Eq. 8 as compared to Eq. 7: the characteristic frequency scales as γ for IC and as γ^2 for synchrotron. We note that the Synchrotron self-Compton with the best-fit parameters of the synchrotron radiation is inconsistent with the Fermi/LAT and MAGIC data.

4 Discussion and Conclusion

The log-parabola particle spectrum can be derived in the assumption that the probability for energetic electrons of remaining in proximity of the acceleration region and not being advected downstream or leak upstream of the shock, P, decreases with γ as $P = g/\gamma^{-q}$, with the normalization constant g and $q > 0$ [8]. As well known, a P independent of energy results in a power-law particle spectrum. For a quasi-perpendicular shock geometry such as the nebula TS, to the lowest order in U_1/v where

Table 1. Best-fit value of the electron log-parabola distribution parameters within the specified range

	s	r	$N_0(10^{44})$	$\bar{\varepsilon}_0 = m_e c^2 \gamma_0$ (TeV)
Synchrotron (5×10^{-6} eV – 10 keV)	2.795 ± 0.003	0.2523 ± 0.0018	$10^3 \pm 90$	1.049 ± 99
IC ($2.2 - 10^4$) GeV	1.3616 ± 0.0054	0.06557 ± 0.0024	1.000 ± 0.009	1.003 ± 0.099

U_1 is the upstream flow speed, we find $q = 2rU_1/c$ (see [9] for detailed discussion). If $U_1/c \simeq 0.2$, we find for the two electron populations $q \simeq 0.026$ (VHE) and $q \simeq 0.1$ (Radio/X-ray). The origin of the two distinct electron populations is not discussed in this short note.

Acknowledgment

We thank D. Mazin, N. Otte and M. Holler for sharing low-energy, VERITAS and HESS II data respectively. The work of FF was supported in part by NASA under grants NNX13AG10G and NNX15AJ71G. FF thanks the organisers of the 6th RICAP held in 2016. FF thanks for the hospitality the Harvard/Smithsonian Center for Astrophysics where part of this work was performed.

References

[1] F. Aharonian, A. Akhperjanian, M. Beilicke, K. Bernlöhr, H.G. Börst, H. Bojahr, O. Bolz, T. Coarasa, J.L. Contreras, J. Cortina et al., *ApJ* **614**, 897 (2004), astro-ph/0407118
[2] J. Aleksić, S. Ansoldi, L.A. Antonelli, P. Antoranz, A. Babic, P. Bangale, J.A. Barrio, J. Becerra González, W. Bednarek, E. Bernardini et al., Journal of High Energy Astrophysics **5**, 30 (2015), 1406.6892
[3] M. Holler, D. Berge, C. van Eldik, J.P. Lenain, V. Marandon, T. Murach, M. de Naurois, R.D. Parsons, H. Prokoph, D. Zaborov et al., ArXiv e-prints (2015), 1509.02902
[4] Meagher, K. for the VERITAS Collaboration, ArXiv e-prints (2015), 1508.06442
[5] G.R. Blumenthal, R.J. Gould, Reviews of Modern Physics **42**, 237 (1970)
[6] R. Schlickeiser, *MNRAS* **398**, 1483 (2009)
[7] R. Buehler, J.D. Scargle, R.D. Blandford, L. Baldini, M.G. Baring, A. Belfiore, E. Charles, J. Chiang, F. D'Ammando, C.D. Dermer et al., *ApJ* **749**, 26 (2012), 1112.1979
[8] E. Massaro, M. Perri, P. Giommi, R. Nesci, *A&A* **413**, 489 (2004), astro-ph/0312260
[9] F. Fraschetti, M. Pohl, submitted, ArXiv e-prints **(2017), 1702.00816**

The DAMPE experiment: first data from space

Ivan De Mitri[1,2,a] (on behalf of the DAMPE Collaboration)

[1] *Dip.to di Matematica e Fisica "E. De Giorgi", Università del Salento, Via per Arnesano, I-73100, Lecce, Italy*
[2] *Istituto Nazionale di Fisica Nucleare (INFN), Sezione di Lecce, Via per Arnesano, I-73100, Lecce, Italy*

Abstract. The DAMPE satellite has been successfully launched in orbit on December 2015. The science goals of the mission include the study of high energy cosmic ray electrons, photons, protons and nuclei in a wide energy range: 10^9 – 10^{14} eV. A report on the mission status will be presented, together with on-orbit detector performance and first data coming from space.

1 Introduction

The DAMPE (DArk Matter Particle Explorer) satellite was launched on December 17, 2015 and is in smooth data taking since few days after. Data from various subdetectors allow an efficient identification of the electron signal over the large (mainly proton-induced) background. As a result, the all-electron spectrum will be measured with excellent resolution from few GeV up to few TeV, thus giving the opportunity to cleary observe a cutoff and/or identify a possible contribution of nearby sources. Moreover the study of cosmic ray (CR) electrons and high energy photons will allow an indirect search for dark matter signatures. Even if primarily optimized to collect electrons and gammas, DAMPE provides good tracking, calorimetric and charge measurements also in case of protons and nuclei. This will allow precise measurement of CR protons and nuclei energy spectra from few GeV up to about hundreds TeV. The high energy region will be explored with better precision compared to previous experiments: spectral indexes for individual species could then be well measured and the observed hardenings around 250 GeV/n could be studied and better quantified.

2 The DAMPE detector

DAMPE consists of a Plastic Scintillator strip Detector (PSD) that serves as anti-coincidence and charge detector, a Silicon-Tungsten tracKer (STK) to reconstruct the direction (and charge) of incident particles, a BGO imaging calorimeter (BGO) of about 32 radiation lengths that measures the energy with high resolution and distinguishes between electrons and protons, and a NeUtron Detector (NUD) that can further increase the hadronic shower rejection power (see Fig. 1).

The PSD is designed to measure the charge (Z) of incident high-energy particles, up to $Z = 26$, by looking at the energy release due to ionization process, that is approximately proportional to Z^2. It has a 825×825 mm^2 active area made by organic plastic scintillators. Only light materials have been

[a] e-mail: ivan.demitri@le.infn.it

 DOI: 10.1051/epjconf/201713602010

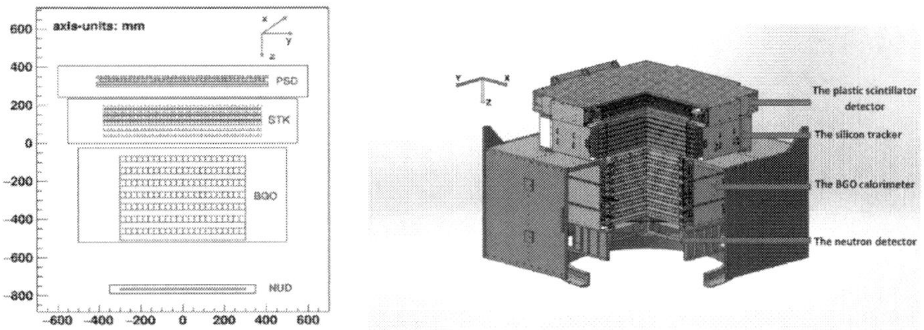

Figure 1. Structure diagram (left) and 3D view (right) of the DAMPE detector.

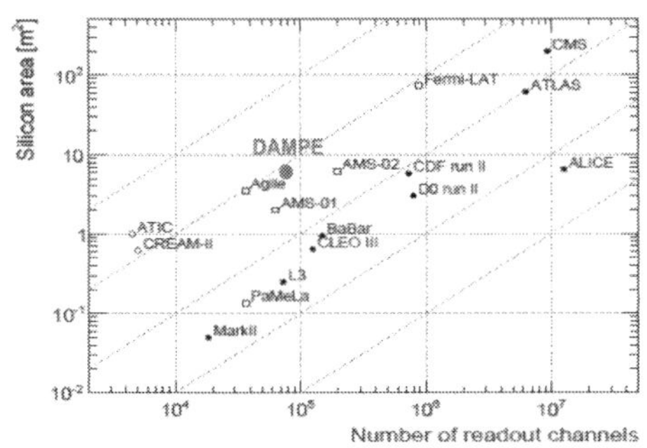

Figure 2. Comparison of the DAMPE STK features with other detectors.

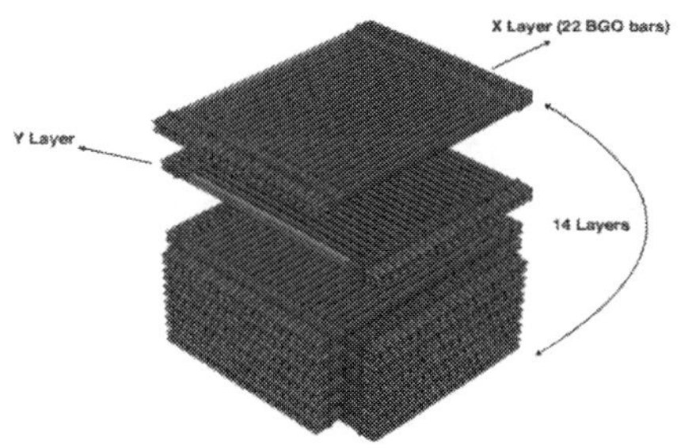

Figure 3. 3D view of the DAMPE BGO scintillator calorimeter.

used in the mechanical structure, in order to avoid initiating electromagnetic showers before the STK. It is arranged in a double layer configuration and has in total 82 detector modules. Each module is made by a long plastic scintillator bar of 825 mm with a $28 \times 10\,\text{mm}^2$ cross section and is read out by two PMTs at each end. The modules are staggered by 8 mm in the layer, and the bars in the two layers are perpendicular, resulting in a overall efficiency of 99.75% . A double dynode readout for each PMT has been implemented in order to cover a wide range, namely from 0.1 to 1400 times the energy released by a Minimum Ionizing Particle (MIP), with an overlap region for calibration.

The Silicon-Tungsten tracKer is devoted to the precise reconstruction of the particle track. It is made by twelve position-sensitive silicon detector planes (six planes for the x-coordinate, six planes for the y-coordinate). Three layers of tungsten are inserted in between the silicon $x - y$ planes (2, 3, and 4) to convert gamma rays in electron-positron pairs. The specifications of STK are given in Table 1 and a comparison with other experiments is shown in Fig. 2 looking at active area and number of channels. The tracking resolution is expected to be 0.2°. The STK analog readout can also provide an independent measurement of the incident particle charge, at least up to medium mass nuclei. It is a compact system where readout and power supply electronics are mounted on the sides [2, 5].

The BGO calorimeter is devoted to measure the energy deposition of incident particles and to reconstruct the shower profile. Moreover the trigger of the whole DAMPE system is based on the signals from the BGO and the shower profile is fundamental to distinguish between electromagnetic and hadronic showers. The calorimeter is composed of 308 BGO crystal bars ($2.5 \times 2.5 \times 60\,\text{cm}^3$ is the volume of each single bar). The crystals are optically isolated from each other and are arranged

Table 1. Main STK specifications.

Active area of each silicon layer	$0.5534\,m^2$
Silicon thickness	$320\,\mu m$
Silicon strip pitch	$121\,\mu m$
Tungsten thickness	1 mm
Depth (rad. lengths)	0.976
Power consumption	82.7 W
Mass	154.8 kg

Table 2. Main BGO specifications.

Active area	$60 \times 60\,cm^2$
Depth (rad. len.)	32
Sampling	> 90%
Long. segmentation	14 layers
Lat. segmentation	~ 1 Molière radius

horizontally in 14 layers each made by 22 bars (Fig. 3). The bars of a layer are ortogonal to those of the following one in order to reconstruct the particle path in both views ($x - z$ and $y - z$). The total vertical depth of the calorimeter is about 32 radiation lengths and 1.6 nuclear interaction lengths. This unprecedented depth ensures that almost 100% of the energy of electrons and γ-rays is deposited in the calorimeter, and about 40% in the case of protons anad nuclei. The high energy limits of the DAMPE measurement ranges (about few TeV for electrons/photons and hundreds TeV for nuclei, depending on the their fluxes) are manly determined by its very large geometric factor, which is about $0.3\,m^2$ sr for hadrons and even larger for electrons/photons. Table 2 summarizes the key parameters of BGO calorimeter. Each crystal element is read out by two PMTs, mounted on both ends of the crystal. The measured light asymmetry allows to further estimate where the energy has been deposited along the BGO bar. In order to cover a very large dynamic range (2×10^6), the signals are collected from three different dynodes covering different ranges on each side.

Typically hadron-induced showers produce roughly one order of magnitude more neutrons than electron-induced showers. The neutrons quickly thermalize in the BGO calorimeter and their activity can be detected by a suitable NeUtron Detector (NUD, see Fig. 1) within few μs, thus providing a further tool to distinguish the types of high-energy showers. The NUD is made by four boron-loaded plastics, each with a set of PMTs and related electronics. Indeed neutrons entering the boron-loaded scintillator undergo a capture process (namely ${}^{10}B + n \rightarrow {}^{7}Li + \alpha + \gamma$), whose probability is inversely proportional to their velocity, within an average capture time that is inversely proportional to the ${}^{10}B$ loading. The scintilators are wrapped with a layer of aluminium film for photon reflection, and anchored in an aluminium alloy framework. The space between plastic scintillators and the framework (1 mm on each side) has been filled with silicone rubber to relieve the vibration during the launch of the satellite.

3 Ground tests and on-orbit operation

An extensive Monte Carlo simulation activity was done during R&D phase in order to find a proper compromise between physics goals and limitations on geometry, power consumption and weight. DAMPE performances were verified by a series of beam tests at CERN with electrons, photons, pions, protons and nuclear fragments. The wide energy range of the electron beams and the high-purity of the proton beams were sufficient to check energy resolution (Fig. 4), linearity and e/p separation. Also a beam of argon fragments was used for testing the detector response to nuclei. Details of the test-beam results as well as the features of the qualification module can be found in [1–6]. Furthermore many tests (namely energy response to MIPs, efficiencies, detector alignment, etc.) have been performed during different stages of the DAMPE assembly using a large samples of CR mouns at ground.

After launch, the spacecraft entered a sky-survey mode immediately and a dedicated-calibration of the detector was performed in two weeks. The calibration included the determination of pedestals,

 DOI: 10.1051/epjconf/201713602010

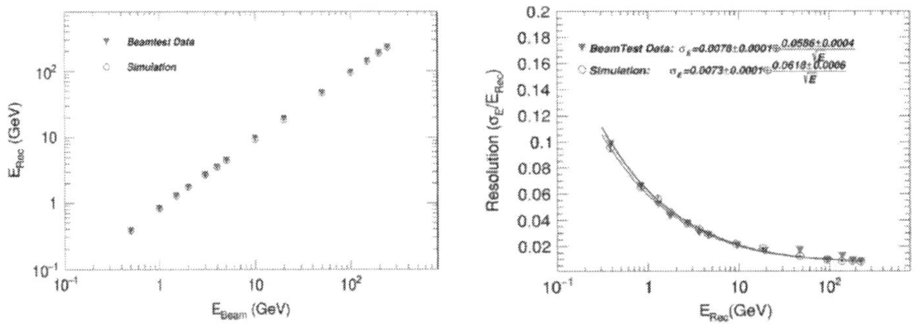

Figure 4. Linearity and resolution of the DAMPE BGO calorimeter for electromagnetic showers: comparison of beam test data with simulations [3].

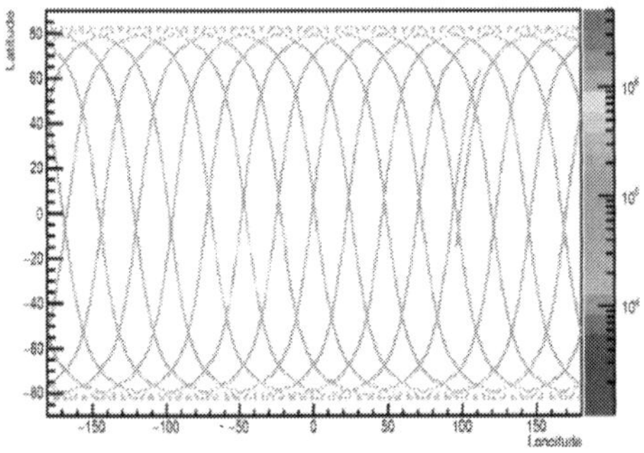

Figure 5. Detected counting rate vs latitude and longitude (arbitary units on the color palette). The South Atlantic Anomaly is clearly visible.

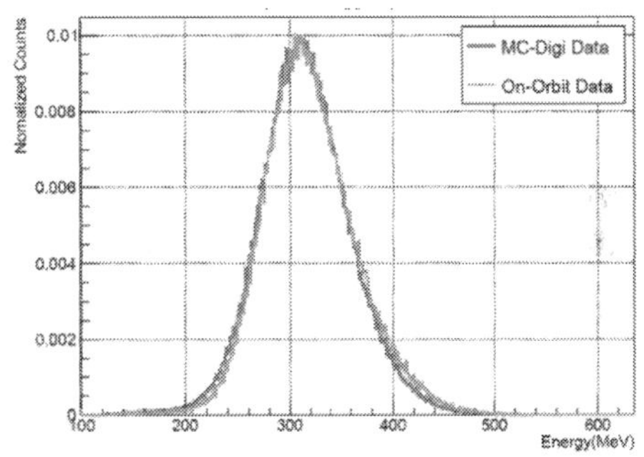

Figure 6. MIP peak in the BGO calorimeter: comparison of simulation with on-orbit data.

response to MIPs, alignment, and timing. By comparing on-orbit data with simulations and on-ground tests we can conclude that the detector is working very well. The satellite is on a solar-synchrous orbit lasting 95 minutes. The pedestal calibration is performed twice per orbit and the global trigger rate is kept at $\sim$ 60 Hz by using different pre-scale for unbiased and low-energy triggers at different latitudes (see Fig. 5 and [2]). Data are packaged with timestamp and transmitted to ground at a rate of about 5 millions events per day corresponding to $\sim$ 15 GB/day, that increases up to $\sim$ 100 GB/day after offline reconstruction.

4 First on-orbit data and expected performances

The DAMPE detector started taking physics data very soon after the launch. The performance parameters (temperature, noise, spatial resolution, efficiency) are very stable and very close to what expected. The absolute energy measurement has been checked by using the geomagnetic cut-off and it results well calibrated. Also the absolute pointing has been successfully verified. Indeed the photon data collected in 165 days were enough to draw a preliminary high-energy sky-map where the main gamma-sources are visible in the proper positions. Fig. 6 shows the MIP peak reconstructed with the energy released in the BGO crystals. It is evident the agreement with the distribution derived from the simulation. The energy released in the PSD allows measuring the particle charge and distinguishing the different nuclei in the CR flux. Figs. 7 and 8 show preliminary results of these measurements for

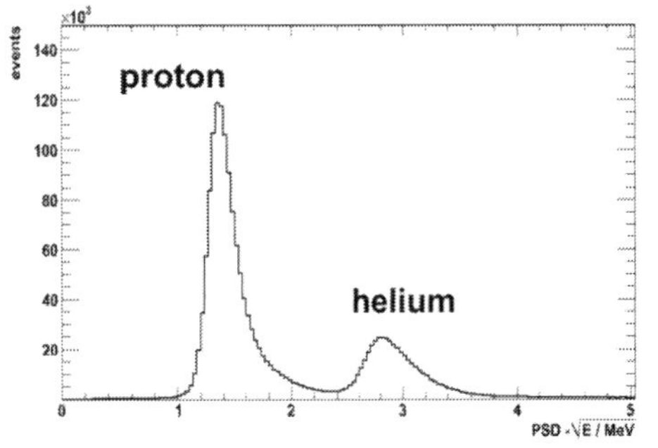

Figure 7. Preliminary distribution for light CR component as measured in the DAMPE PSD.

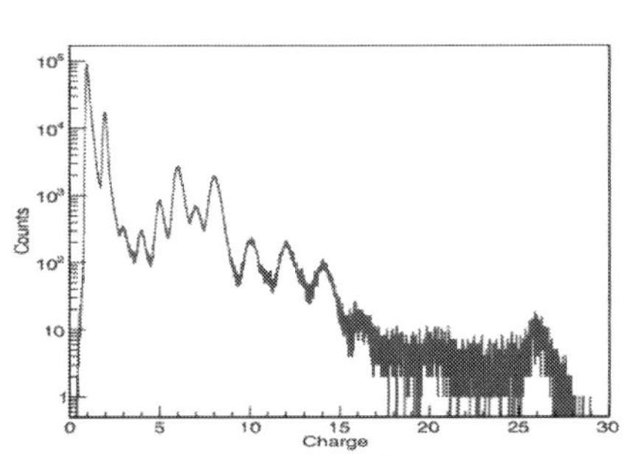

Figure 8. Preliminary Z measurement up to iron with the DAMP PSD.

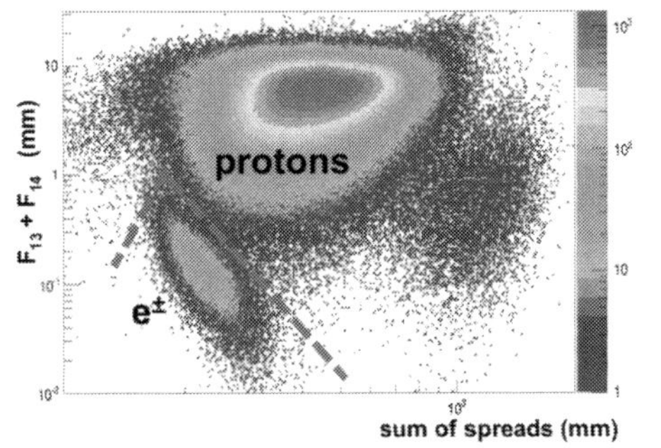

Figure 9. Preliminary result about e/p separation (see text). The background on the right is due to CRs entering the satellite from the sides.

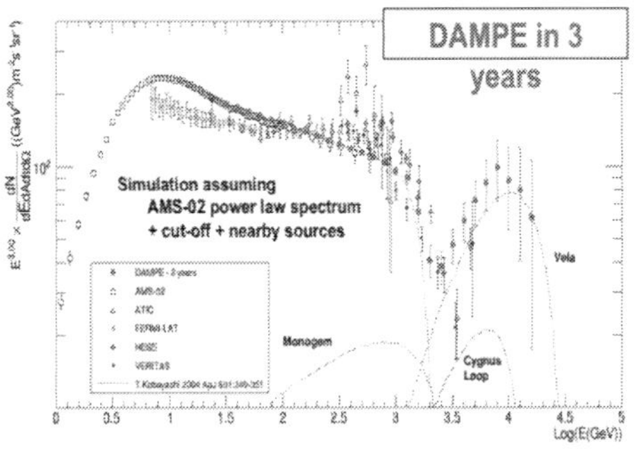

Figure 10. All-electron spectrum. The red dots represent the possible DAMPE measurement in 3 years.

the light component (H and He) and for the full range up to iron ($1 \leq Z \leq 26$), respectively, for a first sample of on-orbit data.

The measurement of the electron (plus positron) flux is one of the main goal of the DAMPE mission. The shower development in the BGO provides the tool to distinguish leptons from hadrons. Among others, a shape parameter has been defined: $F_i = spread_i \times \frac{E_i}{E_{tot}}$, where i is the index of the BGO layer ($1 \leq i \leq 14$), $spread_i$ is the shower width in a given layer, E_i and E_{tot} are the energy on the single i-th layer and on all the calorimenet, respectively. Using the shape parameters on the last BGO layers (13, 14) it is possible to separate leptons from hadrons with a very large rejection power (preliminary results in Fig. 9). The rejection capability will be further enhanced by means of the NUD.

The DAMPE detector is expected to take data for more than 3 years. This time interval is sufficient to deeply investigate many open questions in galactic CR studies. In Fig. 10 the possible DAMPE measurement of the so called "all-electron" spectrum in 3 years is shown. Given the excellent resolution, spectral features like a possible cut-off at about 1 TeV or a flux increase due to nearby astrophysical sources or dark matter signatures [7], could be well detected.

Many experiments [8–11] observed a discrepant hardening of the CR elemental spectra at about 250 GeV/n. This is another interesting topic related to CR origin and propagation. Due to its very large geometric factor, DAMPE will be able to significantly reduce statistical uncertainties in the break region and give a precise measurements on the high energy spectral index of each nuclear

DOI: 10.1051/epjconf/201713602010

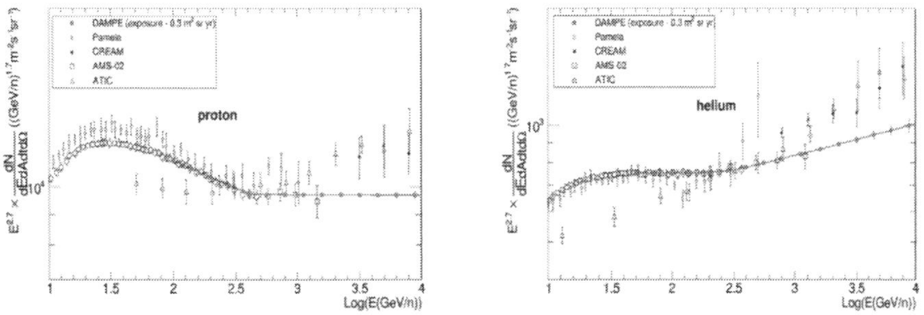

Figure 11. Proton (left) and helium (right) spectra in the range 10 GeV - 10 TeV. The red dots indicate the possible DAMPE measurement (for an exposure of 0.3 m^2 sr yr) assuming the AMS-02 data fit and its extrapolation to high energy.

species (see Fig. 11). Moreover the large exposure will allow extending energy spectra measurements for protons and nuclei up to hundreds TeV. This also apply quantities like the boron-to-carbon ratio, that could be well measured up to about 10 TeV/n.

5 Conclusions

The DArk Matter Particle Explorer (DAMPE) is a space mission supported by the strategic space projects of the Chinese Academy of Sciences with the contribution of Swiss and Italian (namely INFN and MIUR) institutions. The rocket has been successfully launched on December 17, 2015 and DAMPE presently flies regularly on a sun-synchronous orbit at the altitude of about 500 km. Four different detectors are arranged on the satellite: a plastic scintillator array (PSD), a silicon-tungsten tracker (STK), a BGO calorimeter and a neutron detector (NUD). They are devoted to measure the fluxes of charged CRs (electrons, protons and heavier nuclei), to study high energy gamma rays from astrophysical sources and to search for indirect dark-matter signatures. The main technical features of the detectors and a short review of the first preliminary results have been presented.

References

[1] J. Chang, Chinese J. Space Science **34** (2014) 550
[2] G. Ambrosi et al. (DAMPE Collaboration), "The DAMPE mission", to be submitted
[3] Z. Zhang et al., Nuclear Instrum. & Meth. in Phys. Res. **A836** (2016) 98
[4] V. Gallo et al., Proc. of the 34th Intern. Cosmic Ray Conference (2015) 1199
[5] P. Azzarello et al., Nuclear Instrum. & Methods **A 831** (2016) 378
[6] Z. Zhang et al., Nuclear Instrum. & Methods **A 836** (2016) 98
[7] T. Kobayashi et al., The Astrophys. Jour. **601** (2004) 340 ; D. Grasso et al., Astroparticle Physics **32** (2009) 140
[8] A.D. Panov et al., Bull. Russ. Acad. Sci. **71** (2007) 494
[9] H.S. Ahn et al., Astrophys. J. Lett. **714** (2010) L89; Y.S. Yoon et al., Astrophys. J. **728** (2011) 122
[10] O. Adriani et al., Science **332** (2011) 69
[11] L. Accardo et al., Phys. Rev. Lett. **114** (2015) 171103

DOI: 10.1051/epjconf/201713602011

CaloCube: a novel calorimeter for high-energy cosmic rays in space

A. Rappoldi[1,a], P.W. Cattaneo[1,b], O. Adriani[2,3], A. Agnesi[1,4], S. Albergo[5,6], L. Auditore[7,8], A. Basti[10], E. Berti[2,3], G. Bigongiari[9,10], L. Bonechi[2,3], S. Bonechi[9,10], M. Bongi[2,3], V. Bonvicini[11], S. Bottai[3], P. Brogi[9,10], G. Cappello[6], G. Carotenuto[12], G. Castellini[3,13], R. D'Alessandro[2,3], S. Detti[3], M. Fasoli[14,15], N. Finetti[3,16], A. Italiano[6], P. Lenzi[3], P. Maestro[9,10], P.S. Marrocchesi[9,10], M. Miritello[17], N. Mori[2,3], M. Olmi[2,3], G. Orzan[11], L. Pacini[2,3,13], P. Papini[3], M.G. Pellegriti[6], F. Pirzio[1,4], S. Ricciarini[3,13], P. Spillantini[2], O. Starodubtsev[3], F. Stolzi[9,10], J.E. Suh[9,10], A. Sulaj[9,10], A. Tiberio[2,3], A. Tricomi[5,6], A. Trifirò[7,8], M. Trimarchi[7,8], E. Vannuccini[3], A. Vedda[14,15], G. Zampa[11], and N. Zampa[11]

[1] *INFN Pavia, Via Bassi 6, I-27100 Pavia, Italy*
[2] *University of Florence, Department of Physics and Astronomy, via G. Sansone 1, I-50019 Sesto Fiorentino (Firenze), Italy*
[3] *INFN Firenze, via B. Rossi 1, I-50019 Sesto Fiorentino (Firenze), Italy*
[4] *University of Pavia, Dipartimento di Ingegneria Industriale e dell' Informazione, I-27100 Pavia, Italy*
[5] *University of Catania, Department of Physics and Astronomy, via S. Sofia 64, I-95123 Catania, Italy*
[6] *INFN Catania, via S. Sofia 64, I-95123 Catania, Italy*
[7] *University of Messina, Dipartimento di Scienze MIFT, viale F. Stagno d' Alcontres, I-98166 Messina, Italy*
[8] *INFN Messina, c. da Papardo, sal. Sperone 31, I-98166 S. Agata, Messina, Italy*
[9] *University of Siena, Department of Physical Sciences, Earth and Environment, I-53100 Siena, Italy*
[10] *INFN Pisa, via F. Buonarroti 2, I-56127 Pisa, Italy*
[11] *INFN Trieste, via Valerio 2, I-34127 Trieste, Italy*
[12] *Institute of Polymer, Composites and Biomaterials (CNR), piazzale E. Fermi 1, I-80055 Portici (Napoli), Italy*
[13] *IFAC (CNR), via Madonna del Piano 10, I-50019 Sesto Fiorentino (Firenze), Italy*
[14] *University of Milano Bicocca, Department of Materials Science, via Cozzi 55, I-20125 Milano, Italy*
[15] *INFN Milano Bicocca, piazza della Scienza 3, I-20126 Milano, Italy*
[16] *University of L'Aquila, Dipartimento di Scienze Fisiche e Chimiche, Coppito, I-67100 L'Aquila, Italy*
[17] *MATIS (CNR), via S. Sofia 64, I-95123 Catania, Italy*

Abstract. CaloCube is an R&D project borne to develop a novel calorimeter design, optimized for high-energy cosmic ray measurements in space. A small prototype made of CsI(Tl) elements has been built and tested on particle beams. A final version, made of 5x5x18 crystals and with dual readout (two photodiodes for each crystal), to cover the full required dynamic range, is under construction and will be tested at CERN SPS in Summer 2016. The dual readout compensation technique were developed and the feasibility to extract Čerenkov signals from CsI crystals verified.

1 Introduction

The study of cosmic rays can be done either through indirect or direct experiments. The former are commonly achieved by means of ground-based detectors, which detect the secondary particles produced by cosmic rays of very high energy ($E \geq 1$ PeV) interacting with the Earth's atmosphere,

[a]e-mail: andrea.rappoldi@pv.infn.it
[b]e-mail: paolo.cattaneo@pv.infn.it

taking advantage of the multiplication effect due to the shower development. On the contrary, the direct measurement concerns the individual detection of the primary cosmic rays of energy up to 10^{15} eV (given the rapid attenuation of the flux with increasing energy), which is of great interest in the study of cosmic rays produced by galactic sources.

This type of measurement is done using instruments placed abovee the atmosphere (on board of orbiting satellites), and represents a significant technological challenge, because such detectors must have: large geometrical acceptance, high energy resolution and optimal particle identification capability, with a limited weight and low power consumption.

2 The CaloCube concept

The direct measurement of individual protons and nuclei spectra in space at high energy (up to 10 PeV) requires to have: an extremely large acceptance (few $m^2 sr$), a good energy resolution (better than 40%) and mass identification capability, whereas the direct measurement of the electron component above 10 TeV requires: an excellent energy resolution (better than 2%), high h/e rejection (better than 10^5) and large acceptance above 1 TeV. Such stringent requirements can be obtained using a calorimeter in conjunction with a dE/dx detector, and for this purpose the R&D project **CaloCube**[1] was created, with the aim of studying the design and optimization of a calorimeter for measurements of high-energy cosmic rays in space[1]. The proposed solution consists of a finely segmented calorimeter made of

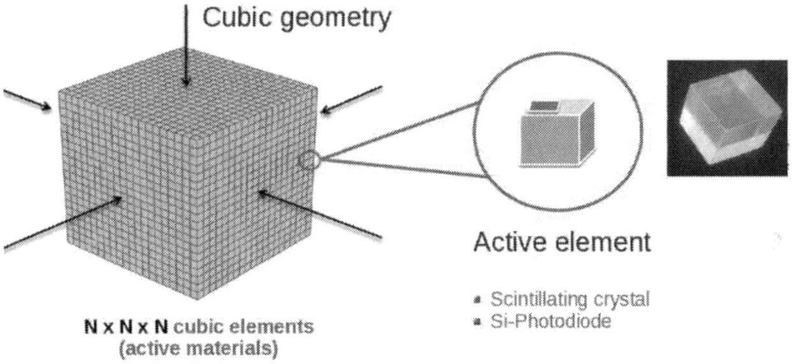

Figure 1. The CaloCube layout is based on a quasi-isotropic geometry with cubic symmetry, where the active volume is made of NxNxN small cubic scintillator crystals, optically isolated from each other, read from one or more solid state detectors (Si-Photodiode).

a large number of active scintillating elements, arranged in a geometry with cubic symmetry in order to have a quasi-isotropic response (see Fig. 1). Each active element, acting as an independent scintillator, is optically isolated from each other, and is read by one or more silicon photodiode (to have a larger dynamic range). The scintillating crystals also work as active absorber (concerning the energy deposition), resulting in a good energy resolution. Moreover, the 3D segmentation of the calorimeter allows to obtain a precise shower imaging[2].

The calorimeter response depends on the adopted geometry and on the material used for the single scintillating crystals. For this reason, an accurate Montecarlo, based on the FLUKA package, has been developed to carefully study the calorimeter operation in different configurations.

A NxNxN cubic geometry, with cubes of about 1 Molière radius size each, has been studied with different scintillating materials, sizes and gaps (distance between the adjacent cubes). Some re-

[1]CaloCube is an R&D project financed by INFN for 3 years (end 2016).

sults are summarized in Table 2, where different configurations, comparable in terms of total weight ($W_{tot} \simeq 2$ tons) and active volume fraction (about 78%) are shown. Among the five materials, LYSO

	CsI(Tl)	BaF$_2$	YAP(Yb)	BGO	LYSO(Ce)
l (cm)	3.60	3.20	2.40	2.30	2.10
Gap (cm)	0.30	0.27	0.20	0.19	0.18
N cubes	20 x 20 x 20	22 x 22 x 22	28 x 28 x 28	27 x 27 x 27	30 x 30 x 30
L (cm)	78.0	76.3	72.8	67.2	68.2
L_{int} (λ_{int})	1.80	2.31	3.09	2.72	3.01
L_{rad} (X_0)	38.88	34.73	24.96	55.54	53.75
GF ($m^2 sr$)	9.56	9.15	8.32	7.10	7.35

Table 1. List of some of the studied calorimeter geometry. Each configuration has been simulated with a FLUKA based Montecarlo, in order to find the best behavior of the calorimeter, having fixed some global parameters, such as the total weight $W_{tot} \simeq 2$ tons.

($Lu_{1.8}Y_{0.2}SiO_5(Ce)$) is the best one for protons, due to the better shower containment, which compensates for the smaller volume (due to its high density of 7.1 g/cm^3).

The collection efficiency of the scintillation light depends on the type of coating used to reflect the light and isolate each crystal from the adjacent ones. Several different materials have been tested measuring the signal induced by a 5.5 MeV α emitted by a ^{241}Am source, and the result (in terms of signal amplitude) is shown in Fig. 2.

3 Test and results

A small mechanical structure has been built, to allow the test of different geometric configurations on an accelerator beam. In particular, a prototype with 135 cubic crystals of 3.6 cm size, arranged in 15 planes of 3 x 3 cubes each, with a gap of 0.4 cm between them, has been implemented (Fig. 3). In this case, CsI(Tl) has been chosen for practical reasons: it is widely available on on the market at an affordable price, it has a very high light yield and its emission spectrum matches very well the spectral response of a large variety of Si photodetectors.

This prototype results to have a shower containment of about 1.5 Molière radius (R_M) and an active depth of 28.4 X_0 and 1.35 λ_{int}, and has been tested at CERN with different particle beams, summarized in Table 3.

Test	Beam	Energy
Feb. 2013	ions Pb + Be	13-30 GeV
Mar. 2015	Ions Ar + Poly	19-30 GeV
Aug./Sep. 2015	μ, π, e	50, 75, 150, 180 GeV

During the last test beam at CERN[2](Summer 2015), the prototype was initially exposed to μ beams, to obtain the response of each crystal to a minimum ionizing particle (m.i.p.), thus determining the individual conversion factor between the deposited energy and the photodiode signal (Fig. 4). These factors are used to equalize the response of all the cubes that compose the calorimeter.

Then an estimate of the energy resolution has been determined, exposing the calorimeter to e$^+$/e$^-$ beams of different energy, and determining the total deposited energy, given by the sum of the equalized signals of all the cubes. A preliminary result is shown in Fig. 5, referring to a beam of 50 GeV electrons. In this case the measured total energy (expressed in m.i.p) is in good agreement with the expectation, and the corresponding resolution is at level of 1%.

[2] *This project has received funding from the European Union's Horizon 2020 research and innovation programme under grant agreement No 654168*

 DOI: 10.1051/epjconf/201713602011

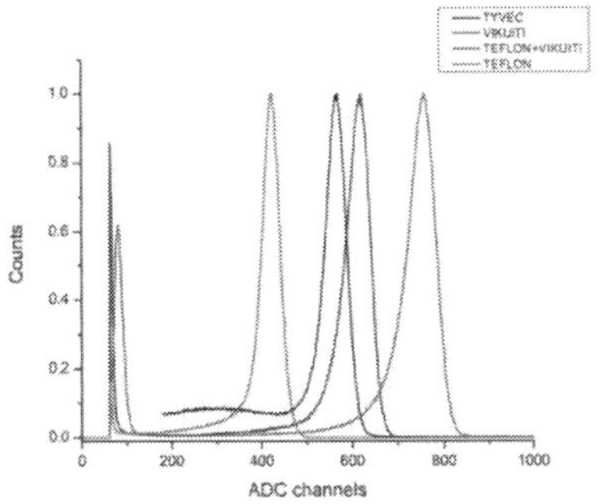

Figure 2. Signal amplitude of a single CsI(Tl) scintillator cubic crystal of size (36 mm)^2 with different coating materials.

Figure 3. The prototype with 3 x 3 x 15 CsI(Tl) crystals used to perform the test beams at CERN in 2015.

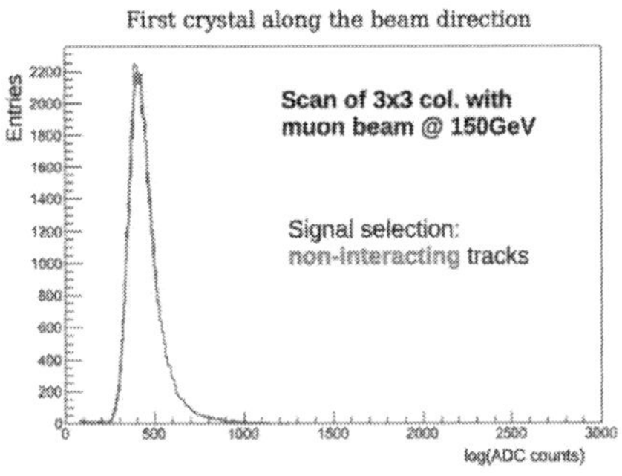

Figure 4. Response of a single crystal to a non-interacting track at 150 GeV (mainly μ). The distribution is well fit with a Landau distribution (red).

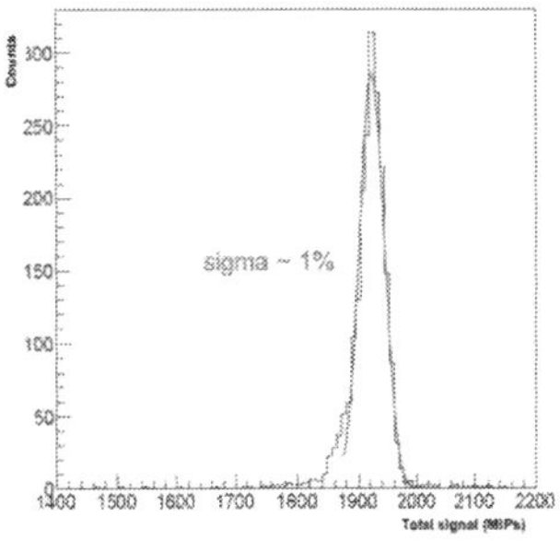

Figure 5. Measured distribution of total energy (expressed in m.i.p.) released with a 50 GeV electrons beam, fit with with the expected distribution (red). The resulting energy resolution is about 1%.

4 Next developments

A new calorimeter prototype, consisting of 5 x 5 x 15 cubes, is in preparation and under test at CERN in the summer 2016. The larger number of cubes (especially in the transverse dimension) allows to have a better event containment, even for hadronic showers. In this prototype the light signal of each crystal will be read out by two photodiodes, with different sensitive area, in order to cover the full expected dynamic range.

Furthermore, some tests are in progress to verify the feasibility of detecting also the Čerenkov light produced by charged particles inside the crystals (CsI is transparent down to 340 nm wavelength), to increase the calorimeter performance. The first results demonstrate that the separation between the scintillation and Čerenkov light can be obtained relying both on timing (prompt vs. delayed signal) and on wavelength (UV vs. green).

References

[1] M. Bongi et al., J. Phys. Conf. Ser. **587**, 012029 (2015)
[2] R. D'Alessandro et al., Nucl. Instrum. Meth. **A824**, 609 (2016)

Precision study of radio emission from air showers at LOFAR

Olaf Scholten[1,2,a], Antonio Bonardi[5], Stijn Buitink[3], Arthur Corstanje[5], Ute Ebert[6,7], Heino Falcke[4,5], Jörg Hörandel[5], Pragati Mitra[3], Katharine Mulrey[3], Anna Nelles[8], Jörg Rachen[5], Laura Rossetto[5], Casper Rutjes[6], Pim Schellart[9], Satyendra Thoudam[10], Gia Trinh[1], Sander ter Veen[1], and Tobias Winchen[3]

[1] *University of Groningen, KVI Center for Advanced Radiation Technology, 9747 AA Groningen, The Netherlands*
[2] *Vrije Universiteit Brussel, Dienst ELEM, IIHE, Pleinlaan 2, 1050 Brussels, Belgium*
[3] *Astrophysical Institute, Vrije Universiteit Brussel, Pleinlaan 2, 1050 Brussels, Belgium*
[4] *Netherlands Institute for Radio Astronomy (ASTRON), 7990 AA Dwingeloo, The Netherlands*
[5] *Department of Astrophysics/IMAPP, Radboud University, P.O. Box 9010, 6500 GL Nijmegen, The Netherlands*
[6] *CWI, Centrum Wiskunde & Informatica, Amsterdam, The Netherlands*
[7] *TU/e, Eindhoven University of Technology, Eindhoven, The Netherlands*
[8] *Physics and Astronomy, University of California, Irvine, CA 92697-4575,U.S.A*
[9] *Department of Astrophysical Sciences, Princeton University, Princeton, NJ 08544, USA*
[10] *Department of Physics and Electrical Engineering, Linnéuniversitetet, 35195 Växjö, Sweden*

Abstract. Radio detection as well as modeling of cosmic rays has made enormous progress in the past years. We show this by using the subtle circular polarization of the radio pulse from air showers measured in fair weather conditions and the intensity of radio emission from an air shower under thunderstorm conditions.

The paper emphasizes the measurement of circular polarization of the radio pulse from air showers as a great technical achievement, without explaining what its source is. It is stated that the main polarization is linear from the transverse current in the shower front (presumably from the opposite bending of electrons and positrons in the geomagnetic field) and the polarization of the Askaryan radiation is radially polarized. Is circular polarization somehow a coherent combination of these two? Whatever the case, its source and significance should be described.

1 Introduction

We will show that even the finest details of radio emission, such as its circular polarization, are well understood. Thus the detecting the radio emission from extensive air showers (EASs) as induced by energetic cosmic rays is understood to the extent that it can be regarded as a very sensitive way to determine shower properties, such as energy [1, 2] and $X_{\rm max}$, the atmospheric (slant) depth where the number of air-shower particles reaches a maximum [3, 4]. These shower properties are used in turn to infer the nature of the primary cosmic ray, in particular its mass [5, 6]. We are now even at the stage

[a] e-mail: Scholten@KVI.nl

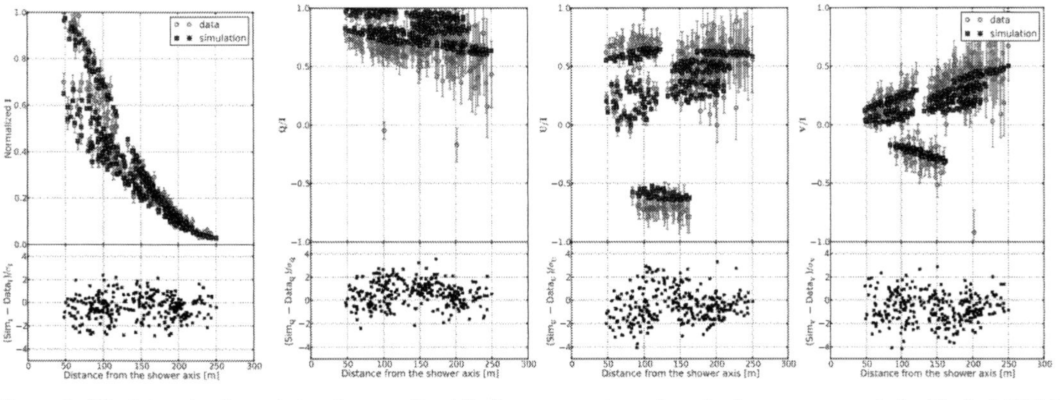

Figure 1. The intensity footprint and normalized Stokes parameters of an air shower as recorded with the LOFAR low-band antennas and projected into the shower plane (open red circles) is compared to the results of a CoREAS simulation (filled blue squares). σ denotes one standard deviation error.

where radio emission can be used to learn about shower development in the atmosphere and thus -for example- strength and direction of atmospheric electric fields [7].

To do so, detailed models have been developed that calculate the radio emission from the EAS based on the motion of individual electrons and positrons in the air shower. Two such microscopic models are CoREAS [8] and ZHAireS [9]. Both successfully reproduce the features of the radio emission [4]. The signal has a dominant linearly-polarized component along the direction of the Lorentz force, $\hat{e}_{\vec{v}\times\vec{B}} \propto \hat{e}_{\vec{v}} \times \hat{e}_{\vec{B}}$, due to the induced transverse current in the shower front [10]. Here the shower direction is given by $\hat{e}_{\vec{v}}$ while $\hat{e}_{\vec{B}}$ denotes the direction of the geomagnetic field. A secondary contribution, also known as Askaryan radiation [11], is due to the build-up of excess negative charge in the shower front [12]. This Askaryan radiation is radially polarized and thus leads to the prediction that the deviation from the main polarization direction depends on the viewing angle, in good agreement with observations [13–17].

2 Circular polarization in fair-weather events

A very complete analysis of the polarization data of radio pulse as measured at LOFAR for fair weather events has been made in Ref. [16, 18], using the Stokes parameters I, Q, U, and V. In terms of the Stokes parameters the linear polarization angle with the $\vec{v}\times\vec{B}$-axis is obtained as $\psi = \frac{1}{2}\tan^{-1}(U/Q)$, while V/I specifies the circular polarization.

In Fig. 1 the measured values of the Stokes parameters for one particular event are show as function of distance to the shower axis. The data are compared with the results of a CoREAS simulation showing an excellent agreement between the two, not only for the intensity I, but in particular also for V/I, the amount of circular polarization. The circular polarization of the radio pulse can be understood by regarded it as the superposition of two separate pulses, one due to geomagnetic and one to charge excess, that have a relative time-delay and a different polarization direction. As a result the polarization vector rotates over the duration of the pulse giving rise to a finite value for V/I.

Since the dominant polarization of the pulse is in the $\vec{v}\times\vec{B}$-direction the ratio of U/I and V/I as given in Fig. 1 can be used to extract the time-delay as explained in detail in Ref. [18]. One observes that at a distance of about 100 m $V/U \approx 0.3$ for antennas at an angle close to 90° with respect to the $\vec{v}\times\vec{B}$ axis (corresponding to extreme values for U and V). Since at 100 m from the axis the signal is dominated by coherent Cherenkov emission the pulse is very narrow in time, thus wide in frequency. In the measured frequency band 30-80 MHz, with mean frequency of 50 MHz, a value $V/U = 0.3$ corresponds to an arrival time difference of approximately $\Delta t = 1$ ns for the two polarization directions [18].

To understand the difference in the timing of the radio pulse emitted through the two mechanisms requires more subtle arguments [18]. For the transverse current contribution the emitted radiation relates to the vector potential as $\vec{E}_{tc} = -d\vec{A}/dt$ while for the charge excess contribution $\vec{E}_{ce} = -dA^0/dr$. Since the different components of the vector potential have a very similar dependence on retarded time ($t_{\rm ret} = Z/c$ where z is the distance along the shower axis to the impact point on Earth) and $dt_{\rm ret}/dt \approx c(R/r)\, dt_{\rm ret}/dr$ [10], where R is the distance to the emission point, we can see that for transverse current emission the lower parts of the shower (small R) are weighted less as compared to charge excess emission. This results in a mean emission height difference of about 1 km translating to an arrival time difference at $d = 100$ m of 1 ns, in agreement with the data.

3 Thunderstorm events

An interesting application of using the measured radio-footprint to infer the structure of the currents in the shower is given in Ref. [7] where this technique is used to determine the structure of the atmospheric electric fields. The polarization direction determines the direction of the fields while the circular polarization tells about the change of orientation of the fields.

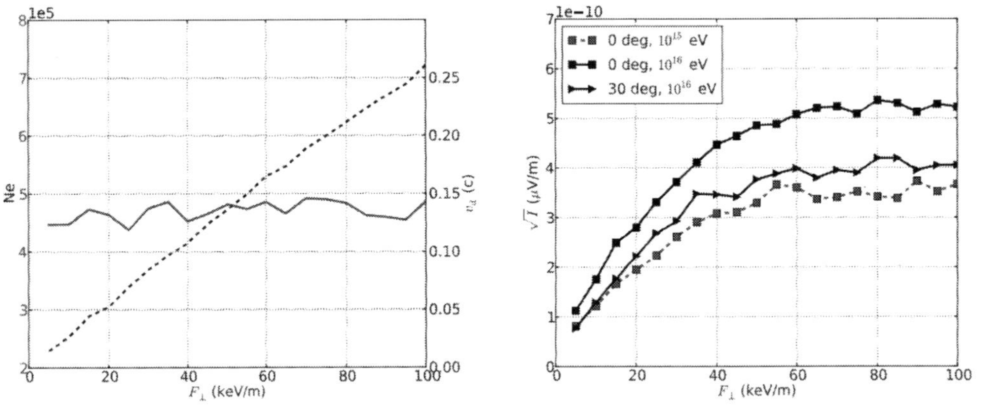

Figure 2. The left hand shows the number of electrons (solid blue line, left axis) and their drift velocity at X_{max} (dashed black line, right axis) of vertical 10^{15} eV showers as a function of the net-transverse forces. On the right the square root of the power $\sqrt{I}$ in the frequency interval of 30-80 MHz at the ring of maximal intensity is shown for vertical 10^{15} eV showers (dashed line) as well as for vertical and inclined 10^{16} eV showers (solid lines) as a function of the net-transverse force. For the 10^{16} eV showers the $\sqrt{I}$ is scaled down by a factor 10.

The amplitude of the radiation (square-root of the intensity) is expected to be proportional to the induced current in the shower front which is the product of the number of charged particles and the drift velocity. As is shown in the left hand plot in Fig. 2, with increasing transverse field the number of charged particles (solid blue line, left axis) remains almost constant while the drift velocity (dashed black line, right axis) shows an almost linear increase. This dependence continues up to a rather large field strength of 100 kV/m. The intensity of the radiation as shown on the right of Fig. 2

shows that only for small fields the thus expected linear increase of the amplitude is observed in the CoREAS simulations while for electric fields exceeding ($E = 50\ kV/m$) the amplitude predicted by the simulation saturates [19].

This counter-intuitive dependency can be understood by investigating the structure of the showers in more detail as is done in Ref [19]. Particularly important in this respect is the structure of the shower front, the pancake. With increasing transverse drift velocity the longitudinal component of the velocity has to decrease as the total velocity of the particles should remain below c, the light velocity. With decreasing longitudinal velocity the particles trail further behind the shower front thus increasing the pancake thickness. Since the radiation of the pancake can only be coherent for wavelength exceeding its thickness [20] the radiation lacks coherence for the frequency interval under consideration. The increase of the current with increasing field strength is thus compensated by the loss of coherence resulting in an amplitude that even drops for very large fields. An immediate consequence of this is that at lower frequencies the amplitude remains sensitive to the field [19].

4 Conclusions

In recent years much progress has been made in the understanding of Radio emission from air showers. Examples of this are subtle circular polarization of air showers in fair weather conditions and the less subtle saturation of the amplitude for showers exposed to rather large thunderstorm electric fields. Because of this high level of understanding, radio emission can be used as a diagnostic tool for the determination of air showers parameters such as X_{max} [3, 6] and atmospheric electric fields [7].

References

[1] A. Nelles *et al.*, J. Cosm. and Astrop. Phys. **05**, 018 (2015); arXiv:1411.7868.
[2] A. Aab *et al.*, Phys. Rev. D **93**, 122005 (2016); arXiv:1508.04267.
[3] S. Buitink *et al.*, Phys. Rev. D **90**, 082003 (2014).
[4] T. Huege, Phys. Rep. **620**, 1 (2016); arXiv:1601.07426.
[5] W.D. Apel *et al.*, Phys. Rev. D **90**, 062001 (2014).
[6] S. Buitink *et al.*, Nature **531**, 70 (2016).
[7] P. Schellart *et al.*, Phys. Rev. Lett **114**, 165001 (2015)
[8] T. Huege, M. Ludwig, and C. James, AIP Conf. Proc. **1535**, 128 (2012).
[9] J. Alvarez-Muñiz *et al.*, Astropart. Phys. **35**, 325 (2012).
[10] O. Scholten, K. Werner, and F. Rusydi, Astropart. Phys. **29**, 94 (2008); K. Werner and O. Scholten, Astropart. Phys. **29**, 393-411 (2008).
[11] G.A. Askaryan, Sov. Phys. JETP **14**, 441 (1962).
[12] K.D. de Vries *et al.*, Astropart. Phys. **34**, 267 (2010).
[13] V. Marin, for the CODALEMA Collaboration, Proc. 32nd ICRC, Beijing, **1**, 291 (2011).
[14] H. Schoorlemmer, for the Pierre Auger Collaboration, Nucl. Instr. and Meth. A **662**, S134 (2012).
[15] A. Aab *et al.*, Phys. Rev. D **89**, 052002 (2014); arXiv:1402.3677.
[16] P. Schellart *et al.*, J. Cosm. and Astrop. Phys. **1410**, 014 (2014); arXiv:1406.1355.
[17] A. Bellétoile *et al.*, Astropart. Phys. **69**, 50 (2015)
[18] O. Scholten *et al.*, Phys. Rev. D **94**, 103010 (2016); arXiv:1611.00758.
[19] T.N.G. Trinh *et al.*, Phys. Rev. D **93**, 023003 (2016), arXiv:1511.03045.
[20] Olaf Scholten, Krijn D. de Vries, Klaus Werner, Nucl. Instr. and Meth. A **662**, S80 (2012).

Exploiting the radio signal from air showers: the AERA progress

B. Revenu[1,a] for the Pierre Auger Collaboration[2]

[1] *Subatech, Université de Nantes, École des Mines de Nantes, CNRS/IN2P3, Nantes, France*
[2] *For the complete author list, see http://www.auger.org/archive/authors_2016_09.html*

Abstract. We present the lastest results and status of the Auger Engineering Radio Array (AERA), located within the Pierre Auger Observatory. AERA, with more than 150 radio stations spread over 17 km^2, is the largest radio detector in the world for extensive air showers above 10^{17} eV. The electric field emitted by secondary electrons and positrons allows us to estimate all characteristics of the primary cosmic ray: arrival direction, energy and mass composition. The performance of AERA together with the analysis methods are described. The final aim of AERA is mainly to improve the composition estimation of ultra-high energy cosmic rays as a standalone detector or in association with other instruments such as a ground particle detector or a fluorescence telescope.

1 Introduction

There are many aspects still not fully understood in the field of ultra-high energy cosmic rays. For the past decade, a clear energy cutoff has been observed in the energy spectrum around 4×10^{19} eV (see [1–3]). Nevertheless, the sources are still unidentified as no significant clustering has been discovered at the highest energies and no departure from isotropic expectations are observed at lower energies at any angular scales [4]. Determining the composition could be the missing key to a much better understanding of the origin of ultra-high energy cosmic rays. Air showers are produced after a first collision of the primary cosmic ray with the atmosphere's constituent nuclei. The secondary electrons and positrons of these showers emit a coherent radio signal in the range 1-300 MHz. This signal allows us to reconstruct the arrival direction and the energy of the cosmic ray, together with the shower's depth of maximum development $X_{\max}$ which is essential to estimate its nature [5–7]. The uncertainty on the $X_{\max}$ is of the same order as that achieved using the fluorescence light (~ 20 g/cm^2 [8]) emitted by the atmospheric nitrogen excited by the air shower. A strong asset of the radio technique is its duty cycle of around 100% compared to 14% with a fluorescence telescope [9].

2 The Auger Engineering Radio Array

The Auger Engineering Radio Array is the largest radio detector of cosmic rays in the world. It is located at the Pierre Auger Observatory [10] and covers an area of 17 km^2. The Observatory contains a surface detector (SD) of 1660 water Cherenkov tanks spread over 3000 km^2, a fluorescence detector (FD) of 27 telescopes at 4 sites around the SD, and the AMIGA instrument with burried

[a] e-mail: revenu@in2p3.fr

scintillators that allows the measurement of the muonic component of air showers. As of the end of 2016, AERA consists of 153 autonomous radio stations running continuously thanks to solar panels and batteries. Radio stations with log-periodic dipole antenna (LPDA) composed the first AERA setup of 24 detectors in 2011, covering 0.6 km^2 and communicating through fibers with the central data acquisition system. In 2013, 100 new stations with butterfly antenna were installed and 29 more in 2015 to complete the actual setup which is presented in Fig 1. The electric field is recorded in the band 30-80 MHz with a sampling rate of 180 MHz or 200 MHz depending on the electronics in use. The detectors run in self-trigger mode together with an external trigger coming from the SD and the

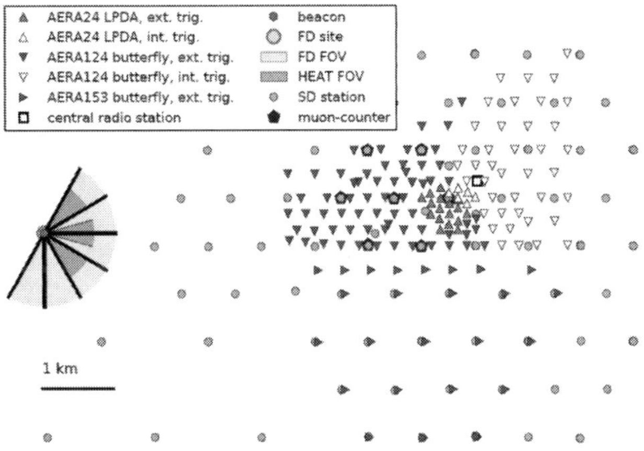

Figure 1. The AERA antennas (triangles) among the other instruments of the Pierre Auger Observatory (water Cherenkov tanks of the SD, fluorescence telescopes and AMIGA muon counters).

FD. The results presented here are those obtained using the external trigger mode.

3 Calibration

The radio stations are calibrated in both time and amplitude. For timing (absolute and relative), we use GPS receivers in conjunction with a reference beacon located at the closest FD site (see Fig 1) and signals emitted by commercial airplanes passing over the AERA site. We obtain an accuracy better than 2 ns [11]. For the amplitude, we first used weather balloon flights equipped with a calibrated transmitter. For cost reasons, we preferred to use a drone carrying the same kind of calibrated light source. The overall accuracy, taking into account all systematic errors, is 9.3% [12], on a LPD antenna. We plan to repeat this calibration procedure to the butterfly antennas.

4 The radio signal from air showers

The electric field emitted by air showers is the superposition of two distincts mechanisms. In the frequency range of interest in AERA (30-80 MHz), the dominant contribution is the geomagnetic effect: the systematic opposite drift of positrons and electrons during their propagation in the geomagnetic field results in an average transverse current leading to an electric field polarized in the direction of $\mathbf{v}\times\mathbf{B}$, where $\mathbf{v}$ is the direction of the shower axis and $\mathbf{B}$ is the geomagnetic field direction. The electric field direction does not depend on the observer's location.

In addtion to the geomagnetic mechanism, the excess of electrons over positrons in air showers leads to another source of electric field, known as the charge-excess contribution. This net negative charge is due to the fast in-flight annihilation of positrons and to the extraction of electrons from the medium. The electric field produced through this mechanism is radially polarized (with respect to the shower axis) and therefore, depends on the observer's location.

A given observer will measure the superposition of both mechanisms as shown in Fig 2. This means that both the amplitude and polarization depend on the ground coordinate with respect to the shower axis. Both mechanisms have been clearly seen in the data thanks to the significant progress of

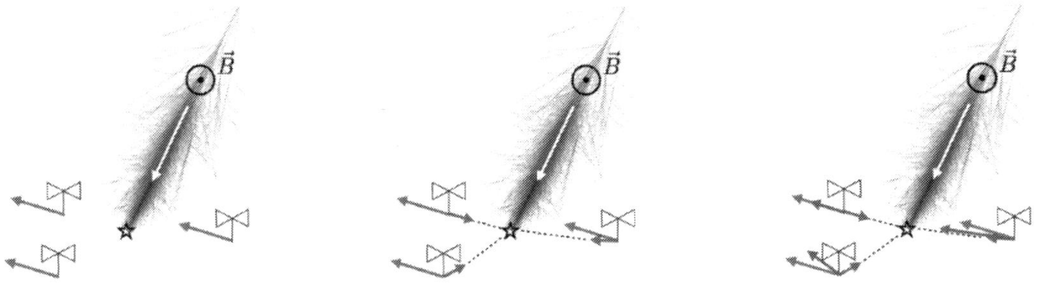

Figure 2. Scheme of an air shower falling between three observers, symbolized by the butterfly antennas. The geomagnetic field **B** is indicated. The electric fields produced by the geomagnetic and charge-excess mechanisms are indicated in green and red respectively. The observer measures the superposition of both electric fields, in blue.

the simulation codes these past few years [13–15]. The impressive agreement between the data and the simulation allows us to use the simulation to extract all characteristics of the primary cosmic ray.

5 Extracting the primary cosmic ray characteristics

As argued in the previous section, the amplitude and polarization of the electric field depend on the observer's location. We cannot describe the electric field using a simple 1D lateral distribution function (LDF). A 2D LDF has been proposed [16], using the code CoREAS [17], as a model of ground distribution of the electric field. This LDF permits us to estimate the amount of energy emitted as radiation in 30-80 MHz from the cosmic ray. We first correct the data recorded by each station participating in an event from the antenna and electronics responses. After this step, we obtain the electric field as a function of time; we can afterwards compute the energy fluence (energy per unit area) using the time integration of the associated Poynting vector. Then we fit the 2D LDF to extract the ground distribution of the energy fluence that we can integrate over space to get the total energy released as radiation. This radiated energy is rescaled by a factor $1/\sin^2\alpha$ to correct for the angular separation α from the geomagnetic field. The final estimator is strongly correlated (quadratic scaling as expected for coherent radiation) with the energy of the primary cosmic ray and we obtain an energy of 15.8 ± 0.7 (stat) $\pm$ 6.7 (sys) MeV for a primary energy of 10^{18} eV and an angular separation from the geomagnetic field of 90° [18, 19].

Concerning the reconstruction of the shower $X_{\max}$, the method is also based on simulations. There are several methods in use in AERA, but as an illustration we can compare the measured amplitudes for an actual event to simulated amplitudes using a simulation code (here SELFAS [20]). We use the fact that showers initiated by light or heavy nuclei have very different electric field distributions at ground level. Very roughly, young showers (initiated by light nuclei) have strong and peaked distributions when old showers (initiated by heavy nuclei) have weaker and wider distributions. Using a set of simulated showers initiated by protons and iron nuclei with the same arrival direction as an actual shower, we obtain the best agreement of the 2D profiles between the data and the simulation for showers in a preferred range of X_{max}. This way, we have an estimation of the best X_{max} together with the confidence interval [7], which in turn gives important constraints on the nature of the primary cosmic ray. Working on a set of air showers detected in coincidence by both the FD and AERA, we can compare in Fig 3 the X_{max} estimated using the radio signal with the FD direct measurement. The correlation is very clear and proves that the radio signal from air showers allows the measurement of the X_{max}.

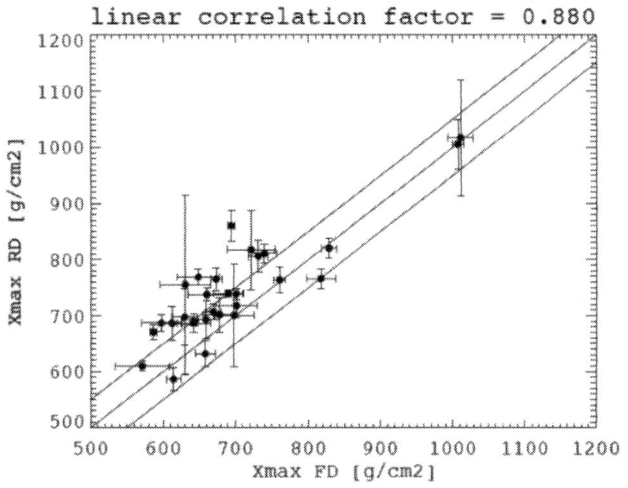

Figure 3. Correlation between the X_{max} estimated using the radio signal and the X_{max} measured by the fluorescence light for showers detected in coincidence by AERA and the Pierre Auger Observatory fluorescence detector.

6 Conclusion

AERA has been running continuously since 2011 and detects air showers also measured by the other instruments of the Pierre Auger Observatory providing an ideal observation of showers with 4 different observables. AERA contains 153 radio stations spread over 17 km^2. The radio signal contains the information needed to reconstruct all characteristics of the primary cosmic rays, in particular their energy and the X_{max} of the showers they initiate when entering our atmosphere. One of the next steps will be to provide the composition of ultra-high energy cosmic rays using the radio signal.

References

[1] J. Abraham et al. (Pierre Auger Collaboration), Phys.Lett. **B685**, 239 (2010), 1002.1975
[2] R.U. Abbasi et al. (HiRes), Phys. Rev. Lett. **100**, 101101 (2008), astro-ph/0703099
[3] T. Abu-Zayyad et al. (Telescope Array), Astrophys. J. **768**, L1 (2013), 1205.5067
[4] P.L. Ghia (Pierre Auger), PoS **ICRC2015**, 034 (2016)

[5] P.A. Bezyazeekov et al. (Tunka-Rex), JCAP **1601**, 052 (2016), 1509.05652
[6] S. Buitink, A. Corstanje, H. Falcke, J.R. Hörandel, T. Huege, A. Nelles, J.P. Rachen, L. Rossetto, P. Schellart, O. Scholten et al., Nature **531**, 70 (2016), 1603.01594
[7] F. Gaté, The Pierre Auger Collaboration, *X_{max} reconstruction from amplitude information with AERA*, in *Proceedings of the ARENA 2016 workshop (Groningen, The Netherlands)* (ARENA, 2016)
[8] J. Abraham et al. (Pierre Auger Collaboration), Phys.Rev.Lett. **104**, 091101 (2010), 1002.0699
[9] J. Abraham et al. (Pierre Auger Collaboration), Nucl.Instrum.Meth. **A620**, 227 (2010), 0907.4282
[10] A. Aab et al. (Pierre Auger), Nucl. Instrum. Meth. **A798**, 172 (2015), 1502.01323
[11] A. Aab et al. (Pierre Auger), JINST **11** (2016), 1512.02216
[12] R. Krause, The Pierre Auger Collaboration, *Detection of High-Energy Cosmic Rays with the Auger Engineering Radio Array*, in *Proceedings of the 14th VCI, Vienna* (Elsevier, 2014)
[13] V. Marin, the CODALEMA Collaboration, *Charge excess signature in the CODALEMA data. Interpretation with SELFAS2*, in *proceedings of the 32nd ICRC, Beijing, China* (ICRC, 2011)
[14] The Pierre Auger Collaboration, Phys. Rev. D **89** (2014)
[15] A. Belletoile, R. Dallier, A. Lecacheux, V. Marin, L. Martin, B. Revenu, D. Torres Machado, Astroparticle Physics **69**, 50 (2015)
[16] A. Nelles et al., JCAP **1505**, 018 (2015), 1411.7868
[17] T. Huege, M. Ludwig, C. James, *Simulating radio emission from air showers with CoREAS*, in *Proceedings of the ARENA 2012 workshop (Erlangen, Germany)* (to be published), AIP Conference Proceedings
[18] A. Aab et al. (Pierre Auger), Phys. Rev. **D93**, 122005 (2016), 1508.04267
[19] A. Aab et al. (Pierre Auger), Phys. Rev. Lett. **116**, 241101 (2016), 1605.02564
[20] V. Marin, B. Revenu, Astropart.Phys. **35**, 733 (2012), 1203.5248

Measurement of the cosmic ray flux with the ANITA experiment

Daniel García-Fernández[1,a], Jaime Alvarez-Muñiz[2], Washington R. Carvalho Jr.[3], Harm Schoorlemmer[4], and Enrique Zas[2]. Based on joint work with the ANITA collaboration

[1] *Subatech, CNRS. École des Mines, Université de Nantes*
[2] *Departamento de Física de Partículas, Universidade de Santiago de Compostela*
[3] *Instituto de Física, Universidade de São Paulo*
[4] *Max Planck Institut für Kernphysik*

Abstract. The ANITA experiment consists on an aerostatic balloon flying over Antarctica and carrying a payload with antennas. Although ANITA was designed to detect the electric field of netrino-induced showers in the ice cap, it has also detected 16 radio pulses coming from extensive air showers, and the ANITA collaboration has used these data to produce the first cosmic ray flux measurement obtained by employing radio as a stand-alone technique. We review the experimental results and its interpretation. We also focus on the simulations and the method used for obtaining the cosmic ray flux.

1 Introduction

The ANITA experiment [1], originally conceived for the detection of the radio emission coming from neutrino-induced showers, detected 16 pulses emitted by cosmic ray showers. Out of those 16 radio pulses, 14 were reflected on the polar ice cap [2].

The arrival times of the pulses coming from different parts of the shower differ, due to reflection, from those an antenna located on the ground would see, and a realistic treatment of this process is key to the understanding of the received electric field.

In this work we discuss the calculation of the electric field via the ZHAireS code [3] and review the experimental results obtained by ANITA.

2 The experiment

The ANtarctic Impulse Transient Antenna (ANITA) (see Fig. 1) is an experiment composed of a balloon with antennas flying over Antarctica at an altitude of $\sim$ 36 km and with a receiving band streching from 200 MHz to 1200 MHz. The choice of a balloon allows to cover a wide fiducial area and helps to reduce the anthropogenic noise. There have been three different ANITA flights (I, II and III) [4], with the fourth scheduled for December 2016.

ANITA I found that the measured electric field for certain events was polarised perpendicularly to both the shower axis and the geomagnetic field, suggesting that the electric field had been created by means of the geomagnetic effect, and therefore its origin was a cosmic ray shower. The radiation

[a] e-mail: garciafe@subatech.in2p3.fr

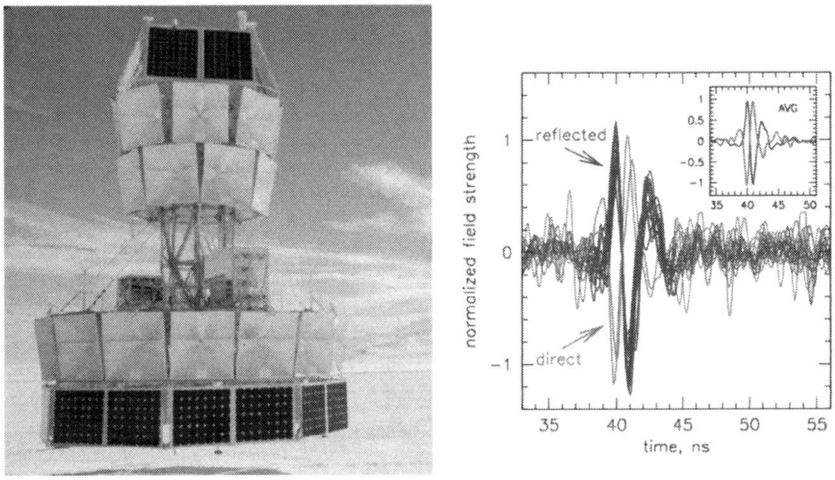

Figure 1. Left: Picture of the ANITA payload. Taken from [1]. Right: voltage as a function of time for the 16 cosmic ray events detected by ANITA. The reflected events possess an opposite polarisation to that of the direct events. Taken from [2].

of 14 of these events came from the ground and had a polarisation opposed to that of the 2 events that came from avobe the horizon, implying a reflection on the ground [2]. The electric field induced by the shower illuminates the ground, gets reflected and propagates through the atmosphere until it arrives at the payload.

3 Simulations

The reflection and the propagation in the atmosphere change the arrival times of the radiation. For these reflected events, it is important to model the atmosphere properly (its sphericity), calculate the arrival times according to the altitude-dependent refractive index and treat the reflection on the ground.

In order to do so, we used the ZHAireS code [5], a combination of the Aires Monte Carlo for particle showers [6] with the ZHS algorithm for the calculation of the electric field [7]. We upgraded the code to a new version called ZHAireS-Reflex [3] which embeds the treatment of the reflection. Our main assumptions were the reflection on a flat surface, the application of the Fresnel coefficients at the interface and the rectilinear propagation of the electric field. A ray-tracing model showed that this rectilinear propagation is valid for the ANITA configuration up to 85° zenith angle showers.

Simulations with ZHAireS-Reflex show the existence of a region where the field is coherent even at 1.4 GHz and that the Cherenkov cone is preserved upon reflection (see Fig. 2, left). The spectrum falls exponentially as a function of the frequency (see Fig. 2, right). Besides, the slope of the spectrum falls is dependent on the off-axis angle, defined as the angle between the observer and the shower axis. Another important result is that the flux of the electric field scales quadratically with the energy of the primary particle. This scaling is a general feature of the electric field produced by air showers that is preserved after reflection.

This suggests a method for measuring the energy of cosmic rays: the slope of the spectrum tells us the off-axis angle, which allows us to estimate the electric field flux at the Cherenkov angle. This is then used for calculating the energy of the cosmic ray.

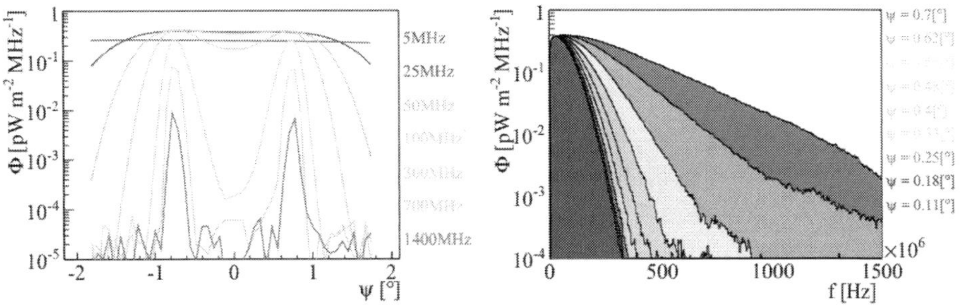

Figure 2. Left: electric field flux as a function of the off-axis angle for several frequencies. Right: electric field flux as a function of the observation frequency for several off-axis angles. See text and [3] for details.

4 Data analysis

The detailed analysis can be found in [8]. When treating the data for the cosmic ray events that ANITA found, the exponential fall-off of the spectrum with frequency was observed. Fits obtained from simulations help us to obtain the electric field at the Cherenkov angle using the amplitude of the measured spectrum at 300 MHz and the constant for the exponential fall-off of the spectrum. Afterwards, the energy of the primary is calculated.

The analysis included a treatment of the defocusing of the electric field due to the curvature of the Earth, and also the loss of coherence due to the roughness of the Antarctic ice. These are also a sort of systematic uncertainties, along with the uncertainties on the shower maximum, the atmospheric refractive index, *etc*, included in the analysis.

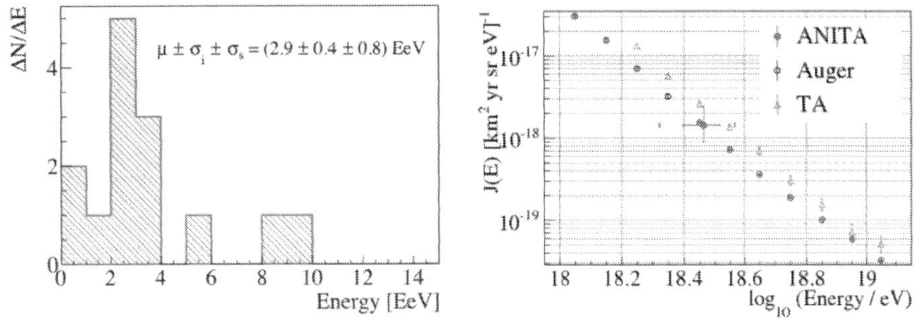

Figure 3. Left: energy distribution of the cosmic ray events detected by ANITA, with the mean energy depicted. Right: flux for the cosmic ray events measured by ANITA and its comparison with the spectra obtained by the Pierre Auger Observatory and Telescope Array. See text and [8] for details.

The energy distribution of the reflected events was obtained, with its mean being $\mu \pm \sigma_i \pm \sigma_s =$ $(2.9 \pm 0.4 \pm 0.8)$ EeV. σ_i corresponds to the statistical uncertainty and σ_s corresponds to the energy scale uncertainty due to the systematic errors of the method and the measurements.

After a calculation of the acceptance of the detector, a point for the cosmic ray flux spectrum was calculated. The result is in agreement within uncertainties with the spectra measured by Pierre

Auger and Telescope Array and it constitutes the first cosmic ray flux measurement using radio as a stand-alone technique.

5 Conclusions

The ZHAireS-Reflex code has been shown to reproduce some of the important features of the cosmic ray events observed by ANITA, such as the spectrum of the electric field and the coherence up to the GHz frequency. As evidenced by the data, a single payload is enough to measure the radio spectrum. The simulations obtained with the ZHAireS-Reflex code were used, together with the ANITA data, for measuring the energy of the primary cosmic rays and the primary cosmic ray spectrum. This is the first analysis that has produced a cosmic ray spectrum using the radio technique exclusively.

Acknowlegdements

We thank Région Pays de la Loire, Ministerio de Economía (FAPESP 2015/15735-1), Consolider-Ingenio 2010 CPAN Programme (CSD2007-00042), Xunta de Galicia (GRC2013-024), Feder Funds Marie Curie-IRSES/EPLA-NET (European Particle physics Latin American NETwork) and 7th Framework Program (PIRSES-2009-GA-246806) for their financial support.

References

[1] P.W. Gorham, *et al.*, Astroparticle Physics **32**, 10 (2009)
[2] S. Hoover, *et al.*, Physical Review Letters **105** (2010)
[3] J. Alvarez-Muñiz, W. Carvalho, D. García-Fernández, H. Schoorlemmer, E. Zas, Astroparticle Physics **66**, 31 (2015)
[4] P. Gorham, *Overview of the Third Flight of the ANITA Long-duration Balloon Payload*, in *Proceedings of the 34th International Cosmic Ray Conference (ICRC)* (2015), 1111
[5] J. Alvarez-Muñiz, W.R.C. Jr., E. Zas, Astroparticle Physics **35**, 325 (2012)
[6] S. Sciutto, <http://www.fisica.unlp.edu.ar/auger/aires/>
[7] D. García-Fernández, J. Alvarez-Muñiz, W.R.C. Jr., A. Romero-Wolf, E. Zas, Physical Review D **87** (2013)
[8] H. Schoorlemmer, *et al.*, *Energy and flux measurements of ultra-high energy cosmic rays observed during the first ANITA flight* (2016)

 DOI: 10.1051/epjconf/201713602015

The FAST Project - A Next Generation UHECR Observatory

Toshihiro Fujii[1,a], Max Malacari[2], Jose A. Bellido[3], Pavel Horvath[4], Miroslav Hrabovsky[4], Jiaqi Jiang[2], Dusan Mandat[5], Ariel Matalon[2], John N. Matthews[6], Pavel Motloch[2], Miroslav Palatka[5], Miroslav Pech[5], Paolo Privitera[2], Petr Schovanek[5], Stan B. Thomas[6], and Petr Travnicek[5]

[1]*Institute for Cosmic Ray Research, University of Tokyo, Kashiwa, Chiba, Japan*
[2]*Kavli Institute for Cosmological Physics, University of Chicago, Chicago, Illinois, USA*
[3]*Department of Physics, University of Adelaide, Adelaide, S.A., Australia*
[4]*Palacky University, RCPTM, Olomouc, Czech Republic*
[5]*Institute of Physics of the Academy of Sciences of the Czech Republic, Prague, Czech Republic*
[6]*High Energy Astrophysics Institute and Department of Physics and Astronomy, University of Utah, Salt Lake City, Utah, USA*

Abstract. We present a concept for large-area, low-cost detection of ultrahigh-energy cosmic rays (UHECRs) with a Fluorescence detector Array of Single-pixel Telescopes (FAST)[b], addressing the requirements for the next generation of UHECR experiments. In the FAST design, a large field of view is covered by a few pixels at the focal plane of a mirror or Fresnel lens. We report on the first results of a FAST prototype installed at the Telescope Array (TA) site, consisting of a single 200 mm photomultiplier tube (PMT) at the focal plane of a 1 m^2 Fresnel lens system taken from the prototype of the JEM-EUSO experiment. We also report on the status of the full-scale FAST prototype soon to be installed at the TA site, comprising a segmented spherical mirror of 1.6 m diameter and a 2×2 PMT camera.

1 The FAST Project

The origin and nature of ultrahigh-energy cosmic rays (UHECRs) is one of the most intriguing mysteries in particle astrophysics [1]. Given their minute flux, less than one per century per square kilometre at the highest energies, a very large area must be instrumented to collect significant statistics. The energy, arrival direction, and mass composition of UHECRs can be inferred from studies of the cascade of secondary particles (Extensive Air Shower, EAS) produced by their interaction with the Earth's atmosphere. Two well-established techniques are used for UHECR detection: 1) arrays of detectors (e.g. plastic scintillators and water-Cherenkov stations) sample EAS particles reaching the ground; 2) large field of view telescopes allow for reconstruction of the shower development in the atmosphere by imaging ultraviolet (UV) fluorescence light from atmospheric nitrogen excited by EAS particles.

The Pierre Auger Observatory (Auger) [2] and the Telescope Array Experiment (TA) [3, 4], the two largest UHECR experiments currently in operation, combine the two techniques, with arrays of particle detectors overlooked by fluorescence detector (FD) telescopes. Auger covers an area of over

[a]e-mail: fujii@icrr.u-tokyo.ac.jp
[b]webpage: www.fast-project.org

3000 km^2 close to the town of Malargüe in the province of Mendoza, Argentina. TA is located near the town of Delta in central Utah, USA, and covers an area of 700 km^2. Significant advances in our understanding of UHECRs have been achieved in the last decade by these experiments [5]. However, these results are limited by statistics at the highest energies. To further advance the field, the next generation of experiments will require an aperture which is larger by an order of magnitude. This may be accomplished by fluorescence detection of UHECR showers from space, as in the proposed JEM-EUSO mission [6], or with a giant ground array. Low-cost, easily-deployable detectors will be essential for a future ground-based experiment.

We present a ground-based FD telescope concept which would fulfill these requirements. The Fluorescence detector Array of Single-pixel Telescopes (FAST) would consist of compact FD telescopes featuring a smaller light-collecting area and far fewer pixels than current-generation FD designs, leading to a significant reduction in cost. In the FAST design, a 30° × 30° field of view is covered by just a few 200 mm PMTs at the focal plane of a mirror or Fresnel lens of $A \sim 1$ m^2. FAST stations, powered by solar panels and with wireless connection, could be deployed in an array configuration to cover a very large area at low cost.

2 A first test of the FAST concept at the Telescope Array site

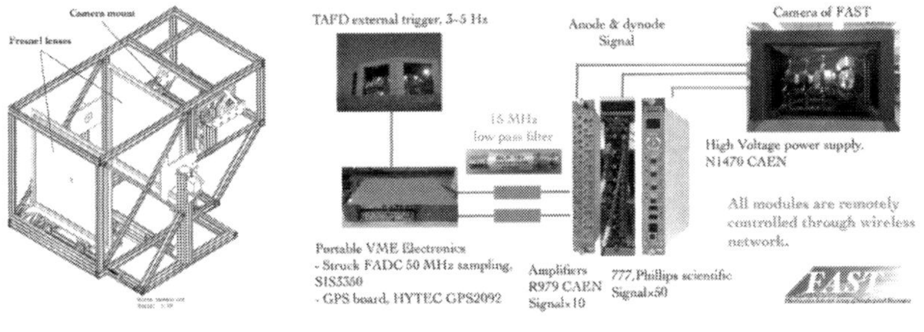

Figure 1. The EUSO-TA telescope optics (left) [7] and the DAQ system of the FAST prototype (right). The dimensions of the of the EUSO-TA optics are approximately 1.8 m × 2.0 m × 2.6 m (H × W × L). The FAST prototype camera, consisting of a single 200 mm PMT and a UV transparent filter, was installed at the focal plane of the telescope.

A first test of the FAST concept was performed profiting from the existing infrastructure of the JEM-EUSO experiment at the TA site in Utah, where a prototype [7] is currently installed for a comprehensive test of the optics and electronics of this space-based detector. The light collecting area (~ 1 m^2) and circular field of view ($\sim 7°$ radius) of the JEM-EUSO prototype telescope (EUSO-TA telescope) are close to the FAST reference design (for a single pixel), providing a perfect test-bed for the FAST concept. Its optical system consists of two 1 m^2 Fresnel lenses, with a UV transparent acrylic plate placed at the diaphragm for protection (Figure 1). For the purpose of the FAST test we installed an 200 mm PMT (mod. R5912-03, Hamamatsu) and an AC-coupled active base (mod. E7694-01, Hamamatsu) at the focal plane of the telescope. A UV band-pass filter (mod. Schott MUG-6 glass) was placed in front of the PMT to reduce the night sky background. The electronics and Data Acquisition System (DAQ) of the FAST prototype was built from commercial modules (Figure 1) [8]. Whenever any of the FD telescopes in the adjacent TA building were triggered by a candidate UHECR shower an external trigger was issued to the FAST DAQ, with a typical rate of ~ 3 Hz.

DOI: 10.1051/epjconf/201713602015

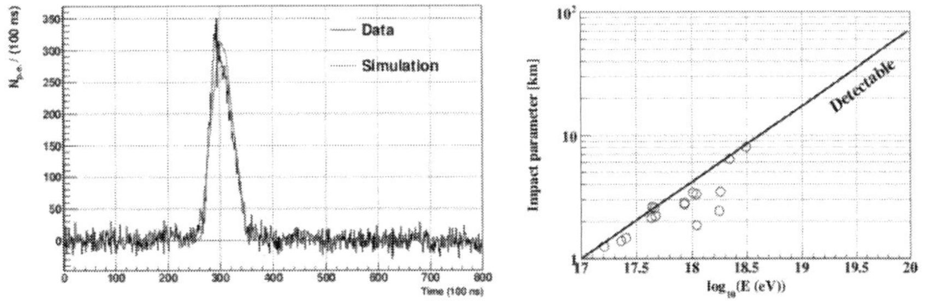

(a) (b)

Figure 2. Measured signal and simulated expectation for a vertical UV laser (a). Correlation between the impact parameter and energy of the 16 cosmic ray shower candidates detected by FAST (b). Both shower parameters were obtained from the TA standard reconstruction. The line indicates the maximum detectable distance consistent with our limited data set [8].

UV laser shots are used for calibration of FD telescopes and atmospheric monitoring [9, 10]. While traveling up through the atmosphere, the laser light scatters on air molecules and aerosol particles into the FD field of view, producing signals similar to a UHECR shower. Figure 2.a shows the measured signal and simulated expectation from a UV vertical laser at a distance of 6 km [11]. The energy of the laser pulse is ~2.2 mJ, approximately equivalent in intensity to a $\approx 10^{19.2}$ eV shower.

A shower signal search was performed, driven by well reconstructed TA FD events which generated an external trigger for the FAST DAQ. We first selected TA FD events with a reconstructed shower geometry passing through the field of view of the FAST prototype. We then searched the corresponding FAST FADC traces for pulses with a maximum signal greater than 5σ. As a result, we found 16 shower candidates in the 83 hour dataset, with an estimated background of < 1 event. Although small, this sample provides an estimate of the sensitivity of the FAST prototype. The correlation between the impact parameter (i.e. the distance of closest approach of the shower axis with respect to the FAST prototype) and the energy of the 16 showers is plotted in Figure 2.b, with shower parameters given by the standard reconstruction of the TA FD [12].

3 The Full-scale FAST prototype

Motivated by these encouraging results using the EUSO-TA telescope optics, we have developed a full-scale FAST prototype as shown in Figure 3. This prototype consists of a segmented spherical mirror of 1.6 m diameter and a UV band-pass filter with a 1 m^2 aperture. Four 200 mm PMTs and active bases are installed at the focal plane of the segmented mirror in a 2×2 matrix, covering a $25° \times 25°$ field of view. The telescope is covered with a shroud to shield the optical system from dust and stray light.

4 Summary and Future Plans

We have presented a novel concept for a next-generation air-shower fluorescence detector, which features just a few pixels covering a large field of view. We confirmed the validity of the FAST concept with test mesurements made using the exiting EUSO-TA telescope prototype optics and a 200 mm PMT. We successfully observed a vertical UV laser along with 16 UHECR showers in time

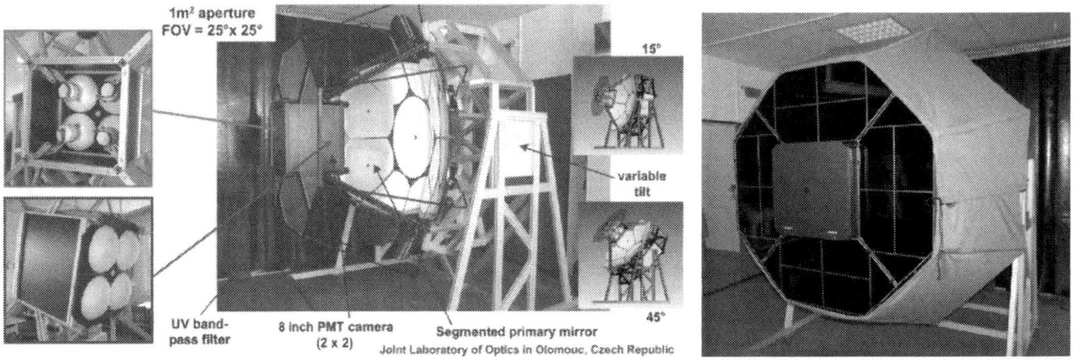

Figure 3. The full-scale FAST prototype being constructed at the Joint Laboratory of Optics in Olomouc, Czech Republic.

coincidence with the TA FD telescopes. Based on these encouraging results, we have developed a full-scale FAST prototype telescope for future installation at the Telescope Array site.

Acknowledgements

This work was supported in part by NSF grant PHY-1412261 and by the Kavli Institute for Cosmological Physics at the University of Chicago through grant NSF PHY-1125897 and an endowment from the Kavli Foundation and its founder Fred Kavli. The Czech authors gratefully acknowledge the support of the Ministry of Education, Youth and Sports of the Czech Republic project No. LG15014, LE13012, LO1305. This work was supported by the Japan Society for the Promotion of Science through the Grant-in-Aid for Young Scientist (A) 15H05443, Grant-in-Aid for JSPS Research Fellow 16J04564 and JSPS Fellowship H25-339, H28-4564.

References

[1] A.A. Watson, Rep.Prog.Phys. **77**, 036901 (2014), 1310.0325
[2] A. Aab et al. (Pierre Auger Collaboration), Nucl. Instrum. Meth. **A798**, 172 (2015), 1502.01323
[3] H. Tokuno et al. (Telescope Array Collaboration), Nucl.Instrum.Meth. **A676**, 54 (2012), 1201.0002
[4] T. Abu-Zayyad et al. (Telescope Array Collaboration), Nucl.Instrum.Meth. **A689**, 87 (2012), 1201.4964
[5] K.H. Kampert, P. Tinyakov, C.R.Phys. **15**, 318 (2014), 1405.0575
[6] Y. Takahashi (JEM-EUSO Collaboration), New J.Phys. **11**, 065009 (2009), 0910.4187
[7] Y. Takizawa et al. (JEM-EUSO Collaboration), Proc. of the 33rd International Cosmic Ray Conference, Rio de Janeiro, Brazil **ID0832** (2013), 1307.7071
[8] T. Fujii et al., Astropart. Phys. **74**, 64 (2016), 1504.00692
[9] B. Fick et al., J.Instrum. **1**, P11003 (2006)
[10] S. Udo et al., Proc. of the 30th International Cosmic Ray Conference, Merida, Mexico **5**, 1021 (2007)
[11] K. Yamazaki et al. (Telescope Array Collaboration), Proc. of the 33rd International Cosmic Ray Conference, Rio de Janeiro, Brazil **ID0526** (2013)
[12] T. Fujii et al. (Telescope Array Collaboration), AIP Conf.Proc. **1367**, 149 (2011)

Recent development on the realization of a 1-inch VSiPMT prototype

F.C.T. Barbato[1,2,a], G. Barbarino[1,2], L. Campajola[1], F. Di Capua[1], C.M.Mollo[2], A. Valentini[3], and D. Vivolo[1]

[1] *University of Naples "Federico II" - Complesso Universitario di Monte Sant'Angelo 80126 Naples*
[2] *INFN Section of Naples - Complesso Universitario di Monte Sant'Angelo 80126 Naples*
[3] *University of Bari "Aldo Moro" - Piazza Umberto I, 1, 70121 Bari*

Abstract. The VSiPMT (Vacuum Silicon PhotoMultiplier Tube) is an innovative design for a revolutionary hybrid photodetector. The idea, born with the purpose to use a SiPM for large detection volumes, consists in replacing the classical dynode chain with a SiPM. In this configuration, we match the large sensitive area of a photocathode with the performances of the SiPM technology, which therefore acts like an electron detector and so like a current amplifier. The excellent photon counting capability, fast response, low power consumption and great stability are among the most attractive features of the VSiPMT. In order to realize such a device we first studied the feasibility of this detector both from theoretical and experimental point of view, by implementing a Geant4-based simulation and studying the response of a special non-windowed MPPC by Hamamatsu with an electron beam. Thanks to this result Hamamatsu realized two VSiPMT industrial prototypes with a photocathode of 3mm diameter. We present the progress on the realization of a 1-inch prototype and the preliminary tests we are performing on it.

1 Introduction

We propose an innovative photodetector for the astroparticle physics experiments: the VSiPMT. In this scenario the Vacuum Silicon PhotoMultiplier Tube represents an appealing solution to use SiPMs from large detection area or volumes, a current big challenge.
This new device is based on the combination of a SiPM with a PMT standard envelope, see fig. 1. In this device, invented in Naples in 2007 [1], the multiplication stage is provided by a special SiPM designed for electrons detection, called SiEM (Silicon Electron Multiplier). The SiEM acts as electron detector and so as current amplifier.
Using a SiEM allows to reach a very high gain totally provided by the pixels working in geiger mode. This has many advantages:

- Excellent photon counting. The photocathode is only a passive intermediary, so the device shows an exellent resolution of the single electron allowing an easy photon counting.

[a]e-mail: barbato@na.infn.it

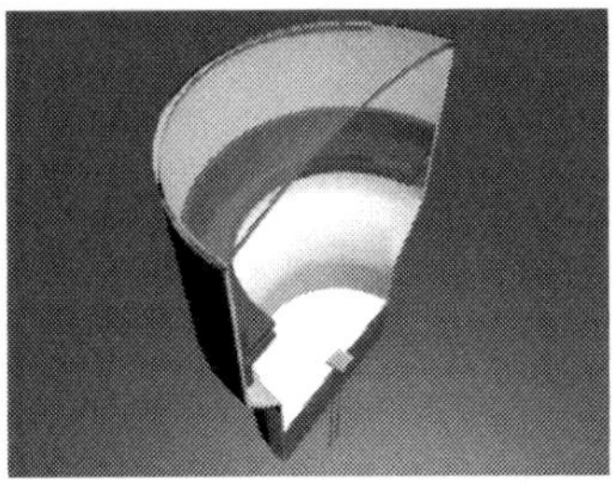

Figure 1. *A cutaway of the VSiPMT showing the interior composition of the device. On the top there is the light entrance window, then a photocathode for the photons conversion into electrons. In the middle there is a focusing ring producing an electric field which accelerates and focuses the photoelectrons on the SiEM surface. Finally, on the bottom there is a special SiPM whose structure is modified to act as an electron detector and current amplifier. Everything is assembled into and hermetically sealed container.*

- High gain with low voltage. Differently from a classical HPD, in this case the high gain is totally realized by the SiEM pixels operating in geiger mode. The average voltage supplied to the photocathode in a VSiPMT is commonly $\sim 3\,kV$.
- Negligible power consumption thanks to the absence of the voltage divider;
- High speed. The absence of a dynode chain means that a sistematical reduction of the VSiPMT TTS with respect to a classical PMT TTS is expected.
- Compactness and simplicity. Only 3 output connection are required.

2 Realization of a 1-inch prototype

The characterization of the first industrial prototype gave results beyond expectation [2]. Anyway an optimization of the focusing was necessary in order to optimize both the dynamic range and the PDE. With this aim we studied an optimized design for a 1-inch prototype.

2.1 The focusing

Since the SiEM is a pixelated device if the focusing is too weak, the photoelectron spot exceeds the size of the SiEM, in this case a fraction of the photoelectrons misses the target and is systematically lost, thus decreasing the overall PDE of the device.
On the other side, a too strong focusing produces a too much squeezed photoelectron beam, therefore the photoelectron spot intercepts only a fraction of the active surface of the SiEM, with a consequent reduction of the linearity.
In order to fully and correctly exploit the VSiPMT features the photoelectrons coming from the photocathode should: have a spot size comparable with that of the SiEM and be uniformely distributed on the SiEM surface.
SimION 8.0-based simulations have been implemented in order to find a configuration that meets this requirements.
The proposed solution is preliminary and consists of a single stage focusing system. Therefore, the device can be schematized as follows:

- flat 25 mm ⌀ photocathode;

- focusing ring 25 mm in external diameter, 10 mm in height and 1.5 mm in thickness;
- $3 \times 3\ mm^2$ SiEM target.

The photocathode and the focusing ring are kept at the same voltage HV, while the SieM is at its nominal operating voltage, i.e. 67.2V.
A mesh of $3 \times 3\ mm^2$ with a $50\mu m$ pixel size is used to simulate the SiEM surface. In the simulations we first set HV to -3kV and then we varied the SiEM distance from the photocathode. The distance that better meets the requirements for the focusing is 16mm, see fig. 2.

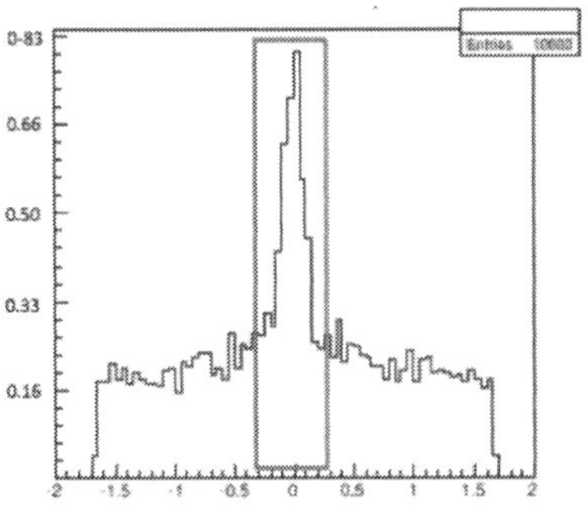

Figure 2. *Electrons distribution on the SiEM surface. The distribution in this position results to be flat in the 99% of the surface and only 1% of the SiEM has an higher probability to be fired by electrons.*

2.2 The photocathode

We chose the CsI as photocathode material since it is not sensitive to the oxigen in the air and thus allows an easy handling in the assembly phase. Since it is an insulating material, we needed a conductive layer to supply voltage. We performed a dedicated study in order to find a material that is conductive as well as transparent, in order to maximize the quantum efficiency of the photocathode. Between all the deposited materials we found the best solution to be a substrate made of Carbon and Nichel that will ensure a high quantum efficiency. We then deposited the CsI photocathode obtaining a QE = 15% with a MgF_2 window and QE = 11% with a quartz window at $\lambda = 170\ nm$,[3][4].

2.3 The SiEM

The selected SiEM is a custom MPPC S10943-3360(X) realized for us by Hamamatsu upon request. It is without epoxy resin and with a p-over-n internal structure. The surface is $3 \times 3\ mm^2$ with a $50\mu m$ pixel size. It has been fully characterized with a light source before being used in the VSiPMT.

3 The preliminary results

Once the several parts that composes the VSiPMT have been chosen, an intermediate phase between the preparation of the parts and the final assembly is necessary.
In view of the final assembly of the VSiPMT, we set up a test bench in the DAFNE Light facility available at Frascati National Laboratories (LNF) aimed at testing the focusing.
The SiEM is mounted on a movable support, the photocathode is lighted by a deuterium lamp. We

 DOI: 10.1051/epjconf/201713602016

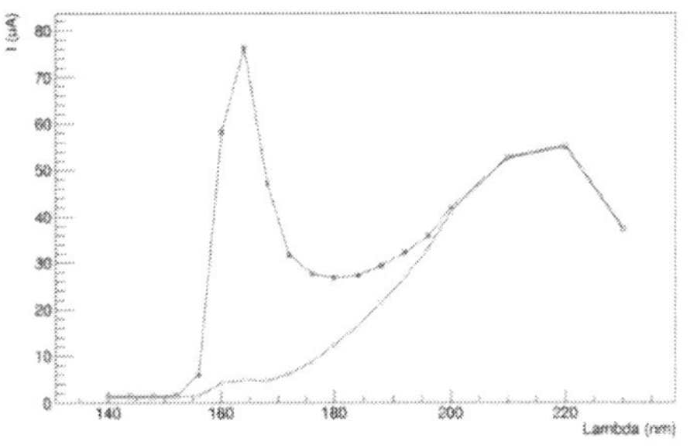

Figure 3. *Output current with HV off (green dots) and HV on (blue dots).*

first measured the output current with HV off and on, in such a way to check the proper operation of the device, see fig. 3. This is a very important result since the difference between the two states of the photocathode (on and off) concentrates in the region of sensitivity of the CsI photocathode, this means that the exceeding current is due to photoelectrons coming from the photocathode.
We finally tested the focusing by varying the SiEM distance from the photocathode. In fig. 4 the SiEM readout current is plotted with respect to the SiEM distance from the photocathode. The results

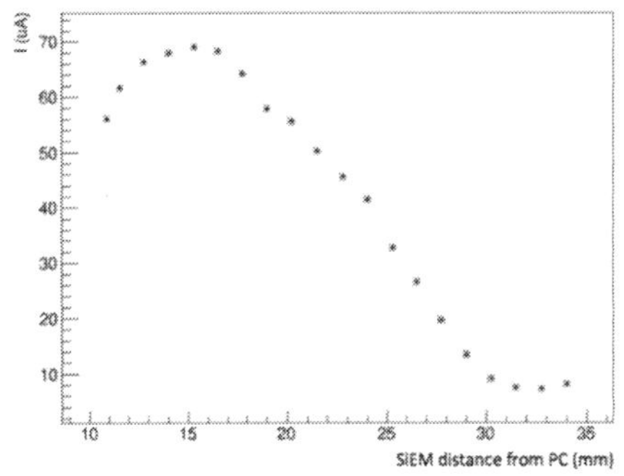

Figure 4. *Plot of the SiEM output current with respect to the distance from the photocathode.*

of the test were good in terms of focusing. It's easy to observe that the position that maximize the readout current (i.e. the number of fired pixels) correspond to the position where the photoelectron distribution on the SiEM surface is expected to be mainly flat from the simulations. These two plots combined together represent a significative milestone in the project since it means that for the first time the sensitive surface of the SiEM has been enlarged of more than 50 times.
In addition, the dark counts rate remains that typical of a single SiEM. This is equivalent to reduce the dark-rate/mm2, that up to now has been one of the main issue for the SiPM manufacturers.

4 Conclusions

The VSiPMT is an innovative hybrid photodetector which combines a photocathode with the SiPM technology. Simulations for an optimized focusing have been done. A 1-inch pre-prototype has been realized in Naples and Bari laboratories and tested at LNF. The results of the tests were in agreement with the simulations. A 1-inch prototype realized by Hamamatsu is currently under test in Naples laboratories.

References

[1] G. Barbarino et al., A new high-gain vacuum photomultiplier based upon the amplification of a Geiger-mode p–n junction, NIM A, **594** 3 326-331 (2008),

[2] G. Barbarino et al., A new generation photodetector for astroparticle physics: The VSiPMT, Astroparticle Physics, **67** 18 - 25 (2015),

[3] F.C.T. Barbato et al., R& D of a pioneering system for a high resolution photodetector: The VSiPMT, NIM A, Available online 31 December 2016, ISSN 0168-9002, http://dx.doi.org/10.1016/j.nima.2016.12.064.

[4] F.C.T. Barbato, Research and development of a pioneering system for single photon detection: the VSiPMT, PhD Theses, JINST Instrumentation Theses Archive, https://jinst.sissa.it/jinst/theses/2016_ JINST_ TH_ 003.pdf

The Pierre Auger Observatory status and latest results

Corinne Berat[1,a] for the Pierre Auger Collaboration[2,b]

[1] *LPSC, Université Grenoble-Alpes, CNRS/IN2P3, Grenoble, France*
[2] *Observatorio Pierre Auger, Av. San Martin Norte, 304, 5613 , Malargue, Argentina*

Abstract. The Pierre Auger Observatory, in Argentina, is the present flagship experiment studying ultrahigh-energy cosmic rays (UHECRs). Facing the challenge due to low cosmic-ray flux at the highest energies, the Observatory has been taking data for more than a decade, reaching an exposure of over 50 000 km^2 sr yr. The combination of a large surface detector array and fluorescence telescopes provides a substantial improvement in energy calibration and extensive air shower measurements, resulting in data of unprecedented quality. Moreover, the installation of a denser subarray has allowed extending the sensitivity to lower energies. Altogether, this contributes to provide important information on key questions in the UHECR field in the energy range from 0.1 EeV up to 100 EeV. A review of main results from the Pierre Auger Observatory is presented with a particular focus on the energy spectrum measurements, the mass composition studies, the arrival directions analyses, the search for neutral cosmic messengers, and the investigation of high-energy hadronic interactions. Despite this large amount of valuable results, the understanding of the nature of UHECRs and of their origin remains an open science case that the Auger collaboration is planning to address with the AugerPrime project to upgrade the Observatory.

1 Introduction

The nature and the origin of ultrahigh-energy cosmic rays are still enshrouded in mystery, even if in last decade measurements have shed light on these puzzling questions. UHECRs are very scarce and their characteristics are inferred from the measurement of extensive air showers (EASs) they produce. The Pierre Auger Observatory [1] located in the province of Mendoza (Argentina) and covering 3000 square kilometres, brings unique capabilities to the UHECR study. It combines two techniques to measure the EAS properties by observing their longitudinal development in the atmosphere as well as their lateral spread at ground level. Charged particles and photons that reach the ground are detected with the surface detector (SD) consisting of 1660 autonomously operated water-Cherenkov detectors (WCDs). The WCDs are arranged on a triangular grid of 1500 m spacing, except for a denser infill area of ~ 30 km^2, where the spacing is 750 m. The surface detector operates 24 hours per day, and provides a huge collecting area. The atmosphere above the SD is observed by the fluorescence detector (FD) which consists of 27 fluorescence telescopes to detect the faint UV light emitted by nitrogen molecules excited by the charged particles from the EAS. The field of view (FoV) of each

[a] e-mail: berat@lpsc.in2p3.fr
[b] e-mail: auger_spokespersons@fnal.gov – Full author list: http://www.auger.org/archive/authors_2016_06.html

telescope is 30° in azimuth, and 1.5 – 30° in elevation, except for three of them, the High Elevation Auger Telescopes (HEAT), whose FoV is 30 – 60° in elevation, allowing the observation of nearby low-energy showers above the denser infill area. The FD can only operate during dark, moonless nights with a field of view free of clouds (duty cycle ~15%). On-line and long-term performances of the detectors and data quality are monitored continuously, and a set of high-quality devices installed in the Observatory array monitor the atmospheric conditions during operation [2]. A 17 km^2 sub-array of 153 radio sensors (Auger Engineering Radio Array) is dedicated to EAS radio detection. The progresses made in this detection technique in particular to identify mass-sensitive radio parameters are reported elsewhere in these proceedings [3].

High-quality data have been collected continuously for about ten years, with a SD annual exposure of ~5500 km^2 sr yr. The longitudinal profile reconstructed by the FD is providing a nearly calorimetric measurement of the primary energy, with total systematic uncertainty of 14% [4]. From the shower lateral distribution reconstructed using the WCD signals, a SD energy estimator is inferred. A high-quality subset of hybrid events recorded by both the SD and the FD is used to calibrate the SD energy estimator with the FD energy measurement, hence providing an almost model-independent energy calibration.

2 Spectrum measurements

The UHECR spectrum is obtained from four different datasets [5], corresponding to an exposure now larger than 50000 km^2 sr yr. Data from the SD-750 m allow measurements down to 10^{17} eV; SD-1500 m vertical data (i.e. zenith angle < 60°) are covering the range from the full trigger efficiency energy threshold of $3 \cdot 10^{18}$ eV up to the highest energies; SD-1500 m horizontal events contribute above $4 \cdot 10^{18}$ eV; hybrid data cover the range between 10^{18}eV and $10^{19.6}$ eV. The four measurements are combined taking into account the systematic uncertainties of the individual measurements. The

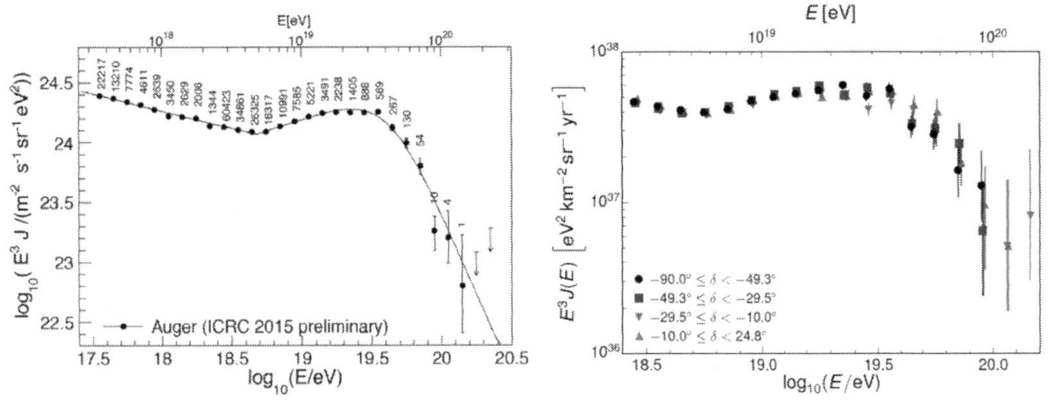

Figure 1. Cosmic-ray energy spectrum measured by Auger ([5]) - Left: from combining four independent measurements (see text) - Right: in four declination bands, using SD-1500 vertical event.

resulting spectrum (Fig. 1-left) flattens from a power law with index (3.29 ± 0.02(stat) ± 0.05(sys)) to one with index (2.60 ± 0.02(stat) ± 0.1(sys)) at E_{ankle} = 4.8 ± 0.1 ± 0.8 EeV. A clear suppression is observed at a significance in excess of 20σ beyond E_s = 42.1 ± 1.7 ± 7.6 EeV, the energy at which the differential flux is reduced to one-half of that expected from the extrapolation of the power law above the ankle. The large number of events and the wide range of declinations δ from −90° to +25°, allow

DOI: 10.1051/epjconf/201713602017

the study of the UHECR flux as a function of δ. Fig. 1-right shows the four spectra obtained with events separated into four declination bands of roughly equal exposure. The agreement between the spectra is within 5% below E_s and 13% above, therefore there is no indication of a δ-dependent flux.

3 Mass composition studies

The most robust observable sensitive to the mass of the primary particle is X_{max}, the depth of maximum of the shower development, directly measured from the longitudinal profile reconstructed with the FD. The most recent analysis includes the data of HEAT, allowing measurement from 0.1 EeV [6], as shown in Fig. 2. Up to 2 EeV the increase of $\langle X_{max}\rangle$ is larger than the one expected for a constant mass composition (see proton/iron simulations), indicating that the mean primary mass is getting lighter. Around $\simeq$ 2 EeV the elongation rate becomes significantly smaller, then the composition is becoming

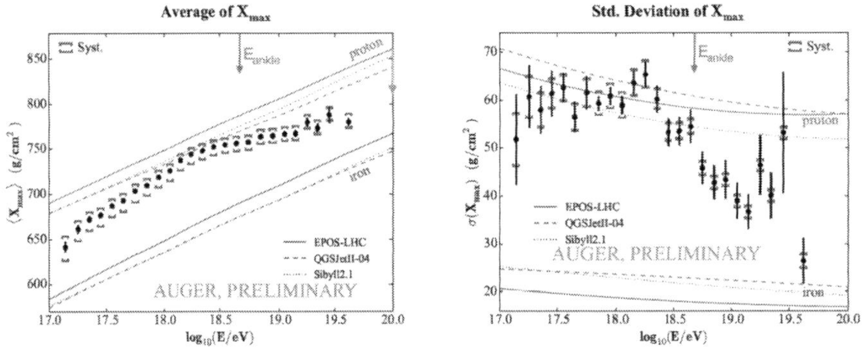

Figure 2. The mean (left) and the standard deviation (right) of the measured X_{max} distributions as a function of energy compared to air-shower simulations for proton and iron primaries.

heavier. The fluctuations of X_{max} start to decrease at around the same energy, confirming the previous observations. These measurements can be interpreted by converting them to $\langle$lnA$\rangle$ [7], with A the atomic mass number, based on simulations using current hadronic interaction models EPOS-LHC [9] and QGSJetII-04 [8]. For both models similar trends with energy are observed for the mean and the variance of lnA. The primary mass is decreasing up to $\simeq$ 2 EeV, the spread of the masses being almost constant: several components are expected, evolving from intermediate to light mass. $\langle$lnA$\rangle$ is increasing at the higher energies, the variance showing a decrease: fewer components are expected, with mass evolving from light to heavy. These behaviours might be an indication that the relative fraction of protons becomes smaller for energies above 2 EeV. The distributions of X_{max} were also interpreted in terms of primary masses [10] based on the QGSJetII-04, Sibyll 2.1 [11] and EPOS-LHC hadronic interaction models. The results suggest also a mixed composition. Around the ankle, a very light composition consisting of proton and Helium only is favoured using QGSJetII-04 and Sibyll 2.1, while for EPOS-LHC, intermediate nuclei (from CNO group) contribute. To get a more direct and robust estimation of the spread of masses in the primary beam, relying less on interaction models, the correlation between X_{max} and S(1000) is studied, where S(1000) is the signal at 1000 m from the shower core, reconstructed with the SD. For single nuclear components this correlation is expected to be $\gtrsim 0$, although for mixed composition, the correlation is negative. For hybrid events with energies of $\log_{10}(E/eV) = 18.5 - 19.0$ and zenith angles $< 65°$ a significant negative correlation was found consistent with a spread of masses σ(lnA) > 1, meaning the composition around the ankle is actually mixed [12]. The hypothesis that below 3 EeV, a fraction of protons mainly of extragalactic origin is dominant and that the ankle corresponds to the proton energy loss through e^+e^- pair production in interactions with the cosmic microwave background (CMB) is then disfavoured.

By simultaneously fitting the spectrum and the X_{max} evolution above 5 EeV, the Auger results can be interpreted assuming a simple astrophysical scenario, as reported elsewhere in these conference proceedings. The best fit supports the hypothesis of a flux suppression partly due to the reach of the source maximum energy, while the second local minimum corresponds to a scenario where the suppression is due to propagation effects (Greisen, Zatsepin and Kuz'min, or GZK cutoff) [13].

4 Neutral cosmic messengers

Both neutrinos and photons are sought for in the flux of UHECRs detected by Auger. The neutrino search is performed by studying very inclined showers and earth-skimming ones [14]. The criteria are based on the characteristics expected for "young" showers initiated by neutrinos, developing deep in the atmosphere, compared to "old" ones from inclined hadronic showers, having their electromagnetic component fully absorbed before reaching the detectors. Photon showers, due to their slower devel-

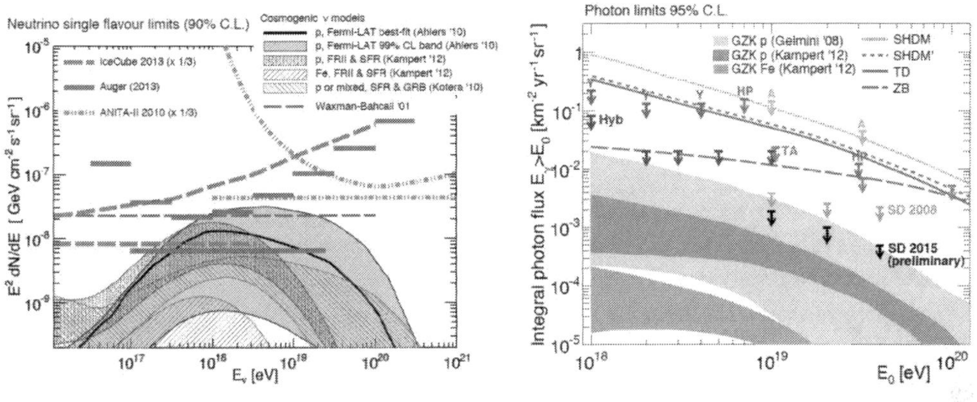

Figure 3. Left: neutrino flux upper limits (at 90% C.L.), in integrated (horizontal lines) and differential forms (see [14] for details). - Right : photon flux upper limits (95% C.L.) (see [15] for details).

opment and the dominance of the electromagnetic component can be distinguished from hadronic showers [15]. SD events on the one hand, and hybrid events on the other hand have been analyzed, to cover the energy range above 1 EeV. Assuming a differential flux $dN(E) = k \cdot E^{-2}$ for both neutrinos and photons, stringent upper limits to their flux are derived. The Auger limits on neutrinos (Fig. 3-left) outperform those from IceCube and ANITA, and also the Waxman-Bahcall limit; in the range $10^{17} - 10^{19}$ eV they are challenging the contribution from cosmogenic-neutrino models. The limits to the integral photon flux are shown in Fig. 3-right. The obtained limits are the most stringent for $E > 10$ EeV and start to constrain the most optimistic predictions of cosmogenic photon fluxes under the assumption of a pure proton composition.

5 Anisotropy searches

The distribution of the arrival directions of ultrahigh-energy cosmic rays is also scrutinised, complementary to the spectrum and mass measurements, to shed light on their origin and nature. The

 DOI: 10.1051/epjconf/201713602017

presence of anisotropies is searched for at small and intermediate angular scales [16] in the distribution of arrival directions of the most energetic cosmic rays above a few tens of EeV, where the magnetic deflections (at least of those with a small charge) may be only a few degrees. Different methods are employed: search for autocorrelation by looking for pairs of events, blind search for localized excesses of events (Fig. 4-left), search for correlations with specific sky regions (galactic center, galactic and super-galactic planes) and with sources from catalogues (2MRS galaxies, Swift-BAT AGNs, ...). The performed tests on the UHECR arrival distribution point out a high degree of

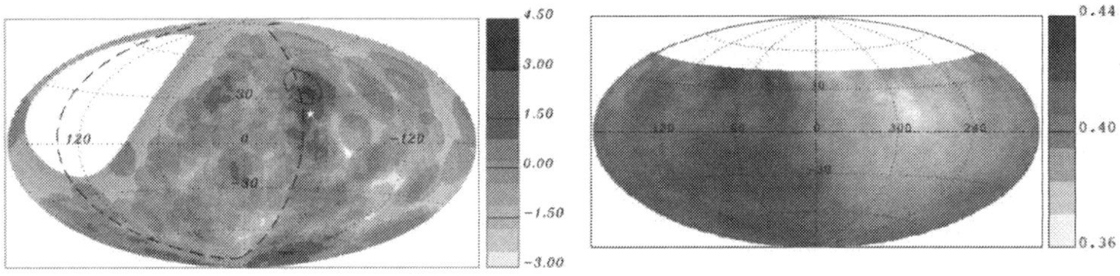

Figure 4. Left: Li-Ma significance map (galactic coordinates) of overdensities in 12°−radius windows for the events with E> 54 EeV. Right: sky map (equatorial coordinates) of flux in km^{-2} yr^{-1} sr^{-1} smoothed in angular windows of 45° for E> 8 EeV.

isotropy, jeopardising the initial expectation of few sources and light primaries.

Large-scale anisotropies could signify a galactic−extragalactic transition which may entail a significant change in the arrival direction distribution. The partial sky coverage limits the harmonic expansion of the cosmic-ray flux distribution, but the large amount of data accumulated by the Pierre Auger Observatory are well-suited to search for dipolar and quadrupolar patterns, from 10 PeV to the highest energies (Fig. 4-left). In the energy band $E > 8$ EeV a dipole component with amplitude $7.3 \pm 1.5\% (p = 6.4 \times 10^{-5})$, pointing to $(\alpha, \delta) = (95° \pm 13°, -39° \pm 13°)$ is found, above isotropic expectations [17].

6 Investigation of high-energy hadronic interactions

The UHECR studies provide a means of investigating hadronic interactions at energies far beyond the reach of the LHC. With Auger hybrid data, using the tail of the X_{max} distribution, the determination of the cross-section of proton-air collisions is achieved in the two energy intervals in $\log_{10}(E/eV)$ from 17.8 to 18 and from 18 to 18.5 [18]. Another approach to investigate high-energy hadronic interactions and to compare measurements to models is to consider the muon content of the EAS, particularly suited to address the multi-particle production in first interactions. The Collaboration examines the muon shower size using different datasets, in particular with inclined events (the electromagnetic component is largely absorbed before reaching the ground) [19] and with hybrid events (comparison of the measured ground signal to the expected one knowing the longitudinal profile) [20]. A muon deficit is observed in simulations compared to EAS data, from 30% to 80% at 10^{19} eV depending on interaction models.

7 Summary and perspectives

A wealth of valuable results is obtained by the Pierre Auger Observatory. The measured all-particle spectrum demonstrates the existence of a flux suppression above 40 EeV, whose origin is still not fully understood. A source effect seems to be favoured when considering the spectrum and the X_{max} evolution together. The image of a mixed composition around the ankle and a heavier composition at the highest energies is strengthened by different composition studies, and by anisotropy searches at small angular scales at the highest energies. The present photon limits also disfavour pure proton composition models. Top-down models were already disfavoured by results on UHE neutrino and photon fluxes, leading to an astrophysical source explanation, but no clear clue on any existing UHECR source has been revealed so far. The hint of dipolar structure in the arrival direction is a step forward in the understanding of the transitions between the galactic and extragalactic components. The comparison of measurements with model predictions remains a delicate operation, knowing that serious hints for deficiencies in UHE interaction models have been observed. The astrophysical scenario resulting from Auger measurements is very complex and cannot at present be understood in terms of a unique interpretation for the sources, propagation and composition of the UHECRs. The great value of these overall results will be reinforced by the knowledge of the nature of the UHECR primaries, event by event. This will be the key to answering the open questions on the highest energy and the suppression region. The Collaboration is undertaking a major upgrade program: AugerPrime (described elsewhere in these proceedings [21]), designed to improve the knowledge on mass composition, mainly by discriminating electromagnetic and muonic shower components from SD-based observables, by having a further and independent measurement. Such additional information will also be valuable to improving the understanding of hadronic interaction models. The upgraded Observatory will take data until 2024, with the objective of clarifying the UHECR puzzle.

References

[1] A. Aab et al., the Pierre Auger Collaboration, Nucl. Instrum. Meth. A 798 (2015) 172 .
[2] P. Abreu et al., the Pierre Auger Collaboration, J. Instrum. 7 (2012) P09001.
[3] B. Revenu for the Pierre Auger Collaboration, these RICAP-2016 proceedings.
[4] V. Verzi for the Pierre Auger Collaboration, Proc. of 33rd ICRC, arXiv:1307.5059 (2013)
[5] I. Valiño for the Pierre Auger Collaboration, Proc. of 34th ICRC, PoS (ICRC 2015) 271; A. Aab et al., The Pierre Auger Collaboration, JCAP 08 (2015) 049.
[6] A. Porcelli for the Pierre Auger Collaboration, Proc. of 34th ICRC, PoS (ICRC 2015) 420.
[7] P. Abreu et al., the Pierre Auger Collaboration, JCAP 02 (2013) 026.
[8] S. Ostapchenko, Phys. Rev. D83 (2011) 014018.
[9] T. Pierog, I. Karpenko, J. Katzy, E. Yatsenko, and K. Werner, Phys. Rev. C92, 034906 (2015).
[10] A. Aab et al., The Pierre Auger Collaboration, Phys. Rev. D90 12, 122006 (2014).
[11] E.-J. Ahn, R. Engel, T. K. Gaisser, P. Lipari, and T. Stanev, Phys. Rev. D80, 094003 (2009).
[12] A. Yushkov for the Pierre Auger Collaboration, Proc. of 34th ICRC, PoS (ICRC 2015) 335.
[13] A. di Matteo for the Pierre Auger Collaboration, Proc. of 34th ICRC, PoS (ICRC 2015) 249.
[14] A. Aab et al., the Pierre Auger Collaboration, Phys. Rev. D 91, 092008 (2015).
[15] C. Bleve for the Pierre Auger Collaboration, Proc. of 34th ICRC, PoS (ICRC 2015) 1103.
[16] A. Aab et al., the Pierre Auger Collaboration, Astrophys. J 804, 15 (2015).
[17] A. Aab et al., the Pierre Auger Collaboration, Astrophys. J 802, 111 (2015).
[18] R. Ulrich for the Pierre Auger Collaboration, Proc. of 34th ICRC, PoS (ICRC 2015) 401.
[19] A. Aab et al., the Pierre Auger Collaboration, Phys. Rev. D 91, 032003 (2015).
[20] A. Aab et al., the Pierre Auger Collaboration, *Testing hadronic interactions at ultrahigh energies with air showers measured by the Pierre Auger Observatory* to be published in Phys. Rev. Lett.
[21] G. Marsella for the Pierre Auger Collaboration, these RICAP-2016 proceedings.

 DOI: 10.1051/epjconf/201713603001

H.E.S.S.-II - Gamma ray astronomy from 20 GeV to hundreds of TeV's

Mathieu de Naurois[1,a] for the HESS collaboration

[1] *LLR - IN2P3/CNRS - Ecole Polytechnique*

Abstract. Since the commissioning of the fifth big telescope in December 2012, H.E.S.S. II is the only hybrid array of Imaging Atmospheric Cherenkov Telescopes operating in the energy range $\sim$ 20 GeV to several hundreds of TeV. The last years have seen a tremendous effort in the design, implementation and optimization of analysis techniques suitable for monoscopic & stereoscopic events. At the same time, a complete redesign of the acquisition scheme resulted in a very significant speed-up of repointing, allowing the big telescope to be on target just $\sim$ 20 s after receiving a Gamma-Ray Burst (GRB) alert notification. With its deep sensitivity, broad energy range, and fast reaction time, H.E.S.S. II provides an unprecedented high-quality view of the Universe at the highest energies, in a multi-wavelength and multi-messenger approach which is currently based on agreements many collaborations including in particular Fermi, IceCube, ANTARES and VIRGO/LIGO.

In the last, we conducted deep observations of several galactic regions of primordial importance, among them are the Galactic Center region and its halo (particularly relevant for dark matter searches), the Crab Nebula, the supernova remnant RXJ 1713.7-3946, the Vela pulsar and several binary systems such as LS 5039 and PSR B1259-63. Outside the Milky Way, the blazars PKS 2155-304 and PG 1553+113 have been extensively monitored, and H.E.S.S.-II forms part of a multi-wavelength campaign of the flaring activity of Mrk 501 in 2014.

Highlights of these observations with H.E.S.S.-II have been presented and discussed at the conference. Moreover, after ten years of H.E.S.S. phase I observations, we are currently preparing a Legacy Release of the H.E.S.S. Galactic Plane Survey. A special edition of Astronomy & Astrophysics is currently under preparation, and will contain many important legacy results from H.E.S.S.-I. Major results from this very deep scan of the Milky Way performed with H.E.S.S.-I, including among others spectacular findings from the Large Magellanic Cloud, have been presented.

1 Introduction

In the recent years the field of very high energy (VHE) γ-ray astronomy has evolved from single source discoveries to large systematic surveys, consisting of several hundred - or even thousands - hours of observation, which allows from the first time population studies to be performed. At the same time, deep exposure have been taken on sources of particular importance (*"Key science projects"*),

[a]e-mail: denauroi@in2p3.fr

establishing new classes of sources as γ-ray emitters. The largest Imaging Atmospheric Cerenkov Telescope, namely the fifth telescope of H.E.S.S.-II (CT5), was inaugurated in September 2012 and is now operating regularly, allowing observation of γ-rays down to 10 GeV.

2 The H.E.S.S. Legacy Survey

The H.E.S.S. Galactic plane survey (HGPS) [1] is a major long-term project, corresponding to $\sim$ 2700 hours of high-quality observations acquired with the H.E.S.S.-I telescope array from 2004 to 2013. Results have already been published on a small ($\sim$ 10%) fraction of the current data set [2, 3]. The region of the Milky Way covered by the HGPS (Galactic longitude between and 65 degrees and Galactic latitude $|b| < 3.5$ degrees) is depicted as a white rectangle in Fig. 1 and compared to the HEGRA (in blue) and VERITAS Cygnus (in green) surveys.

The HGPS data set combines dedicated survey operations (using a fixed-grid pointing strategy) with deep observations of sources of particular interest, leading to non uniform exposure and thus sensitivity across the survey region (Fig. 1, lower panel), the point-source sensitivity being always better than $\sim$ 2% in $0.2 \lesssim E \lesssim 100$ TeV.

A catalogue of 77 cosmic accelerators was derived from the HGPS using a semi-automatic pipeline, out of which 6 are new sources that were previously unknown or unpublished. Less than

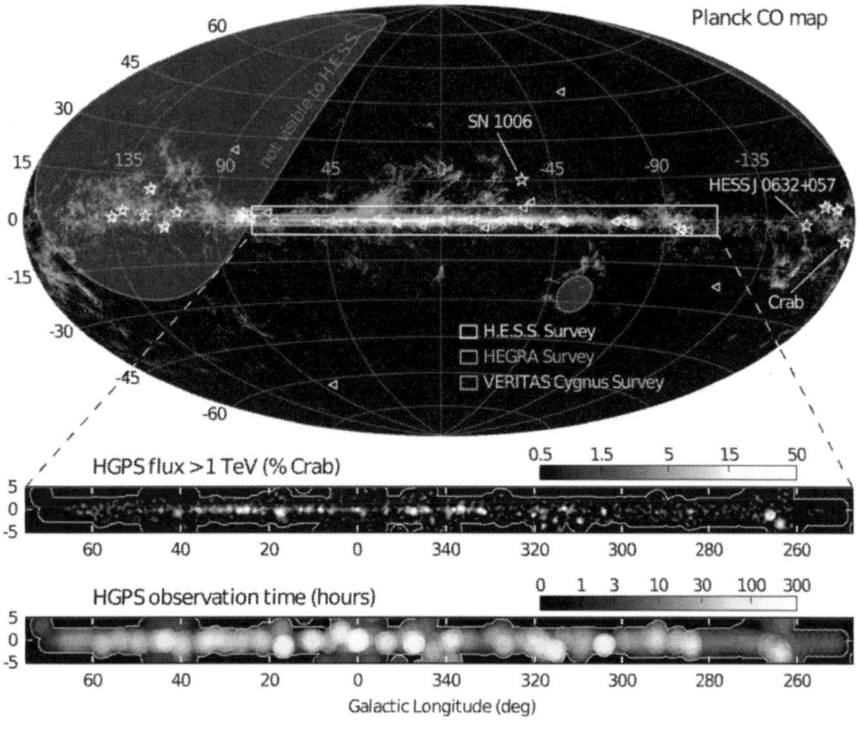

Figure 1. H.E.S.S. Galactic Plane Survey region, flux map and exposure map (from top to bottom). The all-sky image on the top panel shows a Planck CO Map with *Fermi*-LAT identified Galactic 1FHL sources (triangles) and the 15 known Galactic TeV sources (white stars) outside the HGPS region. The HEGRA Galactic Plane Survey [4] and the VERITAS Cygnus survey [5, 6] regions are illustrated in blue and green, respectively. From [1].

half of the source (31) are firmly identified, the largest population consisting of pulsar wind nebulæ (PWN) followed by supernova remnants (SNRs) and binary systems. Most of the remaining sources are confused (several possible counterparts) while a significant fraction (12 sources) have no known counterparts at other wavelengths.

The HGPS is a unique tool for population studies (e.g. PWN [7] or SNR [8]). In particular, it appears that most young and powerful pulsars have a PWN detected in VHE. This allows for the first time to construct a PWN evolution model.

The paper and legacy data will be released in 2017, including FITS maps and a source catalog (morphology & spectra).

3 Very high emission from pulsars

The detection of very high energy pulsations from the Crab pulsar by the MAGIC collaboration came as a surprise to the community [9]. Later detection of pulsed emission above 100 GeV [10, 11], and up to at least 1 TeV [12] implies that the emission takes place in the vicinity of the light cylinder, and triggered an intense theoretical and experimental activity.

Despite its longer period, $P = 89.3$ ms, its larger characteristic age $\tau_c = 11$ kyr and its much smaller spin-down power $\dot{E} = 6.9 \times 10^{36}$ erg s^{-1} [13], the Vela pulsar is, due to its proximity ($d = 287^{+19}_{-17}$ pc [14]) exceptionally bright in radio and in HE γ rays, and is, after the Crab pulsar, one of the best natural candidates for VHE emission. The Vela pulsar is surrounded by a strong wind nebulae, detected in particular by *Fermi*-LAT [15].

The HE γ-ray phasogram is characterized by two main sharp peaks (P1 and P2) and a third peak (P3) in the bridge (Fig. 2, bottom left). The ratio of the peak intensity between P1 and P2, and the location and intensity of P3 vary with energy [16, 17].

The Vela pulsar was observed by H.E.S.S. during 40 hr between 2013 and 2015. A specific reconstruction was developed for monoscopic mode observations, and the resulting VHE phasogram is shown in Fig. 2, top left: the peak P2 is detected with a statistical significance of $> 15\sigma$, and a total

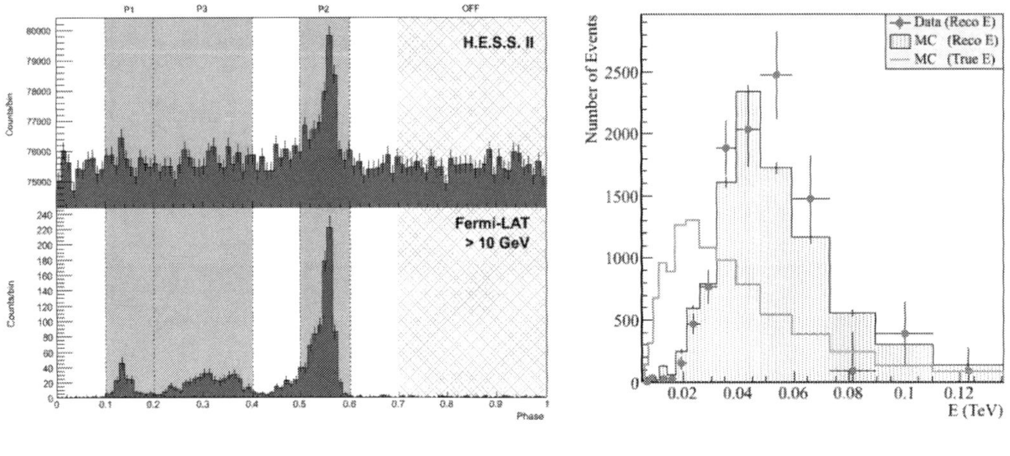

Figure 2. Left: Phase folded distribution of events of the Vela Pulsar with H.E.S.S.-II (top) and *Fermi*-LAT (bottom). **Right:** Energy distribution of events in the P2 Peak compared to Monte Carlo simulations.

excess > 15000 over a huge background, demonstrating that H.E.S.S. operates in a different statistical regime than *Fermi*-LAT.

Detailed comparison with Monte Carlo simulations (Fig. 2, right) validate the analysis pipeline and indicate that γ-rays of energy as low as 10 GeV can be detected by CT5, although with an important reconstruction energy bias. The VHE energy spectrum in the P2 peak is consistent with a steep power law ($\Gamma = 4.1 \pm 0.2_{\mathrm{stat}}$), in very good agreement with that of *Fermi*-LAT.

4 Supernova remnants as sources of cosmic rays

Expanding shock waves in SNRs are believed to be able to accelerate cosmic rays up to multi-TeV energies through the mechanism of diffusive shock acceleration (DSA), see e.g. [18].

In the last years, SNRs have been firmly established as sources of VHE γ-ray emission. At least five SNRs with clear shell-type morphology resolved in VHE γ rays were detected by H.E.S.S., allowing direct investigation of the SNRs as sources of cosmic rays: RX J1713.7-3946 [19, 20], RX J0852.04622 (also known as Vela Junior) [21], SN 1006 [22], HESS J1731-347 [23] and RCW 86 [24]. All of them show a very clear correlation between non-thermal X-ray and VHE γ-ray emissions. Recently, three additional candidates of shell-type SNRs were identified in the HGPS [25], HESS J1534-571, HESS J1614-518 and HESS J1912+101, the latter being the first TeV-only shell candidate, identified through its morphology and without known counterpart.

The SNR RX J1713.7-3946 is one of the brightest Galactic X-ray SNRs, possibly associated with the guest star AD393 which, according to Chinese astronomers, appeared in the constellation Scorpius [26]. It was the first young, shell-type SNR to be resolved in VHE γ rays [19], with a typical shell morphology consistent with X-ray observations. Both leptonic and hadronic scenarios are considered to explain the origin of the emission.

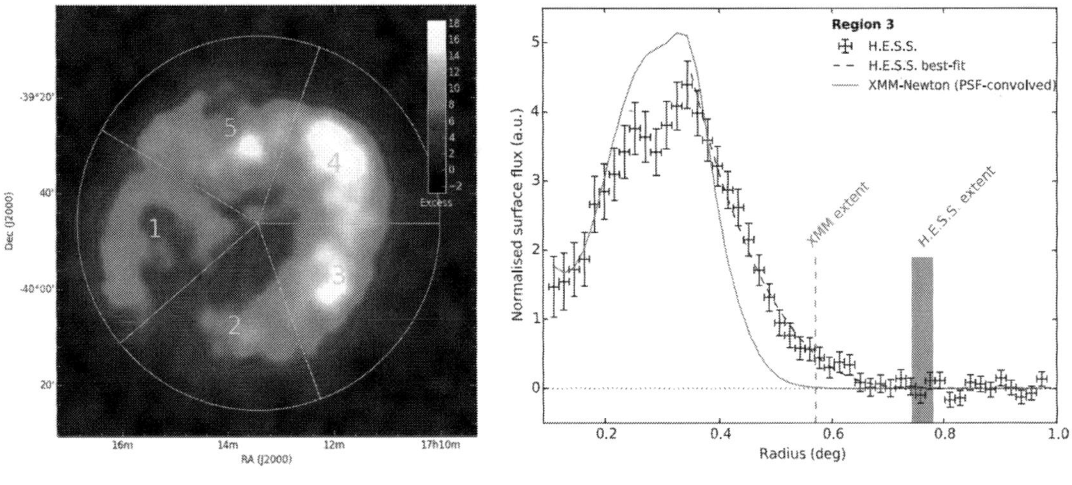

Figure 3. Left: Quadrants used in the H.E.S.S.VHE γ-ray image of RX J1713.7-3946 to investigate possible particle escape. **Right**: Radial profile of γ-ray (black) and X-ray (red) emission in quadrant number 3 [28].

Since the last H.E.S.S. publication [20], the amount of data has been doubled and high-resolution / high-throughput analysis techniques have been developed, e.g. [27], allowing to spatially resolve spectra with unprecedented resolution ($\lesssim 0.05°$). The resulting γ-ray image is shown in Fig. 3, left, and has an energy threshold of ~ 250 GeV. For the first time, maps of physical quantities, such as magnetic field in the case of a leptonic scenario, can be produced. The high statistics collected and the high resolution analysis permits a detailed analysis of the morphology. The γ-ray image from Fig. 3 has been divided in five quadrants for which the radial profiles have been determined. In several regions, such as region 3 (Fig. 3, right) the γ-ray emission appears significantly more extended than the X-ray one (smeared to the H.E.S.S. PSF), indicating for the first time diffusion of particles outside the shell and even escape of high energy particles, most likely of hadronic origin.

5 Galactic Centre

Since the original discovery of TeV emission from the Galactic Centre by H.E.S.S. [29] and diffuse emission from the ridge [30], deep observation of the region have been conducted (> 200 hr), allowing to investigate in more depth the nature and behaviour of the ridge emission [31].

Figure 4 shows the excess map of the galactic centre region, from which a profile of the cosmic ray density along the galactic plane was extracted. This profile shows a $1/r$ dependence, indicating diffusing propagation of cosmic rays injected continuously by a central source. The γ-ray spectrum extracted from the region immediately surrounding the central source shows a power-law spectrum of index 2.3 extending up to 50 TeV without any evidence of a spectral cut-off, giving evidence for a proton injection spectrum extending beyond PeV energies.

The Galactic Centre therefore appears to be the first galactic Pevatron to be confidently detected, and could account for a large fraction of the Galactic cosmic rays around the knee.

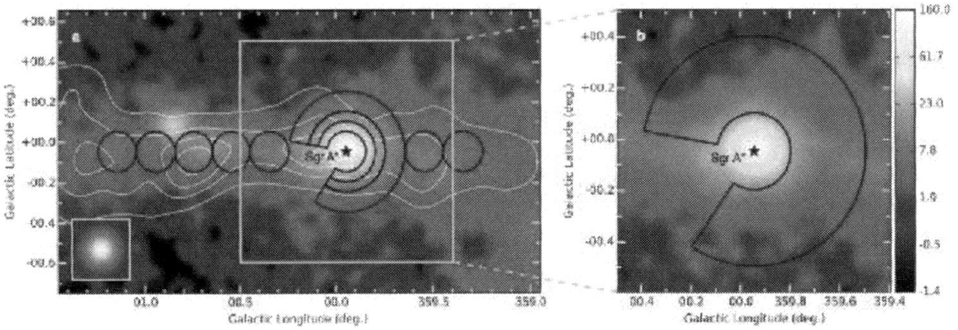

Figure 4. Excess map of the galactic centre region, showing the regions used for the derivation of a radial emission profile (left) and the region used for spectral extraction (right).

6 Active Galactic Nuclei

Several results on active galactic nuclei were presented at the conference, including the first H.E.S.S.-II results, in monoscopic mode, on the well know high synchrotron peaked blazars PKS 2155-304 and PG 1553+113 [32]. The data acquired with CT5 revealed a significant spectral curvature for both sources with respect to a simple power-law spectrum and contributed to close the gap with *Fermi*-LAT.

The discovery of TeV emission from the radio galaxy PKS 0625-354, the sixth source in the class of "γ-ray loud" radio galaxies, is important for modeling and understanding the contribution of non-blazar AGN to extragalactic γ-ray background.

7 Conclusions

The results presented here represents only a small fraction of the recent highlights from H.E.S.S.. Other results were presented at the conference, including the discovery of exceptional sources in the Large Magellanic Cloud (first superbubble in TeV γ-rays, most luminous PWN and a new SNR), results from searches for Dark Matter annihilation and updated measurement of Extragalactic Background Light. A special edition of Astronomy & Astrophysics on the H.E.S.S.-I legacy is currently in preparation, and will be released in 2017.

After more than 10 years of operation, the camera of H.E.S.S.-I are currently been upgraded with a new electronics, inspired from the design of CTA, to optimize the overlap with CT5 in the coming years and reduce the dead-time. Observation with a fully refurbished array will restart early 2017.

References

[1] C. Deil, et al. (H.E.S.S. Collaboration), PoS(ICRC) p. 773 (2015)
[2] F. Aharonian, et al (H.E.S.S. Collaboration), Science **307**, 1938 (2005)
[3] F. Aharonian, et al (H.E.S.S. Collaboration), ApJ **636**, 777 (2006)
[4] F.A. Aharonian, et al., A&A **395**, 803 (2002)
[5] A. Weinstein, et al. (VERITAS Collaboration), ArXiv e-prints (2009), `0912.4492`
[6] A. Popkow (VERITAS Collaboration), PoS(ICRC) p. 750 (2015)
[7] S. Klepser (H.E.S.S. Collaboration), PoS(ICRC) p. 777 (2015)
[8] J. Hahn, et al. (H.E.S.S. Collaboration), PoS(ICRC) p. 860 (2015)
[9] E. Aliu, et al. (MAGIC Collaboration), Science **322**, 1221 (2008)
[10] E. Aliu, et al. (VERITAS Collaboration), Science **334**, 69 (2011)
[11] J. Aleksić, et al. (MAGIC Collaboration), A&A **540**, A69 (2012)
[12] S. Ansoldi, et al. (MAGIC Collaboration), A&A **585**, A133 (2016)
[13] R.N. Manchester, et al., AJ **129**, 1993 (2005)
[14] R. Dodson, et al., ApJ **596**, 1137 (2003)
[15] A.A. Abdo, et al., ApJ **713**, 146 (2010)
[16] A.A. Abdo, et al., ApJ **713**, 154 (2010)
[17] A.A. Abdo, et al., ApJS **187**, 460 (2010)
[18] L.O. Drury, Reports on Progress in Physics **46**, 973 (1983)
[19] F.A. Aharonian, et al. (H.E.S.S. Collaboration), Nature **432**, 75 (2004)
[20] F. Aharonian, et al. (H.E.S.S. Collaboration), A&A **464**, 235 (2007)
[21] F. Aharonian, at al. (H.E.S.S. Collaboration), A&A **437**, L7 (2005)
[22] F. Acero, et al. (H.E.S.S. Collaboration), A&A **516**, A62+ (2010)
[23] F. Acero, et al., PoS(ICRC) (2009), (H.E.S.S. Collaboration)
[24] F. Aharonian, et al. (H.E.S.S. Collaboration), ApJ **692**, 1500 (2009)
[25] G. Pühlhofer, et al. (H.E.S.S. Collaboration), PoS(ICRC) p. 886 (2015)
[26] Z.R. Wang, Q.Y. Qu, Y. Chen, A&A **318**, L59 (1997)
[27] M. de Naurois, L. Rolland, Astropart. Phys. **32**, 231 (2009)
[28] P. Eger, et al. (H.E.S.S. Collaboration), PoS(ICRC) p. 766 (2015)
[29] F. Aharonian, et al. (H.E.S.S. Collaboration), A&A **425**, L13 (2004)
[30] F. Aharonian, et al. (H.E.S.S. Collaboration), Nature **439**, 695 (2006)
[31] A. Abramowski, et al. (HESS Collaboration), Nature **531**, 476 (2016)
[32] D. Zaborov, et al. (H.E.S.S. Collaboration), PoS(ICRC) p. 808 (2015)

A Major Upgrade of the H.E.S.S. Cherenkov Cameras

Iryna Lypova[1,a], Gianluca Giavitto[1,b], Terry Ashton[2], Arnim Balzer[3], David Berge[3], Francois Brun[4], Thomas Chaminade[4], Eric Delagnes[4], Gerard Fontaine[6], Matthias Füßling[1], Berrie Giebels[6], Jean-Francois Glicenstein[4], Tobias Gräber[1], Jim Hinton[7], Albert Jahnke[7], Stefan Klepser[1], Marko Kossatz[1], Axel Kretzschmann[1], Valentin Lefranc[1,5], Holger Leich[1], Hartmut Lüdecke[1], Pascal Manigot[6], Vincent Marandon[7], Emmanuel Moulin[4], Mathieu de Naurois[6], Patrick Nayman[5], Stefan Ohm[1], Marek Penno[1], Duncan Ross[2], David Salek[3], Markus Schade[1], Thomas Schwab[7], Rachel Simoni[3], Christian Stegmann[1], Constantin Steppa[1], Julian Thornhill[2], and Francois Toussnel[5]

[1]*DESY, D-15738 Zeuthen, Germany*
[2]*Department of Physics and Astronomy, The University of Leicester, University Road, Leicester, LE1 7RH, United Kingdom*
[3]*GRAPPA, Anton Pannekoek Institute for Astronomy, University of Amsterdam, Science Park 904, 1098 XH Amsterdam, The Netherlands*
[4]*DSM/Irfu, CEA Saclay, F-91191 Gif-Sur-Yvette Cedex, France*
[5]*LPNHE, Université Pierre et Marie Curie Paris 6, Université Denis Diderot Paris 7, CNRS/IN2P3, 4 Place Jussieu, F-75252, Paris Cedex 5, France*
[6]*Laboratoire Leprince-Ringuet, Ecole Polytechnique, CNRS/IN2P3, F-91128 Palaiseau, France*
[7]*Max-Planck-Institut für Kernphysik, P.O. Box 103980, D-69029 Heidelberg, Germany*

Abstract. The High Energy Stereoscopic System (H.E.S.S.) is an array of imaging atmospheric Cherenkov telescopes (IACTs) located in Namibia. It was built to detect Very High Energy (VHE, >100 GeV) cosmic gamma rays, and consists of four 12 m diameter Cherenkov telescopes (CT1-4), built in 2003, and a larger 28 m telescope (CT5), built in 2012. The larger mirror surface of CT5 permits to lower the energy threshold of the array down to 30 GeV. The cameras of CT1-4 are currently undergoing an extensive upgrade, with the goals of reducing their failure rate, reducing their readout dead time and improving the overall performance of the array. The entire camera electronics has been renewed from ground-up, as well as the power, ventilation and pneumatics systems, and the control and data acquisition software. Technical solutions forseen for the next-generation Cherenkov Telescope Array (CTA) observatory have been introduced, most notably the readout is based on the NECTAr analog memory chip. The camera control subsystems and the control software framework also pursue an innovative design, increasing the camera performance, robustness and flexibility. The CT1 camera has been upgraded in July 2015 and is currently taking data; CT2-4 will upgraded in Fall 2016. Together they will assure continuous operation of H.E.S.S at its full sensitivity until and possibly beyond the advent of CTA. This contribution describes the design, the testing and the in-lab and on-site performance of all components of the newly upgraded H.E.S.S. camera.

[a]e-mail: iryna.lypova@desy.de
[b]e-mail: gianluca.giavitto@desy.de

1 Introduction

The H.E.S.S. experiment is situated at 1800 m hight in the Khomas Highland of Namibia. The first four telescopes (CT1 - 4) were built in 2002 - 2003 (H.E.S.S. phase I). They are arranged in a square with 120 m side length and have 12 m diameter. In summer 2012 the fifth telescope (CT5) with a diameter of 28 m was build in the middle of the array (H.E.S.S. phase II). The larger mirror of the new telescope lowered the array energy threshold from ~100 GeV to ~30 GeV and increased the array trigger rate from (200 - 300) Hz to more than 1.5 kHz. The dead time of the old H.E.S.S. phase I camera was ~450 μs, which is much bigger than the dead time of the modern CT5 (~15 μs) and is not suitable for such high rates. Thus, significant fraction of the triggered events is recorded only by CT5, limiting the performance of the array.

Thereby, one of the main goals of the H.E.S.S. phase I upgrade is to reduce the dead time and to increase the number of recorded stereoscopic events. The second reason for the upgrade is the aging of the electronic components. The cameras of CT1-4 have been operating in the desert environment for more than 10 years leading to an increased rate of failures. To avoid further degradation, the electronics and cabling of the cameras are renewed during the upgrade.

2 Camera components

As it is seen from figure 1 most of the electronic components of the camera were replaced. Only photomultiplier tubes (PMTs), their bases, single photoelectron (ph.e.) and flat fielding units for calibration are the same.

The upgraded camera has a new concept for the ventilation system. Instead of many small fans located inside the camera body now there is a single fan on the back door. It increases air pressure and prevents dust accumulation inside the camera. In addition it can heat or cool air if needed. A new back door design was developed for the upgraded ventilation system. However this increased the weight of the back door and required two pneumatic cylinders to open and close it.

Behind the back door there is a rack with back-end electronics. It includes a power distribution box (PDB) and a drawer interface box (DIB). The PDB provides power to the front-end electronics. The DIB is one of the most important components in the camera and is responsible for the interface of the calibration units, ventilation system, number of different sensors, GPS timestamps for the events. Its most important component is the camera trigger.

2.1 Front-end electronics

The telescope camera contains 960 PMTs arranged in 60 modules (16 PMTs in each) which are called drawers. A drawer includes two analog boards and one slow control board. Each analog board obtains signals from 8 PMTs. The signals are pre-amplified and sent in tree different channels. One of them is a trigger channel and two others are related to readout (low and high gain).

The trigger channel is fed to a comparator. If the signal is high enough (several photoelectrons) then pixel is triggered (level 0 trigger) [1]. The level 0 trigger signals are routed to the DIB, where N-majority trigger is built out of 38 overlapping groups of 64 pixels each. This is the camera-wide trigger (level 1), which helps to reduce night sky background (NSB) and identify useful data (air showers).

At the same time in readout channels signal gets into NECTAr chip [2] where it is sampled, stored and digitised. There are two capacitor arrays for the signal storage. Each of them has a length of 1024 cells, but if camera is triggered only 16 cells will be read.

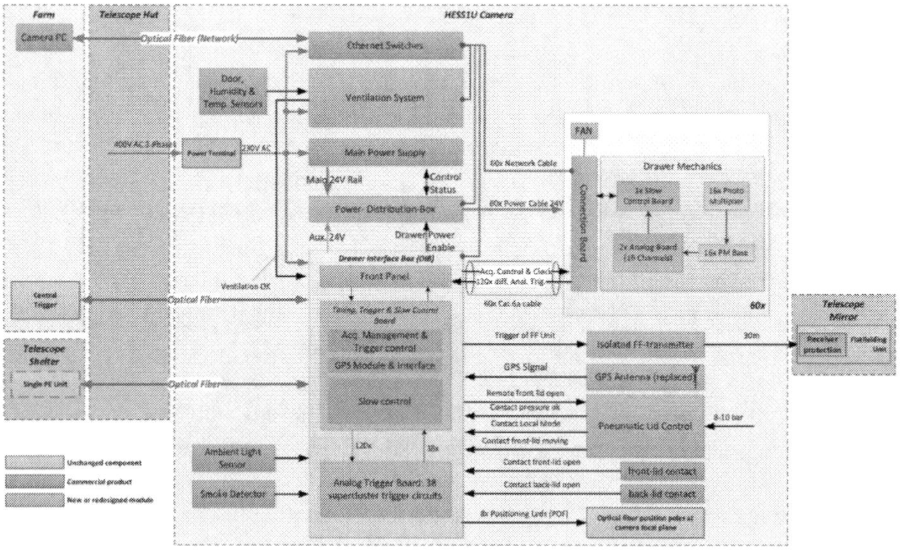

Figure 1: Schematic illustration of upgraded H.E.S.S. camera

2.2 Tests

More than 270 drawers were tested at DESY (Zeuthen) with several test setups. A copy of the camera body (CopyCam, fig. 2, left top) was used for tests of the ventilation system, back door mechanics, pneumatics, centre camera trigger etc. The MiniCam is a dark box for four drawers with PMTs, which offered a possibility to check the function of front-end electronics in realistic conditions.

One more test bench (fig. 2, left bottom) was used to check basic functionality of the electronic components and test the drawers without PMTs. Specially for this setup 8-channel pulse generator was developed. It can generate negative pulses with $\sim$1 ns rising and falling time and amplitudes from $\sim$0.6 mV to $\sim$300 mV. A number of different characteristics were tested here: low gain (LG) and high gain (HG) linearity, cross-talk, pedestal noise, trigger path etc. An example of the linearity test results is illustrated on figure 2 (right). High gain shows linear behaviour from 0 to $\sim$200 ph.e., low gain from $\sim$30 ph.e. to 4200 ph.e. Deviation from linearity do not exceed 2%.

3 Status of the upgrade

The first camera (CT1) was upgraded in summer 2015 and since March 2016 has been included in regular observation. The upgrade of other three cameras started in September 2016 and currently they are in the process of commissioning. Monte-Carlo simulations for the upgraded cameras are ongoing. For that purpose CORSIKA [3] and `sim_telarray` [4] are used. First results showed good match between the simulated trigger rate and the observed one with real upgraded camera.

H.E.S.S. phase I upgrade allows us to keep H.E.S.S. operating in a stable regime and improves the sensitivity of the lowest and highest energies. New fast readout electronics gives us the opportunity to lower the trigger threshold and as a result collect more stereoscopic events at energies $<$100 GeV. In turn, the optimisation of the length of the readout window could improve reconstruction of high energy events. Moreover, the current upgrade of the H.E.S.S. telescopes is a great opportunity to test new electronics, software and algorithms for the future ground based experiments.

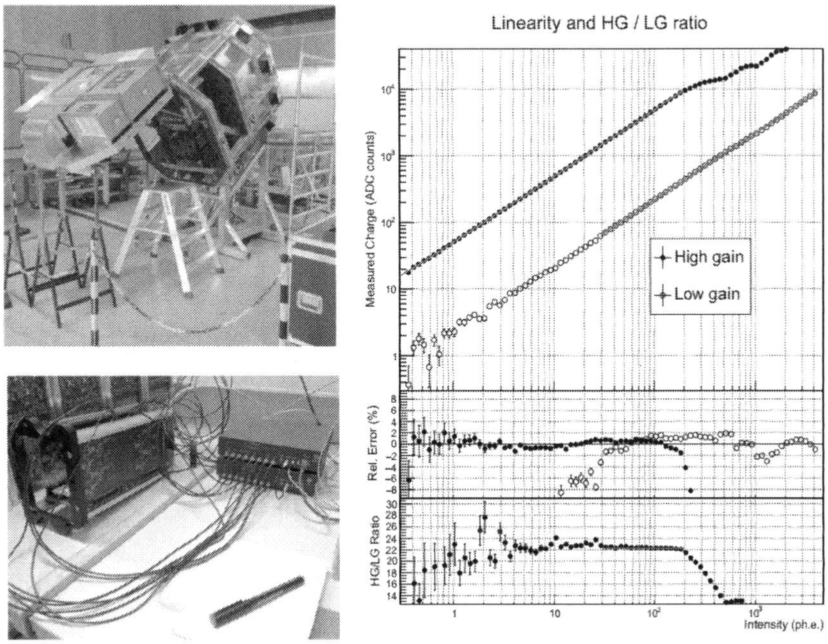

Figure 2: Left: testing setups at DESY (Zeuthen) - CopyCam (top) and test bench with pulse generators (bottom); right: result of linearity tests.

Acknowledgments

The support of the Namibian authorities and of the University of Namibia in facilitating the construction and operation of H.E.S.S. is gratefully acknowledged, as is the support by the German Ministry for Education and Research (BMBF), the Max Planck Society, the German Research Foundation (DFG), the French Ministry for Research, the CNRS-IN2P3 and the Astroparticle Interdisciplinary Programme of the CNRS, the U.K. Science and Technology Facilities Council (STFC), the IPNP of the Charles University, the Czech Science Foundation, the Polish Ministry of Science and Higher Education, the South African Department of Science and Technology and National Research Foundation, and by the University of Namibia. We appreciate the excellent work of the technical support staff in Berlin, Durham, Hamburg, Heidelberg, Palaiseau, Paris, Saclay, and in Namibia in the construction and operation of the equipment.

References

[1] Funk, S. et al., Astropart. Phys. **22**, 285-296 (2004).
[2] Naumann, C. L. et al., NIM A **695**, 44-51 (2012).
[3] Heck D. et al., Wissenschaftliche Berichte FZKA 6019 (1998).
[4] Bernlöhr K., Astropart. Phys. **30**, 149-158 (2008).

The Crab pulsar at VHE

Roberta Zanin[1,a]

[1] *Max-Planck-Institut fur Kernphysik, P.O. Box 103980, D 69029 Heidelberg, Germany*

Abstract.
The last six years have witnessed major revisions of our knowledge about the Crab Pulsar. The consensus scenario for the origin of the high-energy pulsed emission has been challenged with the discovery of a very-high-energy power law tail extending up to ~400 GeV, above the expected spectral cut off at a few GeV. Now, new measurements obtained by the MAGIC collaboration extend the energy spectrum of the Crab Pulsar even further, on the TeV regime. Above ~400 GeV the pulsed emission comes mainly from the interpulse, which becomes more prominent with energy due to a harder spectral index. These findings require γ-ray production via inverse Compton scattering close to or beyond the light cylinder radius by an underlying particle population with Lorentz factors greater than 5×10^6. We will present those new results and discuss the implications in our current knowledge concerning pulsar environments.

1 Introduction

Pulsars are rapidly rotating neutron stars (NS) which form highly magnetized magnetospheres, usually described as magnetic dipoles. The magnetosphere is filled with charged particles, mainly electrons and positrons, extracted from the stellar surface by the induced electric fields. The original model for pulsar magnetospheres foresees that these charged particles satisfy the ideal magnetohydrodynamic (MHD) ($E \cdot B$=0) and the free-force (FF, $\rho E + J \times B$=0) conditions [1]. In this view, the magnetic field lines cannot remained closed beyond the light cylinder (LC, with a radius R_{LC}), where the angular velocity of the NS is equal to the speed of light. Along the open field lines the plasma flows along asymptotically monopole field lines [2] forming the *pulsar wind region*. Nowadays it is commonly accepted that the ideal MHD and FF conditions cannot be fulfilled everywhere inside the magnetosphere. In the past years regions where the electric field is not totally screened by the plasma were invoked as possible sites of particle accelerations. They are referred as to *gaps*. Recently current sheets are taking over though. They are current carrying surfaces where particles can be accelerated via relativistic reconnection and radiate synchrotron emission from optical to γ-ray wavelengths. In particular, an equatorial current sheet was proposed by Contopoulos to ensure the closure of the large-scale currents flowing from the star to infinity and back [3], but only recent particle-in-cell (PIC) simulations of plasma-filled magnetospheres have provided the evidence for its existence, as well as its possible extension down to the NS surface, separating the open and closed field lines (this names them *separatrix*) [4–7]. The equatorial current sheet separates the field lines that originate from the

[a]e-mail: Roberta.Zanin@mpi-hd.mpg.de

two pulsar magnetic poles. When the rotation and magnetic axes are not aligned (oblique pulsars), this current sheet develops corrugation whose amplitude increases linearly with the distance from the star. The equatorial plan is then divided into stripes of magnetic field of opposite polarities, hence the name *stripped wind* [8–10].
In this new paradigma of the pulsar magnetosphere, two classes of pulsars are defined depending on the particle density inside the magnetosphere: the high-multiplicity pulsars where pair production takes place within the magnetosphere, that are the young pulsars with high stellar magnetic fields, and the low-density pulsars where particles are supplied only from the NS surface and are identified with the old pulsars. For the first class of pulsars the FF solution is a good approximation of their magnetosphere characterized by the presence of the separatrices and the equatorial current sheet. They are weakly dissipative [7, 11]. The low-density pulsars exhibit a charged separated magnetosphere with no separatrices since the equatorial sheet is electrostatically supported. They show high dissipation (>40%) beyond the LC [7, 11].

Pulsars are observed from radio frequencies up to γ rays. Radio emission is believed to be produced at low-altitudes, close to the stellar magnetic poles, by a a coherent process either maser amplification or coordinated motion of group of charges. At higher frequencies pulsed emission is incoherent, most likely synchrotron radiation produced by relativistic electron-positron pairs. At high energies, pulsed emission was long attributed to synchro-curvature by particles accelerated in magnetospheric gaps. According to the location of the gap inside the magnetosphere, γ-ray pulsar models were classified into *polar caps*, if close to the magnetic poles, [12], *outer gaps* if at high altitudes [13], or *slot gaps*, if along the last open field lines [14]. In particular, such a synchro-curvature emission has a maximum energy limited by radiation losses or magnetic and γ-γ pair absorption. The recent advances on the high-energy pulsar physics due to discovery of more than 150 γ-ray pulsars by *Fermi*/LAT [15] suggest that pulsed γ rays are produced at high altitudes favouring the *outer gaps* models. This conclusion is mainly drawn by two observational facts: 1) The spectral cutoff at a few GeV is exponential, and often even subexponetial for young pulsars. The smoother spectral break is interpreted as the overlapping of emission beams coming from different regions and/or with different cutoff energies. 2) The majority of the γ-ray pulsars present a double-peaked pulsar profile, which in low-altitude emission models would require a specific geometric configuration (both viewing and inclination angles close to 90°), too special to meet the observed numbers [16]. However, in the context of the new paradigma of the pulsar magnetosphere, the high-energy emission could entirely or partially be of synchrotron origin produced in the equatorial current sheet beyond the light cylinder [17, 18, 26].

2 The Crab pulsar

The Crab pulsar is the second most powerful pulsar known so far with a spin-down luminosity $\dot{E}$ = 3.8×10^{38} erg/s and the only pulsar whose age is known with precision. It is, in fact, the leftover of an historical supernova explosion occurred in 1054 AD which was in detail documented by Chinese astronomers.

Its pulse profile is characterised by two peaks almost aligned at almost all wavelengths, even though their amplitudes change with energy [19, 20]. The main peak at low radio frequencies (0.1–5 GHz), by definition set at phase 0, is usually referred to as P1, whereas P2 or interpulse is located $\sim 0.4°$ away. The emission in between the two peak is called *bridge*.

2.1 Recent results at VHE

All the 160 γ-ray pulsars detected by *Fermi*/LAT show a spectral break at ~few GeV in agreement with the synchro-curvature scenario [15], but one: the Crab pulsar. The spectrum of the Crab pulsar between 0.1-100 GeV is well parametrized by a power-law function with a sub-exponential cutoff (b<1). The best-fit value for the photon index is γ=1.59±0.01, for the energy break is (5.09±0.63) GeV and for the curvature index b=0.43±0.01 [21]. Recent measurements of the Crab pulsar by imaging atmospheric Cherenkov telescopes showed a totally unexpected power-law component emerging above the synchro-curvature cutoff and extending above 100 GeV [22–24]. This new spectral component, that is excluding the spectral break at more than than 6σ [22], is a common feature of both peaks and the bridge. A joint fit of the *Fermi*/LAT data above 10 GeV and the MAGIC ones indicates that P2 is significantly harder than P1 with a difference between the photon indices of 0.5±0.1. Whereas the steeper P1 (Γ=3.5±0.1) spectrum cannot be detected beyond 600 GeV, P2 becomes the dominant component above 50 GeV [23] and extends up to 1.5 TeV without any sign of cutoff [25]. A lower limit on the spectral cutoff is estimated at 700 GeV. On the other hand, the spectrum of the bridge is as soft as P1 and fades out already at 150 GeV [20].

This VHE pulsed emission is very unlikely to be synchro-curvature radiation: the production of 1 TeV photon would require, in fact, a curvature radius of the magnetic field lines one order of magnitude larger than the usual adopted one. If emission up to few hundreds of GeV could be also synchrotron radiation from the equatorial current sheet [17, 26], the emission at higher energies is accredited to be produced via inverse Compton scattering. In this view several models have been put forward. Some of them are listed in the following:

- inverse Compton scattering of relativistic wind electrons off pulsed optical/X-ray photons with magnetospheric origin [27, 28]. This model was proposed to explain the emission up to 400 GeV and can well reproduce the pulse profile by assuming an anisotropic pulsar wind. Nevertheless, it fails in reproducing the spectrum up to 1 TeV. The production of these energetic photons would require electron parent population with a Lorentz factor larger than 5×10^6, hence a continuos acceleration from the LC up to $\sim 100 R_{LC}$ which in turn results in an overestimation of the GeV flux.
- synchrotron-self-Compton scattering off synchrotron photons produced in the current sheet by the same population of synchrotron-emitting electrons. This model reproduces the spectrum up to VHE as a sum of two distinct components: the synchrotron power-law up to few hundreds of GeV and the SSC as an extra bump fading out at a few TeV [17]. However, the model cannot explain the pulse profile.
- inverse Compton scattering off either photons in particle cascading inside the magnetospheric gaps [23, 29, 30].

So far none of the existing model can explain simultaneously the spectrum and the pulse profile with its narrow peaks at VHE. In addition there is no comprehensive theoretical model reproducing all the observable of the pulsed emission i.e. the broadband spectrum with its spectral changes across the pulse-phase in X- and γ rays, and the pulse profile with its energy-dependent peak amplitude and width.

References

[1] P. Goldreich, W.H. Julian, *ApJ* **157**, 869 (1969)
[2] F.C. Michel, *ApJ* **180**, 207 (1973)
[3] I. Contopoulos, D. Kazanas, C. Fendt, *ApJ* **511**, 351 (1999)

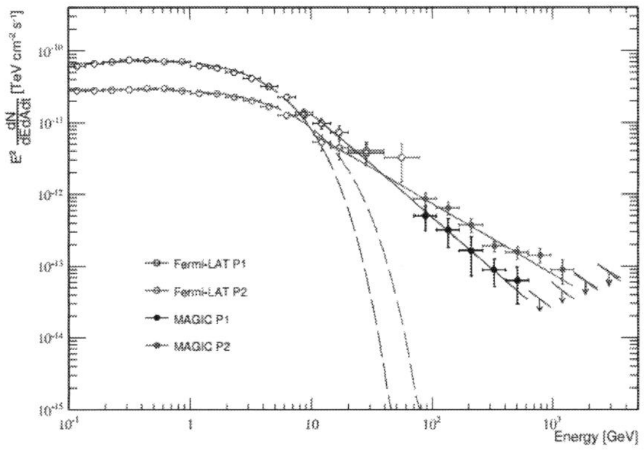

Figure 1. Phase-folded SED of the Crab P1 (black circles) and P2 (blue circles) at HE and VHE (open and filled circles). Taken from [25].

[4] A. Spitkovsky, *ApJ, Letters* **648**, L51 (2006)
[5] C. Kalapotharakos, I. Contopoulos, *A & A* **496**, 495 (2009)
[6] A. Tchekhovskoy, A. Spitkovsky, J.G. Li, *MNRAS* **435**, L1 (2013)
[7] B. Cerutti, A. Philippov, K. Parfrey, A. Spitkovsky, *MNRAS* **448**, 606 (2015)
[8] F.V. Coroniti, *ApJ* **349**, 538 (1990)
[9] S.V. Bogovalov, *AAP* **349**, 1017 (1999)
[10] J.G. Kirk, O. Skjæraasen, Y.A. Gallant, *AAP* **388**, L29 (2002)
[11] I. Contopoulos, *MNRAS* **463**, L94 (2016)
[12] A.K. Harding, E. Tademaru, L.W. Esposito, *ApJ* **225**, 226 (1978)
[13] K.S. Cheng, C. Ho, M. Ruderman, *ApJ* **300**, 500 (1986)
[14] A.G. Muslimov, A.K. Harding, *ApJ* **588**, 430 (2003), astro-ph/0301023
[15] F. collaboration, *ApJ* **208**, 17 (2013)
[16] R.W. Romani, K.P. Watters, *ApJ* **714**, 810 (2010)
[17] I. Mochol, J. Pétri, *MNRAS* **449**, L51 (2015)
[18] B. Cerutti, A.A. Philippov, A. Spitkovsky, *MNRAS* **457**, 2401 (2016)
[19] L. Kuiper, et al, *A & A* **378**, 918 (2001)
[20] MAGIC Collaboration, *A & A* **565**, L12 (2014)
[21] R. Buehler, et al., *ApJ* **749**, 26 (2012)
[22] VERITAS Collaboration, Science **334**, 69 (2011)
[23] MAGIC Collaboration, *ApJ* **742**, 43 (2011)
[24] MAGIC Collaboration, *A & A* **540**, A69 (2012)
[25] MAGIC Collaboration, *A & A* **585**, A133 (2016)
[26] J. Pétri, *MNRAS* **424**, 2023 (2012)
[27] S.V. Bogovalov, F.A. Aharonian, *MNRAS* **313**, 504 (2000)
[28] F.A. Aharonian, S.V. Bogovalov, D. Khangulyan, *Nature* **482**, 507 (2012)
[29] M. Lyutikov, N. Otte, A. McCann, *ApJ* **754**, 33 (2012)
[30] K. Hirotani, *ApJ, Letters* **798**, L40 (2015)

 DOI: 10.1051/epjconf/201713603004

Recent results in γ-ray astronomy with the ARGO-YBJ detector

Tristano Di Girolamo[1,a] on behalf of the ARGO-YBJ collaboration

[1] *Università "Federico II" and INFN - Napoli, Complesso Monte S. Angelo, Via Cintia, Napoli, Italy*

Abstract. The ARGO-YBJ air shower detector has been in stable data taking for five years at the YangBaJing Cosmic Ray Laboratory (Tibet, P.R. China, 4300m a.s.l.) with a duty cycle > 86% and an energy threshold of a few hundreds of GeV. With the scaler mode technique, the minimum threshold of 1 GeV can be reached. In this paper recent results in γ-ray astronomy will be presented, including those from 4.5 years of observations of the blazar Mrk 421 in common with the Fermi satellite.

1 The detector

The ARGO-YBJ experiment is located at Yangbajing (Tibet, P.R. China, 4300 m a.s.l.) and consists of a single layer of Resistive Plate Counters (RPCs) on a total area of about 110×100 m^2. The detector has a modular structure, the basic module being a cluster (5.7×7.6 m^2), made of 12 RPCs. Each RPC is read by 80 strips (6.75×61.8 cm^2) which are the space pixels, logically organized in 10 independent pads (55.6×61.8 cm^2) which are individually acquired and represent the time pixels of the detector [1]. The detector carpet is connected to two different DAQ systems, which work independently: in shower mode, for each event the location and timing of each detected particle is recorded, allowing the reconstruction of the lateral distribution and of the arrival direction; in scaler mode, the counting rate of each cluster is measured every 0.5 s, with little information on the space distribution and arrival direction of the detected particles. The trigger of the shower mode was $N_{pad} \geq 20$ in a time window of 420 ns, with a rate of 3.5 kHz. In the scaler mode, for each cluster four scalers recorded the rate of counts ≥ 1, ≥ 2, ≥ 3 and ≥ 4 in a time window of 150 ns. The corresponding measured rates are, respectively, ~40 kHz, ~2 kHz, ~300 Hz and ~120 Hz [2]. The experiment has been taking data with its full layout from November 2007 to February 2013. The detector pointing accuracy, angular resolution and absolute energy calibration have been determined studying the deficit in the cosmic ray flux due to the Moon [3].

2 Sky survey

The ARGO-YBJ detector surveyed the northern hemisphere, in the declination band from -10° to 70°, at energies above 0.3 TeV. With an integrated sensitivity down to 0.24 Crab unit (depending on the declination) after five years of data taking, six sources were detected with a statistical significance S>5 standard deviations (s.d.), and five excesses are reported as potential (S>4 s.d.) γ-ray emitters. The list of excess regions, with their corresponding significances and TeV associations, is in table 1 [4]. In the rest of this section, a selection of results concerning the detected sources will be presented.

[a] e-mail: tristano.digirolamo@na.infn.it

Table 1. List of ARGO-YBJ excess regions with corresponding statistical significances S and TeV associations. All significances are given for $N_{pad} \geq 20$ except that of ARGO J1841-0332, which is for $N_{pad} \geq 100$ (for $N_{pad} \geq 20$ the significance is S=3.4 s.d.).

ARGO-YBJ Name	S (s.d.)	TeV Association
ARGO J0409-0627	4.8	
ARGO J0535+2203	20.8	Crab Nebula
ARGO J1105+3821	14.1	Mrk 421
ARGO J1654+3945	9.4	Mrk 501
ARGO J1839-0627	6.0	HESS J1841-055
ARGO J1907+0627	5.3	HESS J1908+063
ARGO J1910+0720	4.3	
ARGO J1912+1026	4.2	HESS J1912+101
ARGO J2021+4038	4.3	VER J2019+407
ARGO J2031+4157	6.1	MGRO J2031+41
		TeV J2032+4130
ARGO J1841-0332	4.2	HESS J1843-033

The differential energy spectrum obtained for the Crab Nebula in the range 0.3-20 TeV can be described by the power law $dN/dE = (5.2 \pm 0.2) \times 10^{-12}(E/2\ TeV)^{(-2.63\pm0.05)}$ photons cm^{-2} s^{-1} TeV^{-1}, and it is consistent with results from other experiments [5]. The light curve over five years, with a binning of 200 days, is compatible with a steady emission with a probability of 0.07. An analysis with data collected in 4.5 years of observations in common with Fermi/LAT and divided in bins of 200 days gives a Pearson correlation coefficient $r = 0.56\pm0.22$. Because of the small statistical significance of these results, no flux variability correlated with Fermi/LAT can be claimed.

The blazar Mrk 421 was observed by ARGO-YBJ and the Fermi satellite during a 4.5 year period of common operation time, from August 2008 to February 2013 [6]. Thanks to long-term multiwavelength observations, the variable emission of this source and the correlations among flux variations in different wavebands were investigated. The TeV flux is clearly correlated with the X-ray flux, while only partially with the GeV flux. Seven large flares, of which five in X-rays and two at GeV energies, and one X-ray outburst were identified and used to study the variation of the Spectral Energy Distribution (SED) with respect to two steady phases. The behaviour of the GeV γ-rays allows the classification of SEDs into three different groups. Adopting a simple one-zone synchrotron self-Compton model, we find that in two out of three groups electrons are injected with a power-law spectral index $\sim$-2.2, as expected from relativistic diffuse shock acceleration, while in the remaining group the spectrum is harder ($\sim$-1.8). The variations of the states may be due to environment properties (first two groups) or to the acceleration process itself (last group).

In a location consistent with MGRO J2031+41, Fermi/LAT detected a complex extended source, attributed to the emission by a "cocoon" of freshly accelerated cosmic rays which fill the cavities carved by stellar winds and ionization fronts from young stellar clusters [7]. After reanalysing the complete ARGO-YBJ data set, subtracting the contribution of the overlapping TeV sources and using a larger region to evaluate the excess map (since Fermi/LAT observations revealed a large extended source), ARGO J2031+4157 resulted with an extension $\sigma_{ext} = 1.8° \pm 0.5°$, consistent with that of the Cygnus Cocoon as measured by Fermi/LAT, i.e., $\sigma_{ext} = 2.0° \pm 0.2°$ [8]. The spectrum also shows a good connection with that determined by Fermi/LAT in the 1-100 GeV energy range. Therefore, ARGO J2031+4157 is identified as the counterpart of the Cygnus Cocoon at TeV energies.

3 Diffuse γ-rays from the Galactic plane

The events collected by ARGO-YBJ have been analysed to determine the diffuse γ-ray emission in the Galactic plane at longitudes $25^\circ < l < 100^\circ$ and latitudes $|b| < 5^\circ$ [9]. This analysis was carried out in the energy range connecting the region explored by Fermi/LAT with that investigated by Milagro. In particular, the analysis was focused on two selected regions of the Galactic plane, i.e., $40^\circ < l < 100^\circ$ and $65^\circ < l < 85^\circ$ (the Cygnus region), where Milagro observed an excess with respect to what predicted by current models.

In the Galactic region $40^\circ < l < 100^\circ$, $|b| < 5^\circ$, after masking the discrete sources and subtracting the residual contribution, an excess with a statistical significance of 6.1 s.d. above the background is found. The spectral analysis provides the three fluxes shown in Figure 1 (left), whose fit with a power law gives a spectral index -2.90 ± 0.31, and the corresponding flux at 1 TeV is compatible with the extrapolation of the Fermi/LAT model for the Diffuse Galactic Emission (Fermi-DGE). On the other hand, the first measurement of the diffuse TeV (integral) flux from the Galactic plane made by Milagro revealed a "TeV excess" in the diffuse γ-ray spectrum with respect to expectations [10]. This Milagro measurement, converted into differential flux, is only 34% greater than the value expected from the extrapolation of the Fermi-DGE, and within the experimental uncertainties (see triangle with error bars in left plot of Figure 1). Moreover, considering that the Milagro result does not take into account the contributions from the Cygnus Cocoon (not yet discovered at the time of the measurement) and from overlapping point and extended sources, the discrepancy with the Fermi/LAT predictions is almost cancelled out. Therefore, the full set of measurements with ground-based detectors is in agreement with direct observations by Fermi/LAT, and the evidence of any "TeV excess" is ruled out.

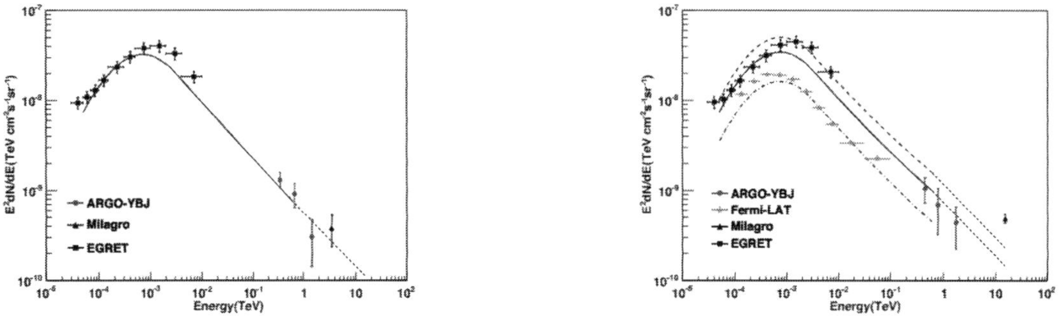

Figure 1. Energy spectrum of the Galactic diffuse γ-ray emission. The EGRET data (squares), showing a "GeV excess", are likely due to instrumental effects. The Fermi/LAT data (stars) were obtained in the region $72^\circ < l < 88^\circ$, $|b| < 15^\circ$. The different lines indicate the energy spectra expected from the Fermi-DGE (with index -2.6, which also rules their extensions) in the different sky regions investigated by the detectors (details are given in [9]). Left: region $40^\circ < l < 100^\circ$, $|b| < 5^\circ$. Right: region $65^\circ < l < 85^\circ$, $|b| < 5^\circ$

In the Galactic region $65^\circ < l < 85^\circ$, $|b| < 5^\circ$, after masking the discrete sources and the Cygnus Cocoon and subtracting the residual contribution, an excess of 4.1 s.d. is left. The SED of γ-ray emission is shown in the right plot of Figure 1 together with the spectra expected from the Fermi-DGE in the different sky regions investigated by the detectors. Milagro measured the diffuse γ-ray emission from the region $65^\circ < l < 85^\circ$, $|b| < 2^\circ$ at a median energy of 15 TeV [11], obtaining the flux reported as a filled triangle in the same plot. For comparison, the long-dashed line shows the expected energy spectrum for this region according to the Fermi-DGE. The Milagro flux results about 75% higher than the Fermi-DGE, suggesting the presence of an excess. The spectral analysis of

ARGO-YBJ data provides the three fluxes shown in the right plot of Figure 1, whose fit with a power law gives a spectral index -2.65 ± 0.44, and the corresponding flux at 1 TeV is about 10% lower than the extrapolation of the Fermi-DGE. These data do not show any excess at energies around 1 TeV which corresponds to the excess found by Milagro. Again, this discrepancy can be explained taking into account that the contribution of all the discrete γ-ray sources was not completely removed from the Milagro data. Finally, the harder spectrum in the Cygnus region compared with that measured in the whole region $40° < l < 100°$, $|b| < 5°$ may suggest the presence of young cosmic rays accelerated by a nearby source.

4 Search for Gamma Ray Bursts in scaler mode

In scaler mode, the energy threshold for photons is about 1 GeV, lower than the highest energies detected by satellite experiments. Moreover, the modular structure of the ARGO-YBJ detector allowed the collection of data during the different mounting phases. Therefore a search for emission from Gamma Ray Bursts (GRBs) in coincidence with satellite detections started in November 2004, when the Swift satellite was launched [12]. Until February 2013 a sample of 206 GRBs was analysed, 24 of them with known redshift z. This is the largest sample of GRBs investigated with a ground-based detector at high energies. Since no significant signal was found in the data, for each GRB fluence upper limits in the 1-100 GeV energy range were determined at 99% c.l. assuming two different power law spectra: a) the index measured by satellite detectors in the keV-MeV energy range; b) the conservative differential index -2.5. For case a), when double power law spectral features have been identified, the higher energy index has been used. Therefore, we obtain ranges of upper limits between the values corresponding to the two spectral assumptions, while a single value results if the low energy spectrum is a cutoff power law, and thus only case b) is considered. For the set of 24 GRBs with known redshift, the fluence upper limits are as low as $\sim 3 \times 10^{-5}$ erg/cm^2 and are the only ones set at GeV energies. For a subset of these GRBs, the upper limits show that the low-energy spectrum cannot be extended to the GeV region and some additional features occur in the keV-MeV range. For GRB090902B the upper limit can be compared with the fluence extrapolated from Fermi/LAT observations in the same energy range [13], which results lower by a factor ~3. More results and details about this search are given in [14].

References

[1] G. Aielli et al., *NIM A* **562** 92 (2006)
[2] G. Aielli et al., *Astropart. Phys.* **30** 85 (2008)
[3] B. Bartoli et al., *PRD* **84** 022003 (2011)
[4] B. Bartoli et al., *ApJ* **779** 27 (2013)
[5] B. Bartoli et al., *ApJ* **798** 119 (2015)
[6] B. Bartoli et al., *ApJS* **222** 6 (2016)
[7] M. Ackermann et al., *Science* **334** 1103 (2011)
[8] B. Bartoli et al., *ApJ* **790** 152 (2014)
[9] B. Bartoli et al., *ApJ* **806** 20 (2015)
[10] T. Prodanović, B. D. Fields & J. F. Beacom, *Astropart. Phys.* **27** 10 (2007)
[11] A. A. Abdo et al., *ApJ* **688** 1078 (2008)
[12] G. Aielli et al., *ApJ* **699** 1281 (2009)
[13] A. A. Abdo et al., *ApJ* **706** L138 (2009)
[14] B. Bartoli et al., *ApJ* **794** 82 (2014)

First year results from the HAWC observatory

Sabrina Casanova for the HAWC Collaboration[1,2,a]

[1] *Institute of Nuclear Physics Krakow, ul.Radzikowskiego 152, 31-342 Poland*
[2] *Max Planck Institute for Nuclear Physics Heidelberg, Saupfercheckweg 1, 69117 Germany*

Abstract. The High Altitude Water Cherenkov Observatory is an all-sky surveying instrument sensitive to gamma rays and cosmic rays from 100GeV to 100TeV. With its 2sr instantaneous field of view and a duty cycle of > 95%, HAWC is carrying out an unbiased survey of the Northern sky and is monitoring known flaring sources and searching for transients. HAWC operation began mid-2013 with a partially-completed detector. The array was terminated in 2015. We here summarize the status of the observatory, and highlight its first scientific results, resulting from the first year of data taking after completion of the detector. In particular, we will present the HAWC map of the sky at tens of TeV.

1 Introduction

The High Altitude Water-Cherenkov Telescope (HAWC) is an air-shower array facility, designed to study TeV gamma rays by observing the air-shower particles, which, produced by the TeV photons in the high atmosphere, arrive in the detector. HAWC data taking started in August 2013 while still under construction. HAWC construction was terminated at the beginning of 2015. The complete detector consists of 300 Water Cherenkov detectors (WCDs) spread on a 22000 square meter area. Every WCD is a 4.5 m high tank made of steel with a diameter of 7.3m, and contains 200,000 liters of purified water. At the bottom of the tank there are three 8-inch photomultipliers (PMTs) arranged in a triangular layout, plus one high quantum-efficiency 10-inch PMT at the center of this triangle. The photomultipliers detects the Cherenkov light produced by the air shower particles in each WCD unit. HAWC is continuously recording signals from each WCD and seeking for multiple coincident signals or events. The data are reconstructed event per event, each event having between 20 and 50 channels hit. The shower direction is reconstructed by collecting and comparing the arrival times of the shower particles in the different WCDs hit by the shower. The detector is calibrated by means of a laser calibration system which delivers light pulses to all the WCD in the array using optical fibers. This laser signal guarantees the timing and charge calibration of the individual PMTs, which are crucial for an accurate reconstruction of the air shower. The shower fronts are fitted with sub-nanosecond residuals, resulting in an angular resolution as good as 0.2 degree for the highest energy events. The charge deposits on each PMT channel is used for energy estimations and for gamma/hadron discrimination through algorithms based on core location and the pattern of the energy deposition in the array

[a] e-mail: sabrinacasanova@gmail.com

- including muon identification. HAWC registers over 20,000 cosmic rays per second, generating 2 TB of data every day.

Its wide field-of-view, with an instantaneous aperture of 2sr, covers more than 15% of the sky so that the detector is exposed to two-thirds of the sky during a 24-hour period. This makes HAWC an ideal survey and monitoring observatory, complementary to the pointing instruments at TeV energies, the imaging air Cherenkov telescopes (IACTs), such as H.E.S.S., MAGIC, VERITAS, and the up-coming Cherenkov Telescope Array (CTA). While the IACTs reach their best sensitivity at about 1 TeV, the HAWC observatory is particularly sensitive to photons of energy above several TeVs. In fact, HAWC sensitivity curve to steady sources peaks at 10 TeV and furthermore is constant for extended sources up to three degrees across [1]. These two key features make HAWC an ideal instrument to discover and study Galactic gamma-ray sources produced by the interactions of the highest energy cosmic rays. In the following section we will review the HAWC map of the sky at tens of TeV obtained by the HAWC Collaboration after collecting one year of data.

2 Scientific results

Primary science goals of the HAWC observatory are searching for the origin of the highest energy cosmic-rays, studying transient phenomena, such gamma-ray binaries and gamma-ray bursts [2], monitoring active Galactic Nuclei, searching for dark matter signals [3] and studying more exotic topics such as primordial black holes [4].

A year after completion of the observatory, the HAWC Collaboration announced a new map of the sky in very high energetic (VHE) gamma rays [5]. The sky map in Fig.1 shows several astrophysical objects along the plane of our galaxy, many of which have not been observed previously. Among the extended galactic sources observed by HAWC it is worth mentioning the extended emission discovered close to two famous middle-age closeby pulsars, Geminga, and PSR B0656+14, associated with the Monogem ring (see Fig.2). The two HAWC sources are likely pulsar wind nebulae powered by ultrarelativistic electrons accelerated at the termination shocks, where the powerful pulsar winds encounter the interstellar medium. The HAWC discovery of extended TeV emission from the directions of both Geminga and the pulsar PSR B0656+14 offers a unique chance to investigate the VHE particle populations in our Galactic neighbourhood, shedding light on the origin of these particles and understanding the local particle transport at the highest energies, which is not constrained by any cosmic ray measurements.

Another region of great astronomical interest mapped by HAWC is the Cygnus region, the brightest region in GeV and TeV gamma-rays in the Northern sky (see Fig.2). The region was previously mapped at TeV energies with the Milagro instrument [6–8] and one of the two Milagro sources in Cygnus, MGRO J2019+37, was later spatially resolved by Veritas in two sources, which still lack a clear counterpart at lower energies [9]. The study of the non-thermal radiation from the Cygnus region, dominated by one of the brightest and closest star forming region in our Galaxy, Cygnus X, is crucial for understanding the origin of cosmic rays in the Galaxy and the relationship between star formation and acceleration of particles. Detailed studies of the different TeV sources observed by HAWC are currently under way.

The new non-thermal map of the universe to the highest energies also includes two famous active galactic nuclei, Markarian 501 and Markarian 421, the closest BL Lac objects known, whose lightcurves are shown in Fig.3 and Fig.4, respectively. Thanks to the high significances of the detections, 33 σ for MrK421 and 23 σ for Mrk501 it was possible to extract their daily flux level. The HAWC light curve of Mrk421, including 387 selected transits, represents the first daily monitoring of a TeV blazar with the sensitivity to actually detect flaring on single transits. Frequent high states of

 DOI: 10.1051/epjconf/201713603005

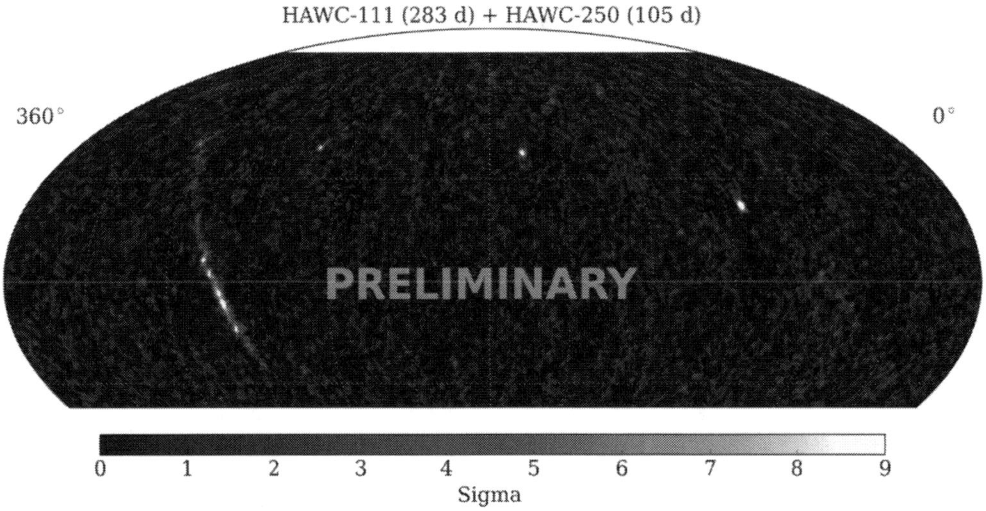

Figure 1. HAWC first year all-sky map in celestial coordinates. Several astrophysical objects along the plane of our galaxy, many of which have not been observed previously, are visible. The three point sources are Mrk 501, Mrk 421 and the Crab Nebula. The HAWC results were announced in a press release during the April 2016 APS meeting [5].

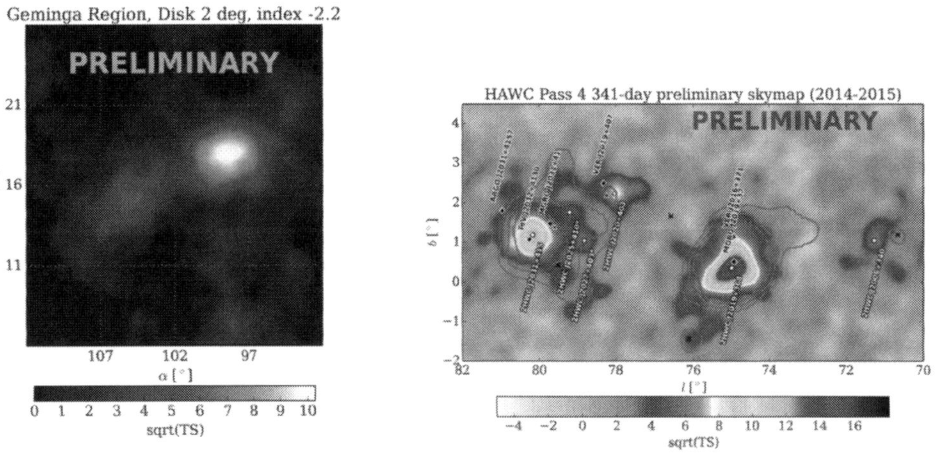

Figure 2. *(Left Panel)* HAWC excess map of the Geminga and PSR B0656+14 regions. *(Right Panel)* Excess map of the Cygnus region, dominated at lower energy by one of the brightest and closest star forming region in our Galaxy, Cygnus X, where a high density in gas, pulsars and supernova remnants create an incredible environment for stars to be born.

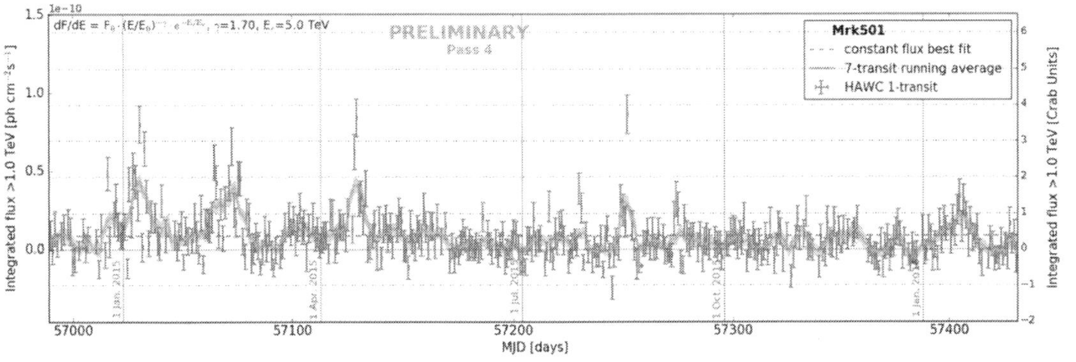

Figure 3. The lightcurve of the active Galactic Nucleus, Markarian 501 between November 26 2014 and February 12 2016 is shown. The fluxes are calculated assuming a spectral index of 1.7 and an exponential cut-off at 5 TeV. Episodes of flaring activity, with twice the Crab flux are evident.

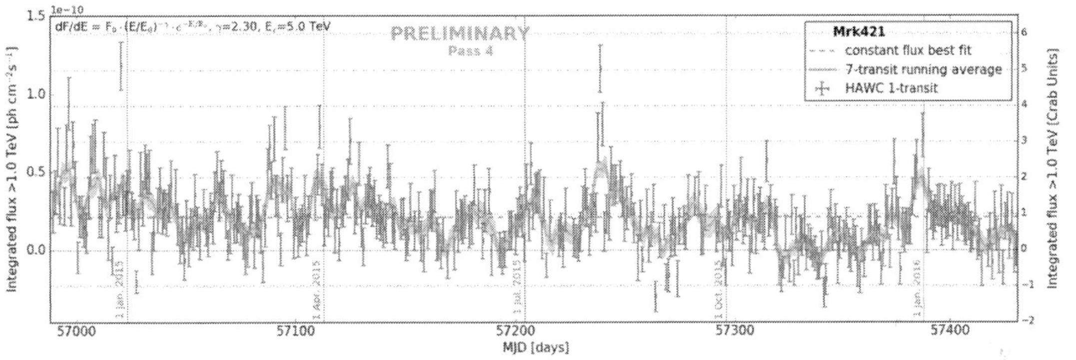

Figure 4. The lightcurves of the active Galactic Nucleus, Markarian 421 HAWC light curve of the BL Lac object Mrk 421 spans from November 2014 to February 2016 and shows several flaring states, in which the TeV flux of Mrk 421 was stronger than that of the Crab.

fluxes larger than 1 or 2 Crab have been observed. Recently, HAWC, FACT and Swift-XRT reported the joint observation of a four day activity of Mrk 421, which reached 2 Crab units above 1 TeV [10]. Mrk501, whole lightcurve is presented in Fig.4, was observed recently by HAWC while it was extremely bright, the two-day flare starting on April 6th 2016 reaching a flux 2.2 times that of the Crab [11]. This confirms that HAWC has the capability to provide timely alerts to the astronomical community.

Gamma-Ray Bursts (GRBs) have been detected up to several GeV by Fermi-LAT. However none of the TeV ground-based experiments has ever detected one yet. The main problem is that they are unpredictable, and due to the reduced duty cycle of the IACTs, it is difficult to detect them. The large duty cycle of HAWC ($> 90\%$), together with its large field of view makes HAWC the ideal experiment to detect GRBs at TeV energies. HAWC is expected to detect a rate as high as 1.6 GRB per year. One of the brightest GRB in the last years happened while a small fraction of the detector

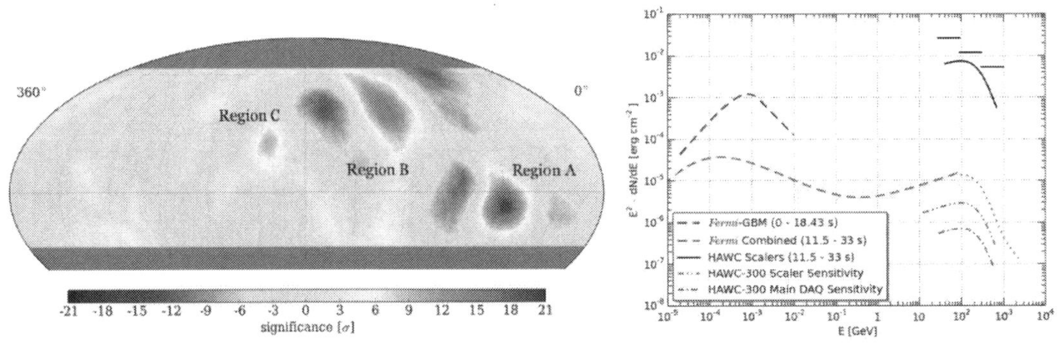

Figure 5. *(Left Panel) HAWC cosmic-ray skymap. The large-scale structures have been subtracted to emphasize the three regions with enhanced cosmic-ray flux.(Right Panel) HAWC upper limits on GRB130427A. Together with the complete HAWC sensitivity and the flux measured by Fermi. The sensitivity for the complete detector (HAWC300) is shown for comparison.*

(HAWC30) was taking data, and only the scaler data acquisition system (DAQ) was active. The burst was not detected, and the corresponding upper limits were derived (see right panel of Fig. 5 [12]).

The anisotropy of the arrival direction of the cosmic-rays, which was first detected by Milagro [13], and later on confirmed by other experiments like IceCube [14], has been also observed with HAWC [15]. The left panel of Figure 5 shows the cosmic ray sky map obtained with 113 days of HAWC-100 (HAWC detector with 100 WCD) data. It contains 5×10^{10} cosmic-ray events with a median energy of 1.9 TeV, and a median angular resolution of 1.2 deg. The map shows the significance computed by using a 10 deg smoothing angle. There are 3 regions with an excess of cosmic-rays above 5 σ level. Region A and B were already reported by Milagro experiment, and region C was recently announced by the ARGO-YBJ collaboration [16]. The origin of the cosmic-ray anisotropy is unclear. HAWC will shed some light on its origin by measuring the spectrum, flux, and variability of the cosmic-ray anisotropy.

3 Conclusions and outlook

Thanks to its nonpointed observations of photons up to energies of tens of TeV, its large field of view, and its high duty cycle HAWC is the perfect instrument for surveying and monitoring the TeV gamma-ray sky. One year after completion of the array, the HAWC collaboration has presented the first map of the Galactic Plane at multi-TeV energies. HAWC continuously monitors the VHE sky. Recently it has detected Markarian 501 in a flaring state and promptly launched an alert to the astronomical community.

Acknowledgments

We acknowledge the support from: US National Science Foundation (NSF); US Department of Energy Office of High-Energy Physics; The Laboratory Directed Research and Development (LDRD) program of Los Alamos National Laboratory; Consejo Nacional de Ciencia y Tecnologia (CONACyT), Mexico; Red de Fısica de Altas Energıas, Mexico; DGAPA-UNAM, Mexico; and the University of Wisconsin Alumni Research Foundation. Sabrina Casanova acknowledges the support from the Polish National Science Center under the Opus Grant UMO-2014/13/B/ST9/00945.

References

[1] A.U. Abeysekara, R. Alfaro, C. Alvarez, J.D. Álvarez, R. Arceo, J.C. Arteaga-Velázquez, H.A. Ayala Solares, A.S. Barber, B.M. Baughman, N. Bautista-Elivar et al., Astroparticle Physics **50**, 26 (2013), 1306.5800

[2] A.U. Abeysekara, J.A. Aguilar, S. Aguilar, R. Alfaro, E. Almaraz, C. Álvarez, J.d.D. Álvarez-Romero, M. Álvarez, R. Arceo, J.C. Arteaga-Velázquez et al., Astroparticle Physics **35**, 641 (2012), 1108.6034

[3] A.U. Abeysekara, R. Alfaro, C. Alvarez, J.D. Álvarez, R. Arceo, J.C. Arteaga-Velázquez, H.A. Ayala Solares, A.S. Barber, B.M. Baughman, N. Bautista-Elivar et al., @@stylePhys. Rev. D**90**, 122002 (2014), 1405.1730

[4] A.A. Abdo, A.U. Abeysekara, R. Alfaro, B.T. Allen, C. Alvarez, J.D. Álvarez, R. Arceo, J.C. Arteaga-Velázquez, T. Aune, H.A. Ayala Solares et al., Astroparticle Physics **64**, 4 (2015), 1407.1686

[5] M. Hui, *The HAWC Galactic Plane Survey*, in *APS April Meeting Abstracts* (2016)

[6] A.A. Abdo, B. Allen, D. Berley, E. Blaufuss, S. Casanova, C. Chen, D.G. Coyne, R.S. Delay, B.L. Dingus, R.W. Ellsworth et al., @@styleApJ**658**, L33 (2007), astro-ph/0611691

[7] A.A. Abdo, B. Allen, T. Aune, D. Berley, E. Blaufuss, S. Casanova, C. Chen, B.L. Dingus, R.W. Ellsworth, L. Fleysher et al., @@styleApJ**688**, 1078-1083 (2008), 0805.0417

[8] A.A. Abdo, U. Abeysekara, B.T. Allen, T. Aune, D. Berley, E. Bonamente, G.E. Christopher, T. DeYoung, B.L. Dingus, R.W. Ellsworth et al., @@styleApJ**753**, 159 (2012), 1202.0846

[9] E. Aliu, T. Aune, B. Behera, M. Beilicke, W. Benbow, K. Berger, R. Bird, A. Bouvier, J.H. Buckley, V. Bugaev et al., @@styleApJ**788**, 78 (2014), 1404.1841

[10] A. Biland, D. Dorner, R. Lauer, J. Wood, B. Kapanadze, A. Kreikenbohm, The Astronomer's Telegram **9137** (2016)

[11] A. Sandoval, R. Lauer, J. Wood, The Astronomer's Telegram **8922** (2016)

[12] A.U. Abeysekara, R. Alfaro, C. Alvarez, J.D. Álvarez, R. Arceo, J.C. Arteaga-Velázquez, H.A. Ayala Solares, A.S. Barber, B.M. Baughman, N. Bautista-Elivar et al., @@styleApJ**800**, 78 (2015), 1410.1536

[13] A.A. Abdo, B.T. Allen, T. Aune, D. Berley, S. Casanova, C. Chen, B.L. Dingus, R.W. Ellsworth, L. Fleysher, R. Fleysher et al., @@styleApJ**698**, 2121 (2009), 0806.2293

[14] R. Abbasi, Y. Abdou, T. Abu-Zayyad, M. Ackermann, J. Adams, J.A. Aguilar, M. Ahlers, M.M. Allen, D. Altmann, K. Andeen et al., @@styleApJ**746**, 33 (2012), 1109.1017

[15] A.U. Abeysekara, R. Alfaro, C. Alvarez, J.D. Álvarez, R. Arceo, J.C. Arteaga-Velázquez, H.A. Ayala Solares, A.S. Barber, B.M. Baughman, N. Bautista-Elivar et al., @@styleApJ**796**, 108 (2014), 1408.4805

[16] B. Bartoli, P. Bernardini, X.J. Bi, I. Bolognino, P. Branchini, A. Budano, A.K. Calabrese Melcarne, P. Camarri, Z. Cao, R. Cardarelli et al., @@stylePhys. Rev. D**88**, 082001 (2013), 1309.6182

EPJ Web of Conferences **136**, 03006 (2017) DOI: 10.1051/epjconf/201713603006
RICAP16

HAWC High Energy Upgrade with a Sparse Array

V.Joshi[1,a] for the HAWC collaboration[2]

[1] *Max Planck Institut für Kernphysik, Heidelberg, Germany*
[2] *For a complete author list, see www.hawc-observatory.org/collaboration/*

Abstract. The High Altitude Water Cherenkov (HAWC) gamma-ray observatory has been fully operational since March 2015. To improve its sensitivity at the highest energies, it is being upgraded with an additional sparse array called outrigger array. We will discuss in this contribution, the different outrigger array components, and the simulation results to optimize it.

1 HAWC and the Motivation for Outriggers

HAWC is situated in central Mexico at an altitude of 4100 m above the sea level. It has a wide field of view of 2 sr and operational energy range of 0.1-100 TeV. It consists of 300 Water Cherenkov Detectors (WCDs) in the main array encompassing a surface area of 20000 m^2. The main array WCDs comprised of cylindrical steel water tanks of diameter 7.3 m and height 4.5 m with 4 Photo Multiplier Tubes (PMTs) (three 8" and one 10") in each one of them. HAWC detects the Cherenkov light produced in the water by particles generated in an atmospheric air shower.

When the energy of the primary particle is of the order of tens of TeV, the footprint of the shower becomes comparable to the size of the main array. Therefore, most of the recorded showers are not contained in the array, which causes challenges to constrain the shower properties. To address these challenges the construction of the outrigger array around the main array has started. It will increase the fraction of well-reconstructed showers above multi-TeV energies. The outrigger array will help in determining the position of the core of the shower falling outside the main array and it will also improve the determination of the primary particle's direction and energy.

2 Outrigger Array

The outrigger array [1] consists of 350 cylindrical tanks of diameter 1.55 m and height 1.65 m (see Figure 1a). Each tank has one Hamamatsu R5912 8" PMT at the bottom of the tank. The outrigger array will be deployed in a circular symmetric way around the main HAWC array with a mutual separation of 12 m to 18 m (see Figure 1b).

To trigger and readout, the system electronics developed for the FlashCAM [2] will be used. FlashCAM is a readout electronics, which has been developed for the cameras of the medium-size telescopes of the Cherenkov Telescope Array. The reason for using the FlashCam readout for outrigger

[a] e-mail: vikas.joshi@mpi-hd.mpg.de

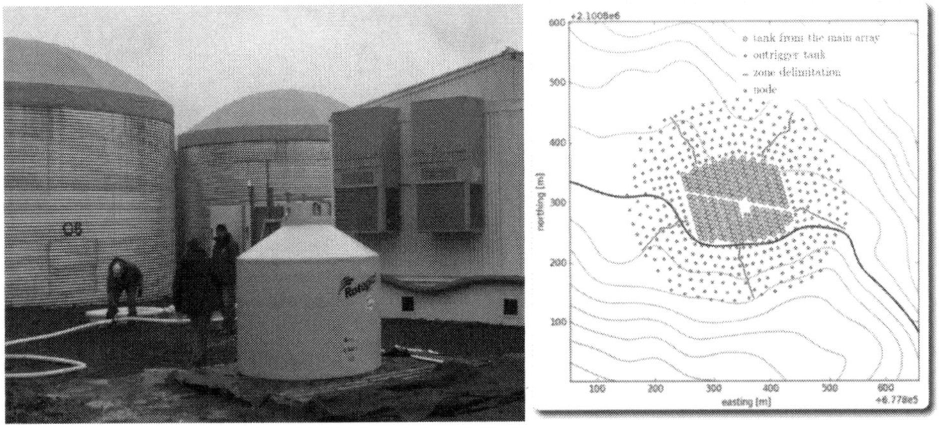

Figure 1. **a.** Outrigger tank and main array tanks. **b.** Outrigger array surrounding the main HAWC array. The red lines shows the different sections of the outrigger array.

array is that each PMT of the outrigger array is equivalent to a pixel of an Imaging Atmospheric Cherenkov Telescopes (IACTs) camera. The outrigger array is divided into five sections with 70 outriggers in each of them. One such section will contain a readout and trigger electronics, which we named as the: Flash Adc electronics for the Cherenkov Outrigger Node (FALCON). A node will contain 3 Flash-ADC boards, each of them can digitize 24 channels with a sampling speed of 250 MHz with a 12-bit accuracy. It also allows a flexible digital multiplicity trigger as well as the readout of full waveforms, with settable length (typically 40 samples i.e. 160 ns) which can be used for charge extraction and signal timing information.

3 Simulations

We performed extensive simulations in order to optimize the outrigger array. This can be further divided into two parts:

1. Simulations to study the effect of different PMT options and tank colors.

2. Simulations to develop a likelihood fit method in order to fit the shower core and to constrain the shower energy and the depth of the shower maximum.

3.1 Simulations for PMT Options and Tank Colors

In order to choose the size of the PMT, different PMT sizes have been simulated in combination with different tank wall colors. Here we present the results for the 3" and 8" PMT with tank wall colors black and white. We have focused on the following figures of merit:

1. Average number of Photo-Electrons (PEs) observed at a given distance from the shower core.

2. RMS of the distribution of the time difference between neighboring tank pairs for the arrival time of the first PE.

It can be seen from the Figure 2 that one gets 10 times more PEs with the 8" PMT in comparison to the 3" PMT and the effect of the white wall color in the contrast of black wall color is 20% increase in the number of PEs observed. Furthermore, the white wall color is more diffusive than the black wall color, and the loss of the timing information by using the white wall color it can be more than 20% (see Figure 3) in comparison to the black wall color. It can be concluded that we don't gain much in the average number of PEs observed by using the white wall color and we lose considerably in our timing information. We decided that black wall color tanks (less diffusive) with 8" PMT seems to be the appropriate choice.

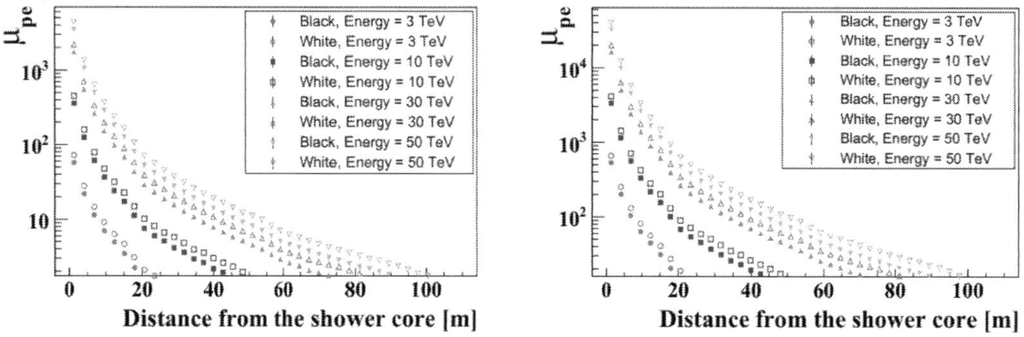

Figure 2. Average number of PE (μ_{pe}) observed for 3" PMT (left) and 8" PMT (right) with black and white tanks as a function of distance from the shower core for different energies.

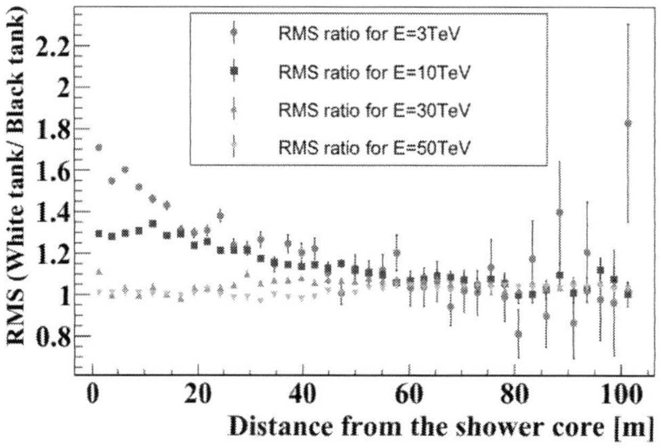

Figure 3. Ratio of the RMSs (see Section 3.1) for white/black tanks for 8" PMT as a function of distance from the shower core for different energies (E).

3.2 Simulations for Likelihood Core Fit Method

To constrain the core location of the multi-TeV γ-ray showers falling outside the main HAWC array a likelihood core fitter is being developed. In Figure 4 we can see that a core resolution of < 10 m is achieved by just using the outriggers for energies > 10 TeV and for zenith angle up to 30°. In addition, this likelihood method also constrains the shower energy and depth of the shower maximum. In the next step, this likelihood fit method for the outriggers will be merged with the one for the main array to ultimately improve the core resolution for multi-TeV showers.

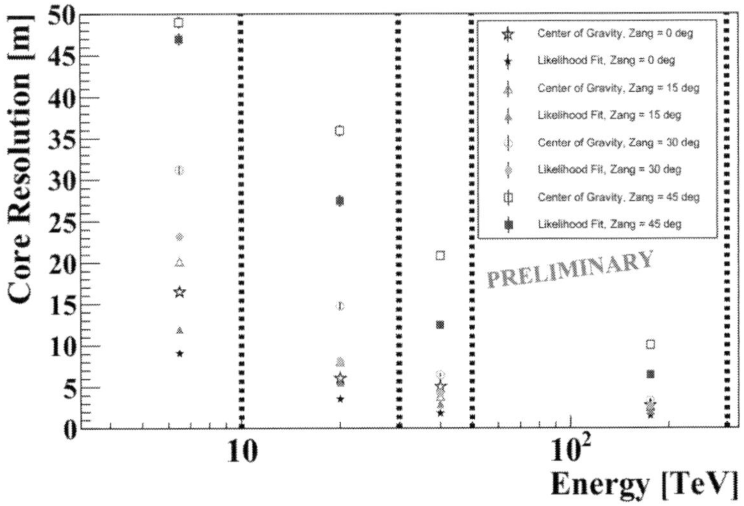

Figure 4. The core resolution obtained with a likelihood fit in comparison with the center of gravity of the signal for different zenith angles (Zang). The vertical dashed lines represent the binning in the energy range. The points in each of these energy bins correspond to the 68% containment of the core resolution distribution.

4 Current Status of the Outrigger Array

The deployment of the outrigger array has already started. FALCON electronics is being used to take the data from the first set of outriggers installed at the HAWC site. Integration of the FALCON readout with the central DAQ is ongoing and will be finished soon. A complete outrigger array will be fully operational by the end of the next year.

Acknowledgements

We acknowledge the support from: the US National Science Foundation (NSF); the US Department of Energy Office of High-Energy Physics; the Laboratory Directed Research and Development (LDRD) program of Los Alamos National Laboratory; Consejo Nacional de Ciencia y Tecnologa (CONACyT), Mexico (grants 260378, 55155, 105666, 122331, 132197, 167281); Red de Fsica de Altas Energas, Mexico; DGAPA-UNAM (grants IG100414-3, IN108713, IN121309, IN115409, IN113612); VIEP-BUAP (grant 161-EXC-2011); the University of Wisconsin Alumni Research Foundation; the Institute of Geophysics, Planetary Physics, and Signatures at Los Alamos National Laboratory; the Luc Binette Foundation UNAM Postdoctoral Fellowship program.

References

[1] A. Sandoval, Proc. of the 34rd ICRC, The Hague, The Netherlands, September (2015), astro-ph.IM:1509.04269.

[2] G. Pühlhofer. C. Bauer, F. Eisenkolb, D. Florin, C. Föhr, A. Gadola, G. Hermann, C. Kalkuhl, J. Kasperek, T. Kihm, J. Koziol, A. Manalaysay, A. Marszalek, P. J. Rajda, W. Romaszkan, M. Rupinski, T. Schanz, S. Steiner, U. Straumann, C. Tenzer, A. Vollhardt, Q. Weitzel, K. Winiarski, K. Zietara, and f. t. CTA consortium, Proc. of the 33rd ICRC, Rio de Janeiro, Brazil, July (2013), astro-ph.IM:1307.3677 .

Observations of the gamma-ray emission from the Quiescent Sun with Fermi Large Area Telescope during the first 7 years in orbit

S. Rainó[1,2,a], N. Giglietto[1,2], Moskalenko[3], E. Orlando[3] and A.W. Strong[4]

[1] *Dipartimento Interateneo di Fisica "M.Merlin" dell'Universitá e del Politecnico di Bari, I-70126, Bari, Italy*
[2] *Istituto Nazionale di Fisica Nucleare, Sezione di Bari, I-70126, Bari, Italy*
[3] *W. W. Hansen Experimental Physics Laboratory, Kavli Institute for Particle Astrophysics and Cosmology, Department of Physics and SLAC National Accelerator Laboratory, Stanford University, Stanford, California 94305, USA*
[4] *Max-Planck Institut für extraterrestrische Physik, 85748 Garching, Germany*

Abstract. The high energy gamma-ray emission from the quiescent Sun is due to the interactions of cosmic ray (CR) protons and electrons with matter and photons in the solar environment. Such interactions lead to two component gamma-ray emission: a disk-like emission due to the nuclear interactions of CR protons and nuclei in the solar atmosphere and a space extended emission due to the inverse Compton (IC) scattering of CR electrons off solar photons in the whole heliosphere. The observation of these two solar emission components may give useful information about the evolution of the solar cycle by probing two different CR components (proton and electrons) in regions not directly accessible by direct observations. We present the results of the observations of the Sun with Fermi-LAT in the first 7 years on orbit, with the exception of the flaring periods. Significantly large photon statistics and improved processing performance with respect to previous analysis allow us to explore both components of the emission in greater details and perform better comparisons of data with current models of the IC component. This allows us to probe CR electrons in the inner heliosphere which is not possible by other methods. Moreover, the longer period of observations allows us to study the variations of the emission between the maximum and the minimum of the solar cycle.

1 Introduction

The high energy gamma-ray emission from the quiescent Sun is due to the interactions of cosmic ray (CR) protons and electrons with matter and photons in the solar environment. Such interactions lead to two component gamma-ray emission: a disk-like emission due to the nuclear interactions of CR protons and nuclei in the solar atmosphere and a space extended emission due to the inverse Compton (IC) scattering of CR electrons off solar photons in the whole heliosphere [1] [2] . While the IC emission is brightest in the region within a few degrees from the Sun, even at larger elongation angles it can be comparable in intensity to the isotropic (presumably extra-galactic) gamma-ray background

[a] e-mail: silvia.raino@ba.infn.it

DOI: 10.1051/epjconf/201713603007

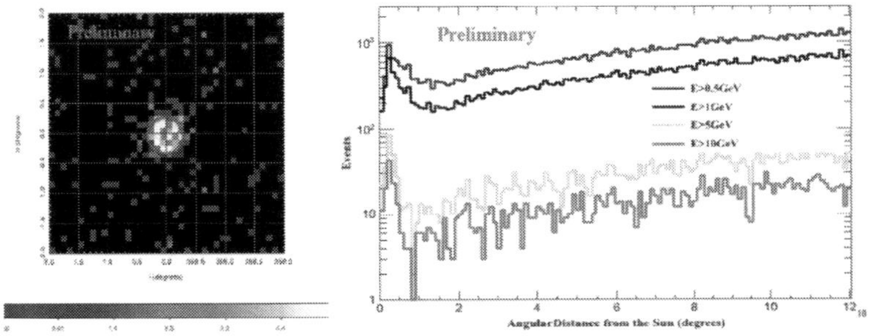

Figure 1. Left: Photons counts map of the solar emission components above 10 GeV. Right: Sun events as a function of the angular distance from the Sun, for different energy thresholds. From top to bottom: E_{min}=0.5 GeV, 1 GeV, 5 GeV, 10 GeV.

[3]. The flux for both components of the CR-induced emission is expected to change over the solar cycle due to the change of the heliospheric flux of the Galactic CRs in anticorrelation with the variations of the solar activity. The observation of these two solar emission components may give useful information about the evolution of the solar cycle by probing two different CR components (proton and electrons) in regions not accessible by direct measurements.

2 Data Selection and background

In our analysis we have evaluated the gamma-ray flux from the Sun using the data collected by the Fermi LAT in the first 7.5 years of operation from 2008 August 4 up to 2015 December 31. The data set used is much larger than the previously published [4], moreover the data are reprocessed with the new Pass 8 reconstruction and event-level analysis that is a comprehensive revision of the entire analysis chain and provides improved processing performance with respect to previous analysis: wider energy range, larger acceptance, better Point Spread Function (PSF), better background rejection, better control of systematic uncertainties [5] . We select events in an energy range between 100 MeV and 30 GeV in a region of interest (ROI) of 20° centered on the Sun position. To reduce the contamination by the albedo gamma-ray emission coming from CR interactions in the Earth's upper atmosphere our selection is refined by selecting events with zenith angles <100°. To reduce the systematic uncertainties due to bright gamma-ray emission from the Galactic plane we exclude the data taken when the Sun was within 30° of the plane. We further excluded the periods when the Sun was within 20° from the Moon or any other bright celestial source with integral flux F $\geq 2\times10^{-7}$cm^{-2}s^{-1} above 100 MeV, as reported in the 3FGL catalog [6]. Since solar flares gamma rays are known to be produced directly from nuclear interactions of accelerated protons and heavier ions we remove all flaring periods observed from the considered data set in order to focus on the quiet component of the sun emission . Because the Sun is a moving source, the analysis of its emission requires a special treatment, so the data are selected in a moving frame centered on the instantaneous solar position. Fig. 1 shows a Sun-centered counts map of the solar emission components above 10 GeV and the Sun events as a function of the elongation angle for different energy thresholds (from top to bottom, E_{min}=0.5 GeV, 1 GeV, 5 GeV, 10 GeV). The peak corresponds to the disk events in the centre of the reference frame (θ=0, where θ is the elongation angle). The correct evaluation of the background in the region around the Sun is of considerable importance for the analysis of the weak extended IC emission. The latter is expected to decrease with $\sim 1/\theta$ [1]. The background is mainly due to the diffuse

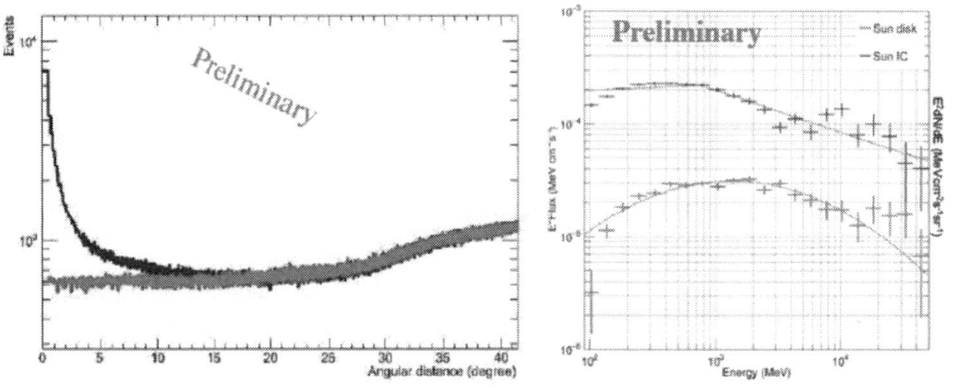

Figure 2. Left: Right: Number of events vs. angular distance from the Sun, for the full data sample (red) and for the off-time source sample (blue). Right: Spectral energy distribution of the two solar components: Sun IC (top, black line), Sun disk (bottom, red line).

Galactic and isotropic (presumably extra-galactic) gamma-ray emission averaged along the ecliptic and to weak point sources. The background is estimated with the trial off-time source method where an off-time source follows the path of the real source but at different times and at 90° distance (same path at different times). Fig. 2 left shows the distribution of the events for both background and sun samples centered on the solar position, as a function of the angular distance from the fake-sun and the sun respectively. While for the solar centered data set the number of events increases considerably for small elongation angles, the fake-Sun profile is flat. The two distributions overlap at distances larger than 20° where the signal becomes insignificant.

3 7-year analysis results

The full data sample has been modeled fitting the background determined with the off-time source method and masked near the solar disk, with the two emission components from the Sun. The disk emission is modeled as a point-like source with a Log Parabolic energy spectrum, while for IC emission a model undependent approach is used. The extended IC emission is made up of a parametric energy spectrum and a generic $1/\theta$ radial dependence fitted on data, where θ is the angular distance from the Sun. Then, the Sun-centered maps are analyzed using a binned maximum likelihood technique developed in Fermi-LAT Science Tools (version v11r03p0). The analysis performed gives a significant detection and separation of the disk and of the IC extended component. Fig.2 right shows the spectral energy distribution (SED) for all solar components evaluated in the first 3 years of *Fermi*-LAT data taking. The disk component is well fitted by a Log Parabola with an integral flux above 100 MeV of $(1.93 \pm 0.07)\times10^{-7}$ph cm^{-2}s^{-1} while we have fitted the IC component with a broken power law resulting in an integral flux above 100 MeV of $(2.02 \pm 0.02)\times10^{-6}$ph cm^{-2}s^{-1}sr^{-1}. As already found in [4] the highest energy points show an excess with respect to model prediction. This result is being carefully investigated on the entire data set period in order to clarify this discrepancy.

4 Modulation

Gamma-ray flux measurements depend on the 11-year solar cycle. The maximum in the solar activity corresponds to the minimum CR flux and vice versa. Solar activity reached its maximum in 2014 after a period of minimum in the first years of Fermi-LAT data taking. The effect of solar modulation has already been observed analyzing the lunar gamma-ray emission [7]. We are investigating the effects of the solar activity on the fluxes of the two components of the Sun emission. On the basis of the first results, the disk component shows a clear trend in anticorrelation with the solar activity, as expected. The analysis on the IC component is still being investigated but we expect to find a dependence with the solar modulation as well.

5 Conclusions

We have analyzed a 7.5-year data sample of Fermi-LAT observations centered on the solar position using the new Pass8 processing performance. We have used a simplified model that takes into account the background separately evaluated and two emission components from the Sun. The first component from the Sun is a pointlike emission from the disk resulting from the interaction of cosmic rays protons and nuclei with the solar atmosphere . The second contribution comes from the IC scattering of CR electrons and positrons scattering off solar photons in the heliosphere and appears as an extended halo around the Sun. The behavior of the IC spectrum as a function of distance from the Sun is as expected. The solar modulation effect is currently still under investigation, though the first results confirm a clear anticorrelation trend.

Acknowledgements

The Fermi LAT Collaboration acknowledges generous ongoing support from a number of agencies and institutes that have supported both the development and the operation of the LAT as well as scientific data analysis. These include the National Aeronautics and Space Administration and the Department of Energy in the United States, the Commissariat á l'Energie Atomique and the Centre National de la Recherche Scientifique / Institut National de Physique Nucléaire et de Physique des Particules in France, the Agenzia Spaziale Italiana and the Istituto Nazionale di Fisica Nucleare in Italy, the Ministry of Education, Culture, Sports, Science and Technology (MEXT), High Energy Accelerator Research Organization (KEK) and Japan Aerospace Exploration Agency (JAXA) in Japan, and the K. A. Wallenberg Foundation, the Swedish Research Council and the Swedish National Space Board in Sweden. Additional support for science analysis during the operations phase from the following agencies is also gratefully acknowledged: the Istituto Nazionale di Astrofisica in Italy and the Centre Nationale d'Études Spatiales in France.

References

[1] Moskalenko, I.V. et al. 2006, ApJL, 652, L65; Orlando, E., & Strong, A.W., 2006, arXiv:astro-ph/0607563 (Ap&SS, 309,359)
[2] Orlando, E. & Strong A.W., 2008, A&A, 480, 847
[3] Abdo, A.A., et al., 2010, Physical Review Letters, 104, 101101
[4] Abdo, A.A., et al., 2011, ApJ 734, 116
[5] *http://fermi.gsfc.nasa.gov/ssc/data/analysis/documentation/Cicerone/Cicerone_LAT_IRFs/IRF_overview.html*
[6] The Fermi LAT Collaboration, arXiv:1501.02003, accepted for publication on ApJS
[7] M.Ackermann, et al. Phys. Rev D 93, 082001 (2016)

The TAIGA timing array HiSCORE - first results

M. Tluczykont[6,a], N.Budnev[2], I.Astapov[9], N.Barbashina[9], A.Bogdanov[9], V.Boreyko[10], M.Brückner[8], A.Chiavassa[4], O.Chvalaev[2], O.Gress[2], T.Gress[2], O.Grishin[2], A.Dyachok[2], S.Epimakhov[6], O.Fedorov[2], A.Gafarov[2], N.Gorbunov[10], V.Grebenyuk[10], A.Grinuk[10], D.Horns[6], A.Kalinin[10], N.Karpov[1], N.Kalmykov[1], Y.Kazarina[2], S.Kiryuhin[2], R.Kokoulin[9], K.Kompaniets[9], A.Konstantinov[1], E.Korosteleva[1], V.Kozhin[1], E.Kravchenko[11], M.Kunnas[6], L.Kuzmichev[1,2], Yu.Lemeshev[2], B.Lubsandorzhiev[3], N.Lubsandorzhiev[1], R.Mirgazov[2], R.Mirzoyan[5,2], R.Monkhoev[2], R.Nachtigall[6], E.Osipova[1], A.Pakhorukov[2], M.Panasyuk[1], L.Pankov[2], A.Petrukhin[9], V.Poleschuk[2], E.Popova[1], A.Porelli[8], E.Postnikov[1], V.Prosin[1], V.Ptuskin[7], G.Rubtsov[3], A.Pushnin[2], V.Samoliga[2], P.Satunin[7], Yu.Semeney[2], A.Silaev[1], A.Silaev (junior)[1], A.Skurikhin[1], M.Slunecka[10], A.Sokolov[11], C.Spiering[8], L.Sveshnikova[1], V.Tabolenko[2], B.Tarashansky[2], A.Tkachenko[10], L.Tkachev[10], D.Voronin[2], R.Wischnewski[8], A.Zagorodnikov[2], V.Zurbanov[2], D.Zhurov[2], and I.Yashin[9]

[1] *Skobeltsyn Institute of Nuclear Physics MSU, Moscow, Russia*
[2] *Institute of Applied Physics, ISU, Irkutsk, Russia*
[3] *Institute for Nuclear Research of RAN, Moscow, Russia*
[4] *Dipartimento di Fisica Generale Universiteta di Torino and INFN, Torino, Italy*
[5] *Max-Planck-Institute for Physics, Munich, Germany*
[6] *Institut für Experimentalphysik, Universität Hamburg, Germany*
[7] *IZMIRAN, Moscow Region, Russia*
[8] *DESY, Zeuthen, Germany*
[9] *MEPhI: National Research Nuclear University MEPhI (Moscow Eng. Phys. Institute), Moscow, Russia*
[10] *JINR, Dubna, Russia*
[11] *Budker Institute of Nuclear Physics SB RAS, Novosibirsk State University, Novosibirsk, Russia*

Abstract. Observations of gamma rays up to several 100 TeV are particularly important to spectrally resolve the cutoff regime of the long-sought Pevatrons, the cosmic-ray PeV accelerators. One component of the TAIGA hybrid detector is the TAIGA-HiSCORE timing array, which currently consists of 28 wide angle (0.6 sr) air Cherenkov timing stations distributed on an area of 0.25 km^2. The HiSCORE concept is based on (non-imaging) air shower front sampling with Cherenkov light. First results are presented.

1 Introduction

The current generation of Imaging Air Cherenkov Telescopes (IACTs, e.g. H.E.S.S., MAGIC, and VERITAS) allowed the detection of more than 150 sources of very high energy (E>100 GeV) gamma rays [1]. The IACT technique is also used for the planned Cherenkov Telescope Array CTA [2]. Given the quick drop in flux with increasing energy, very large instrumented areas are required to access the

[a]e-mail: martin.tluczykont@physik.uni-hamburg.de

ultra high energy (UHE, E>10 TeV) regime. A cost effective method for the instrumentation of large areas with a wide field of view is the (non-imaging) shower-front timing technique. The weak point - a poor gamma-hadron separation towards low energies - can be compensated by a hybrid combination with a small number of small sized IACTs, such as currently implemented by the TAIGA (Tunka Advanced Instrument for Gamma ray and cosmic-ray Astrophysics) experiment [3, 4].

The main astrophysical motivation of TAIGA is the measurement of the cutoff regime of known Galactic sources. Some of these sources might be cosmic-ray Pevatrons, i.e. cosmic-ray accelerators reaching up to the knee energies (around 3 PeV proton energy). For more details see [5, 6].

2 First results from the TAIGA timing array HiSCORE

The experiment: TAIGA currently consists of 28 timing stations distributed over an area of 0.25 km^2 in the Tunka valley in Siberia (see Fig. 1). This array is an implementation of the HiSCORE wide-angle timing array concept [7]. Each of the 28 existing TAIGA-HiSCORE stations consists of four 8 inch-photomultiplier tubes (PMTs), equipped with light concentrators, resulting in 0.5 m^2 light collection area per station, and a field of view of 0.6 sr. The DAQ is based on fast GHz signal sampling [8]. Two independent time-calibration methods were sucessfully employed, reaching the required sub-ns accuracy [8, 9]. The primary event parameters are reconstructed using signal timing and the amplitude [9–11]. Additionally, a small IACT (4.75 m dish diameter) is currently in commissioning and will be operated using a new hybrid approach, combining imaging and timing (see e.g. [12] and [13, 14]). The sensitivity of the current timing array alone is shown as a solid line in Figure 2. For comparison,

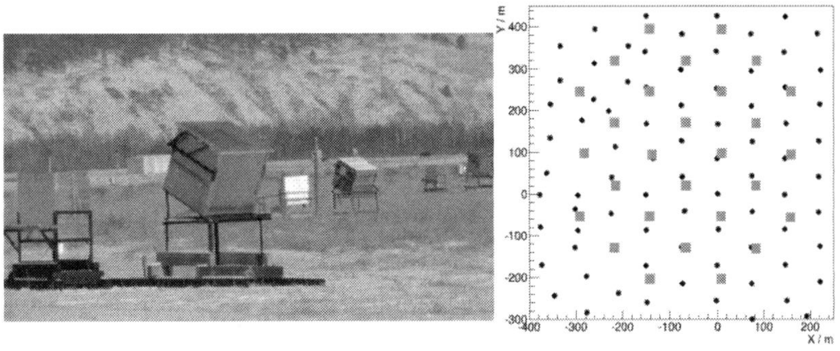

Figure 1. *Left:* photograph of 3 TAIGA timing array stations and their electronic boxes. *Right:* TAIGA timing array layout since 2014. The timing stations (previously HiSCORE) are shown in blue along with the Tunka-133 cosmic-ray array (black).

also the sensitivity of the combined TAIGA array is shown. More details on TAIGA and the new hybrid observation mode are given in [12].

Data-MC verification: from October 2015 to February 2016, 250 h of observation time were taken, containing 10^7 air shower events. The individual station trigger rate (8 to 12 Hz), as well as the 4-station-coincidence array rate (10 to 18 Hz) could be verified using MC simulations, which repor-duce the observed rates using a single station threshold of about 250 p.e. This translates to an energy threshold of 50 TeV. Further studies could verify the validity of the observed station multiplicity per event, as well as the angular resolution. The latter was studied using the *chessboard* method, dividing the array in subarrays, each providing a reconstructed event direction. The angular difference α between the reconstructed directions depends on the reconstruction quality and the number of stations used in each subarray. Applying this analysis to both, real data and (proton) MC simulations, the reconstruction quality of the current setup could be verified (see Fig. 3). The value of the angular resolution for gamma rays is 0.4° at threshold and below 0.2° above 100 TeV, as predicted in [11].

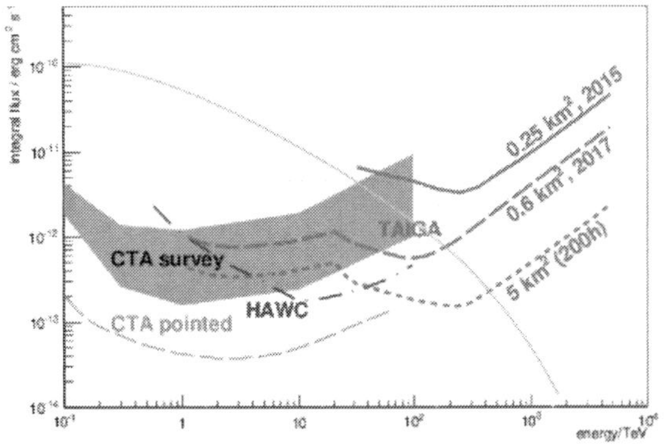

Figure 2. The sensitivity (after 200 h for the timing array and 50 h for the TAIGA IACTs) of the current 0.25 km^2 TAIGA timing array (HiSCORE) and the planned hybrid stage, consisting of an extension of the timing array to 0.6 km^2 in combination with two IACTs. In dotted red, the sensitivity is shown for 5 km^2, including 10 IACTs. TAIGA curves are based on simulations and reconstruction methods presented in [7, 10, 11, 13]. A parametrization of the Crab Nebula flux by [15] is shown as a solid grey line. Figure adapted from [6], see references therein.

Crab Nebula data: using a standard ring background estimation method on blinded real data, the field of view was found to be free of artifacts of positive or negative fluctuations in excess of what is statistically expected for the test region of $10° \times 10°$. Performing the same analysis on the position of the Crab Nebula, a weak excess was found. Up to now the analysis was done without cut optimizations nor gamma-hadron separation. The observations are compatible with reasonable Crab Nebula flux extrapolations to higher energies and an instrumented area of 0.25 km^2. We expect to reach better sensitivity optimizing the analysis, increasing the area as planned, and combining the timing array with IACTs.

A first source, the ISS: a recent serendipitous discovery, opens up opportunities for calibration and possibly different physics. Three intervals of $\sim$ 1 s duration with extremely high array trigger rate (few kHz as compared to the usual 15 Hz array rate) were found from November to February. These data are compatible with a 4 kHz periodic light source, and are best described by a plane-wave time front shape, strikingly different from air shower events with a curved shower front. The reconstructed event directions are coincident with the trajectory of the International Space Station (ISS), passing almost vertically over the TAIGA observation site (see Fig. 3). As the strong light source, we identified the CATS-LIDAR [16] onboard the ISS, emitting $\sim$1 mJ at a wavelength of 532 nm at 4 kHz, and pointing almost vertically downwards at a few-degree inclination. This ISS light source is an interesting object for TAIGA calibration and atmospheric light scattering studies. Also other air Cherenkov installations like IACTs could benefit from this light source. A detailed analysis is underway.

3 Summary

As a part of the hybrid TAIGA array, the TAIGA timing array, based on the HiSCORE concept, is currently being operated with a total instrumented area of 0.25 km^2. First studies of real data and comparisons to MC simulations show that the detector is reasonably well understood. While observations are compatible with expectations, in the current stage, no significant excess can be expected from the Crab Nebula. An unexpected result was achieved very recently, with the detection of a LIDAR onboard the ISS, which can be used for calibration measurements or other studies in future.

Acknowledgements

We acknowledge the support of the Russian Federation Ministry of Education and Science (agreements N 14.B25.31.0010, zadanie No 3.889.2014/K), the Russian Science Foundation (grant 15-12-

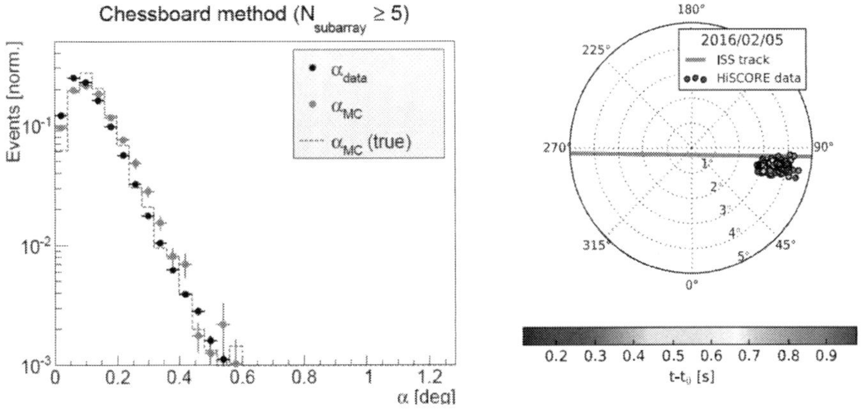

Figure 3. *Left:* angular resolution study with the chessboard method (see text), applied to data (black) and MC (red). *Right:* HiSCORE events (circles) reconstructed for the ISS passage on 5.2.2016 (color coded for registration time). The ISS track is calculated from NORAD ISS-TLE.

20022) the Russian Foundation for Basic research (15-02-10005, 15-02-05769, 16-02-00738, 16-29-13035), the Helmholtz association (grant HRJRG-303), and the DFG (grant TL 51-3).

References

[1] S.P. Wakely, D. Horan, International Cosmic Ray Conference **3**, 1341 (2008)
[2] B.S. Acharya, M. Actis, T. Aghajani, G. Agnetta, J. Aguilar et al., Astr.Part.Phys. **43**, 3 (2013)
[3] R. Mirzoyan, 40th COSPAR Scientific Assembly **40** (2014)
[4] N. Budnev et al., Journal of Physics: Conference Series **718**, 052006 (2016)
[5] M. Tluczykont, D. Hampf, D. Horns et al., Advances in Space Research **48**, 1935 (2011)
[6] M. Tluczykont, N. Budnev, I. Astapov et al., Verlag Deutsches Elektronen-Synchrotron, Connectige Neutrino Physics and Astronomy pp. 135–142 (2016)
[7] M. Tluczykont, D. Hampf, D. Horns, D. Spitschan, L. Kuzmichev, V. Prosin, C. Spiering, R. Wischnewski, Astroparticle Physics **56**, 42 (2014)
[8] O. Gress, I. Astapov, N. Budnev et al., NIMA, dx.doi.org/10.1016/j.nima.2016.08.031 (2016)
[9] A. Porelli, D. Bogorotskii, M. Brückner et al., J. of Phys: Conf. Ser. **632**, 012041 (2015)
[10] S.F. Berezhnev, D. Besson, N.M. Budnev et al., NIMA **692**, 98 (2012)
[11] D. Hampf, M. Tluczykont, D. Horns, NIMA **712**, 137 (2013)
[12] L. Kuzmichev et al., These proceedings (2017)
[13] M. Kunnas, I. Astapov, N. Barbashina et al., Journal of Physics: Conference Series **632**, 012040 (2015)
[14] M. Tluczykont, I. Astapov, N. Barbashina et al., Journal of Physics: Conference Series **632**, 012042 (2015)
[15] M. Meyer, D. Horns, H.S. Zechlin, Astronomy and Astrophysics **523**, A2 (2010)
[16] http://cats.gsfc.nasa.gov/

DOI: 10.1051/epjconf/201713603009

The First *Fermi*-LAT SNR Catalog and Cosmic Ray Implications

F. de Palma[1,2,a], T. J. Brandt[3], and J.W. Hewitt[4]on behalf of *Fermi*-LAT Collaboration

[1] *Istituto Nazionale di Fisica Nucleare, Sezione di Bari*
[2] *Universitá Telematica Pegaso, Piazza Trieste e Trento, 48, 80132 Napoli, Italy*
[3] *NASA/Goddard Space Flight Center*
[4] *University of North Florida*

Abstract. While supernova remnants (SNRs) are widely thought to be powerful cosmic-ray accelerators, indirect evidence comes from a small number of well-studied cases. Here we systematically determine the gamma-ray emission detected by the *Fermi* Large Area Telescope (LAT) from all known Galactic SNRs, disentangling them from the sea of cosmic-ray generated photons in the Galactic plane. Using LAT data we have characterized the 1-100 GeV emission in 279 regions containing SNRs, accounting for systematic uncertainties caused by source misattribution and instrumental response. We classified 30 sources as SNRs, using spatial overlap with the radio emission position. For all the remaining regions we evaluated upper limits on SNRs' emission. In the First *Fermi*-LAT SNR Catalog there is a study of the common characteristics of these SNRs, such as comparisons between GeV, radio and TeV quantities. We show that previously satisfactory models of SNRs' GeV emission no longer adequately describe the data. To address the question of cosmic ray (CR) origins, we also examine the SNRs' maximal CR contribution assuming the GeV emission arises solely from proton interactions. Improved breadth and quality of multiwavelength (MW) data, including distances and local densities, and more, higher resolution gamma-ray data with correspondingly improved Galactic diffuse models will strengthen this constraint.

1 The SNR catalog

Using three years of LAT data starting from 279 SNRs' radio spatial information [1] we detected 102 candidates with a significance above 5σ. 36 of them passed our spatial association probability threshold. For those sources we evaluated their spectral and spatial characteristics with systematic and statistical uncertainties. The latter are obtained summing in quadrature the uncertainty in the effective area and the effects of changing the interstellar emission model (IEM), which accounts for γ-rays produced by CR interactions with interstellar gas and radiation fields in the Milky Way. We used eight alternative IEMs from [5] for more details see appendix B of [2] and [3].

Of these 36, 30 SNRs show significant emission for all alternative IEMs and are classified as likely GeV SNRs. Of the remaining six, four were identified as sources which are not SNRs; while the other two candidates were considered marginal candidates due to their dependence on the IEM. Of the 30 sources likely to be GeV SNRs, 17 show evidence for extension ($TS_{ext} > 16$). Only sources

[a]e-mail: francesco.depalma@ba.infn.it

associated with SNRs G34.7−0.4 and G189.1+3.0 are fit with logP spectra instead of the usual power-law (PL) spectra. Of the classified candidates, four extended and 10 point SNRs are new. Other 14 sources are marginally classified. The observed photon flux range goes from $\sim 7 \times 10^{-10}$ $ph\ cm^{-2} s^{-1}$ to $\sim 7 \times 10^{-8}$ $ph\ cm^{-2} s^{-1}$, while the PL index range goes from 1.5 to $\sim$ 4.0.

We, also, evaluated the chance of spatial coincidence probability using a mock catalog for both the candidates and the marginally classified candidates. We reported also 245 bayesian [4] upperlimits (UL) for all the non detected radio SNRs.

2 Multiwavelength correlations

We have studied the relationship between the newly measured GeV indices and radio and TeV spectral indices from literature, as shown in Fig. 1. Nearly all candidates have γ-ray photon indices that are softer than predicted given their radio spectra, regardless of the possible GeV emission mechanisms. The three young SNRs in blue are consistent with a single underlying particle population, and it has been suggested inverse Compton emission (dashed line in the figure) at GeV energies. The observed soft GeV spectra relative to the radio has several potential explanations. The underlying leptonic and hadronic populations may have different PL indices. The emitting particle populations may may follow a broken PL or even differing spectral shapes. Or, there may be different zones with different properties dominating the emission at different wavelengths.

Six of the ten SNR candidates have TeV indices softer than their GeV indices, while three have GeV and TeV indices consistent with each other, within statistical and systematic errors, as shown in Fig. 1. Such a change of the spectral index from GeV to TeV suggests that another particle population may dominate at higher energies or that the emission mechanism may change between the GeV and TeV regimes.

3 Constraining Cosmic Ray acceleration

SNRs have long been held the most promising candidate sources of Galactic CRs, capable of supplying the flux observed at Earth if their efficiency in accelerating CR protons and nuclei stands between $\sim 5 - 10\%$. In this section we would like to understand if our detected SNR population is compatible with the observed CRs flux and spectra. In the following, we assume that the γ-ray emission from SNRs probed with LAT entirely arises from the interaction of CR protons and nuclei with the surrounding ISM or circumstellar medium through the production and subsequent decay of π^0. Given that two other emission mechanisms involving accelerated leptons, namely non-thermal bremsstrahlung and inverse Compton scattering, could also contribute in the γ-ray domain, the constraints derived from the LAT measurements should be considered as upper limits on the CR energy content in SNRs. As seen in [2] and references therein the expected γ-ray flux is nearly independent of the CR maximal energy as long as $E_{\rm CR,max} \gtrsim 200\,$GeV and $\Gamma_{\rm CR} \gtrsim 2$. In this case, it can conveniently be approximated using the following expression:

$$F(1 - 100\,\mathrm{GeV}) \approx f(\Gamma_{\rm CR}) \times \frac{\epsilon_{\rm CR}}{0.01} \times \frac{E_{\rm SN}}{10^{51}\,\mathrm{erg}} \times \frac{n}{1\,\mathrm{cm}^{-3}} \times \left(\frac{d}{1\,\mathrm{kpc}}\right)^{-2} \times 10^{-9}\,\mathrm{cm}^{-2}\,\mathrm{s}^{-1} \qquad (1)$$

where $f(\Gamma_{\rm CR})$ is such that $f(2.0) = 2.06, f(2.5) = 1.07$, and $f(3.0) = 0.34$. Using in the previous equation the γ-ray flux, density and distance of each SNR we can determine its maximal CR energy content $E_{\rm CR}$ as shown in Fig. 2. Distances and effective densities are taken from the literature when available, see [2]. We use the canonical value of 10^{51} erg for $E_{\rm SN}$.

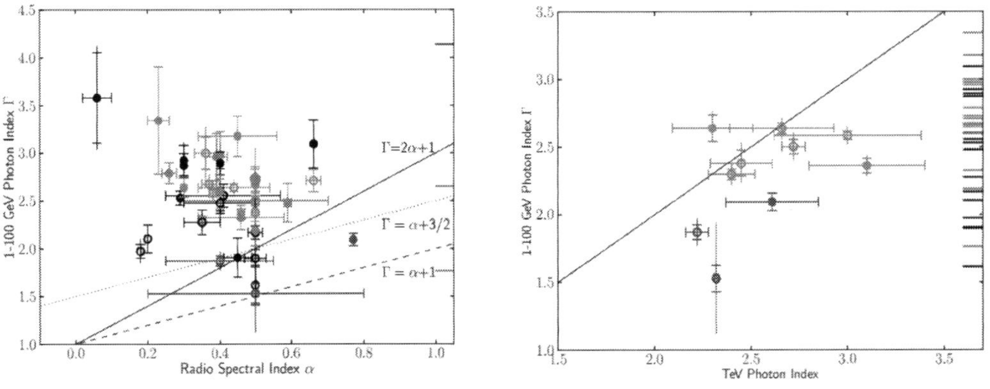

Figure 1: Open circles indicate extended SNRs while filled circles indicate point-like sources. All SNRs that passed classification criteria are shown as black unless also classified as young non-thermal X-ray SNRs (blue) or as interacting with molecular clouds (red). Marginally classified candidates which did not pass classification but still had both fractional overlaps > 0.1 are grey. If they are also young or interacting, they are outlined in blue or red, respectively. Statistical error bars have caps; error bars without caps present the systematic error. Ticks along the right hand side of the right plot show the 1 – 100 GeV photon indices of those SNRs without reported radio or TeV spectral indices. *Left:* Comparison of radio spectral index, α, and GeV photon index, Γ. The expected correlations for π^0 decay or $e^{\pm}$ bremsstrahlung (solid) and inverse Compton (dashed) are plotted. *Right:* GeV index compared to published TeV spectral indices. The line corresponds to equal index values.

As is clearly visible for the first two subclasses of SNRs, the estimates and upper limits on the CR energy content span more than three orders of magnitude, from a few $\times 10^{49}$ erg to several $\times 10^{52}$ erg. For the interacting SNRs that lie above the $\epsilon_{CR} = E_{CR}/E_{SN} = 1$ dashed line, the densities experienced by the CR particles in the molecular cloud interaction region are likely much larger than those used in this analysis. Interacting candidates' lying above this limit suggests that they are likely the sites of hadronic interactions in dense environments. In contrast, most of the young SNRs lie at or below this luminosity limit, suggesting that IC processes may contribute to their measured luminosity. New information about distances and densities can give us further insight on the possibility of known SNRs to provide the observed galactic CRs. The usual assumption of $\epsilon_{CR} = 0.1$, required for the Galactic SNR population to supply the CR flux observed at Earth, is compatible with the results of the *Fermi*-LAT SNR catalog, for more details see [2].

4 Conclusion

In this work we have studied all known radio SNRs in the γ-ray band and the correlation of our results with spectral and spatial information from radio and TeV bands. This was done in order to understand if the underlying CR population, possibly responsible of these emission, is the same and has the same spectral characteristics. Within the limits of existing MW data, our observations generally support previous findings of changes in spectral slope at or near TeV energies and a softening and brightening in the GeV range with age, yet we see indications that new candidates and new MW data may provide evidence of exceptions to this trend.

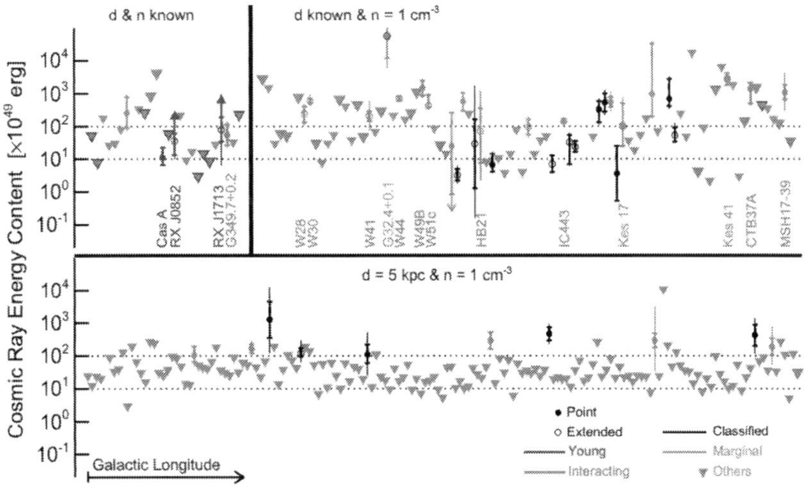

Figure 2: Estimates of the CR energy content for all Galactic SNRs, divided into three categories each sorted by Galactic longitude: SNRs with existing distance and density estimates (upper left panel); SNRs with known distances (upper right panel); and SNRs with unknown distance and density (lower panel).The two dashed lines indicate a CR energy content of 10 and 100% of the standard supernova (SN) explosion energy.

We find, also, that the limits on CR energy content span more than three decades, including many interacting candidates for which the densities in the interaction regions are much greater than the nominal density assumed in the calculation, and young candidates with efficiencies below the nominal $\sim$ 10%, consistent with possible leptonic emission predictions (e.g. IC). The contribution from all SNRs, particularly those with flux upper limits, constrains the energy content put into CRs from the known SNRs to less than 10%, particularly in regions of well characterized IEM background. We find that the candidates and upper limits are generally within expectations if SNRs provide the majority of Galactic CRs.

Acknowledgments

The *Fermi*-LAT Collaboration acknowledges support for LAT development, operation and data analysis from NASA and DOE (United States), CEA/Irfu and IN2P3/CNRS (France), ASI and INFN (Italy), MEXT, KEK, and JAXA (Japan), and the K.A. Wallenberg Foundation, the Swedish Research Council and the National Space Board (Sweden). Science analysis support in the operations phase from INAF (Italy) and CNES (France) is also gratefully acknowledged.

References

[1] Green, D. A. 2009, Bulletin of the Astronomical Society of India, 37, 45

[2] Acero, F. et al. 2016, ApJS, 224, 8

[3] de Palma, F. et al. arXiv:1304.1395

[4] Helene, O. 1983, Nuclear Instruments and Methods in Physics Research, 212, 319

[5] Ackermann, M. et al. 2012, ApJ, 750, 3

The Fermi-GBM Gamma-Ray Burst Catalogs: The First Six Years

E. Bissaldi[1,2,a], P. N. Bhat[3], and H.-F. Yu[4,5] for the Fermi-GBM Team

[1] *Dipartimento Interateneo di Fisica, Università e Politecnico di Bari, Bari, Italy*
[2] *INFN – Sezione di Bari, Bari, Italy*
[3] *CSPAR, University of Alabama in Huntsville, Hunstville, AL, USA*
[4] *Max Planck Institute for Extraterrestrial Physics, Garching, Germany*
[5] *Excellence Cluster Universe, Technical University of Munich, Munich, Germany*

Abstract. Since its launch in 2008, the Fermi Gamma-ray Burst Monitor (GBM) has triggered and located on average approximately two gamma-ray bursts (GRBs) every three days. Here we present the main results from the latest two catalogs provided by the Fermi-GBM science team, namely the third GBM GRB catalog [1] and the first GBM time-resolved spectral catalog [2]. The intention of the GBM GRB catalog is to provide information to the community on the most important observables of the GBM detected bursts. It comprises 1405 triggers identified as GRBs. For each one, location and main characteristics of the prompt emission, the duration, the peak flux and the fluence are derived. The GBM time-resolved spectral catalog presents high-quality time-resolved spectral analysis with high temporal and spectral resolution of the brightest bursts observed by Fermi GBM in a shorter period than the former catalog, namely four years. It comprises 1491 spectra from 81 bursts. Distributions of parameters, statistics of the parameter populations, parameter-parameter and parameter-uncertainty correlations, and their exact values are obtained.

1 The third GBM trigger catalog

From July 12th, 2008 to July 11th, 2014, GBM triggered on 3350 transient events. The left panel of Figure 1 shows the quarterly statistics over the aforementioned period, highlighting different sources with different colors. The total number of detected GRBs is 1404. From this number we can derive the six year average GRB rates, namely (0.662 ± 0.018) day^{-1} or (242 ± 7) yr^{-1}. The GRB sky distribution in celestial coordinates is also shown in Figure 1. There are 1176 long GRBs ($T_{90} > 2$ s, black dots) and 228 short GRBs (blue asterisks). The isotropic distribution of long and short GRB arrival directions is evident. The lower fraction of GBM short GRBs (20.7%) compared to the one measured by BATSE (24%) is due to an excess of long events detected by GBM's longer timescale trigger algorithms.

1.1 The GRB catalog analysis – Main steps

1) Burst localization The distribution of systematic uncertainties in GBM positions is well represented (68% c.l.) by a Gaussian of standard deviation 3.7° with a non-Gaussian tail that contains

[a] e-mail: elisabetta.bissaldi@ba.infn.it

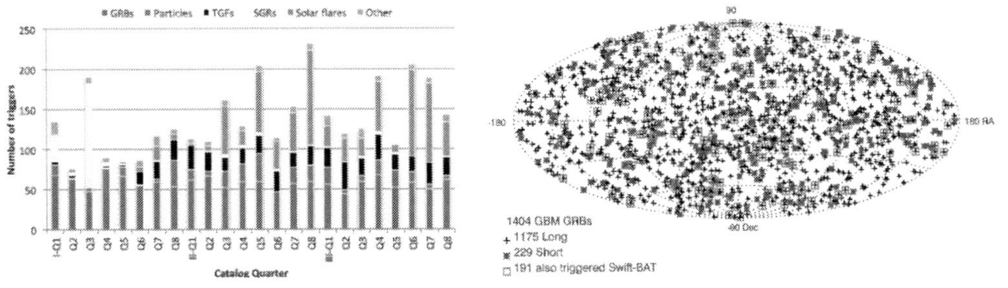

Figure 1. On the *left*: Quarterly trigger statistics over the first six years of the GBM mission. The different types of events triggering GBM are classified as shown at the top. On the *right*: Sky distribution of GBM-triggered GRBs during the first six years in celestial coordinates (RA and Dec). Crosses indicate long GRBs and asterisks indicate short GRBs. Also shown are the GBM GRBs simultaneously detected by Swift (red squares).

about 10% of GBM-detected GRBs and extends to approximately 14° [3]. Moreover, probability maps are produced for every burst through convolution of the statistical uncertainty with the best current model for the systematic errors. The maps reflect the occasional noncircular shape of the statistical uncertainty region as well as its area.

2) Detector Response Matrix (DRM) production DRMs are generated using the General Response Simulation System [4] and are routinely delivered to the Fermi Science Support Center[1].

3) GRB Duration, Peak Flux and Fluence computation GRB T_{50} and T_{90} durations are computed through an automatic batch fit routine implemented within the spectroscopy software package RMFIT[2]. They represent the intervals between the times where the burst has reached 25%/5% and 75%/95% of its maximum fluence in the 50–300 keV energy range. Fluences are computed in two energy ranges, namely 50–300 keV and 10–1000 keV. Peak fluxes in the same energy ranges are given for three timescales (64, 256, and 1024 ms).

[1] `http://heasarc.gsfc.nasa.gov/FTP/fermi/data/gbm/bursts`
[2] `http://fermi.gsfc.nasa.gov/ssc/data/analysis/user/`

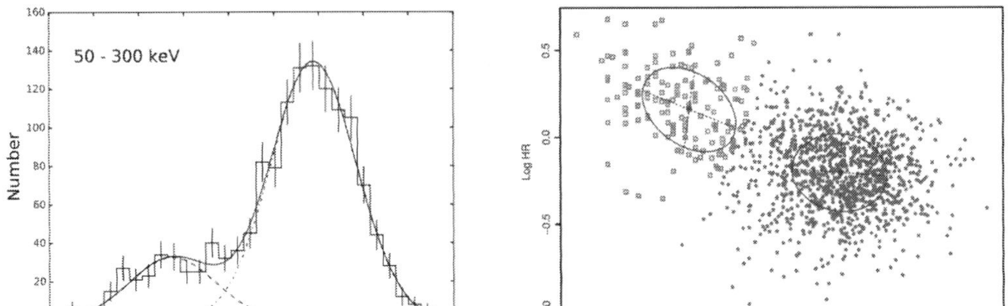

Figure 2. On the *left*: Distribution of GRB T_{90} durations in the 50–300 keV energy range. Also, shown separately are the lognormal fits to long and short GRBs. On the *right*: Classification based on the hardness–duration diagram. Colors indicate their group membership (red: on average short/hard, blue: on average long/soft). Ellipses show the best-fitting multivariate Gaussian models.

1.2 The GRB catalog – main results

The left panel of Figure 2 shows the distributions of GBM durations. This is consistent with a bimodal distribution and it was verified by various independent analyses, namely trough (1) a model-based clustering method with lognormal model components [5]; (2) Monte Carlo simulations; and (3) a Bayesian Dirichlet mixture model [6]. The median T_{90} durations are 26.62 s and 0.58 s for long and short bursts, respectively. The right panel of Figure 2 shows the hardness–duration diagram. It is consistent with two clusters: the short/hard group and the long/soft group. This was verified by independent analysis as was done for the duration distribution.

2 The first GBM time-resolved spectral catalog

Since the time-resolved spectral analysis requires bright bursts with sufficiently high signal-to-noise spectra, we applied following selection criteria to the bursts detected by GBM in the first four years (954 GRBs): energy fluence $f > 4 \times 10^{-5}$ erg cm^{-2} and/or peak photon flux $F_p > 20$ ph s^{-1}cm^{-2} (in either 64, 256, or 1024 ms), both in the 0.01–1 MeV energy range. Moreover, we required the presence of at least five time bins in the light curves when binned with signal-to-noise ratio = 30. 81 bursts satisfy these criteria (with only 1 short GRB 120323A). In the end, we obtained 1802 time-resolved time bins and spectra, which were analyzed with RMFIT considering four different empirical fit models, namely the Band function, the smoothly broken power law (SBPL), the cutoff power law (COMP) and the simple power law (PL). Among all fitted spectra, we further selected the "BEST"

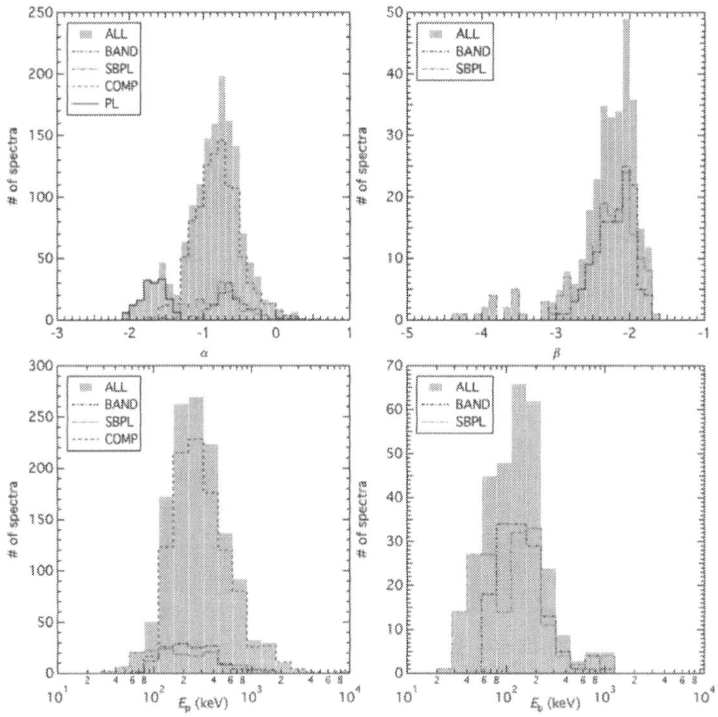

Figure 3. Distributions of the BEST sample spectral parameters: the low-energy spectral index α, the high-energy spectral index β, the peak photon energy E_p and the break energy E_b.

model sample adopting following criteria: that the relative error $\sigma_Q/Q \leq 0.4$ for each parameter Q of a model and that $\sigma_\alpha \leq 0.4$ and $\sigma_\beta \leq 1.0$ for models that have two PL indices. 1491 spectra satisfy these criteria, and the distributions of the BEST sample spectral parameters are shown in Figure 3.

2.1 Discussion of the main results

We can summarize the main results of the time-resolved spectral analysis as follows. **(1)** The preferred model (in 69% of cases) is the COMP one. A possible reason might be a bias due to poor count statistics at high energies. However, GRBs in the LAT FoV which remain undetected show that upper limits are usually inconsistent with the GBM fit Band function's β, extrapolated to the LAT energy range. This might indicate a real manifestation of GRB physics. **(2)** There are no significant deviations of the distributions of fit parameters from those observed in the Fermi GBM GRB time-integrated spectral catalogs. The comparison of averaged time-resolved parameters to time-integrated ones shows that averaged time-resolved α and E_p are harder than time-integrated ones. **(3)** We observed the widening of time-integrated spectra with respect to the time-resolved ones (see also [7, 8]). This effect might be caused by the time-integrated analysis, and this might lead to incorrect physical interpretations. **(4)** We performed a search for plausible blackbody components in time-resolved spectra, adding simulations on individual bursts. Unfortunately, only three GRBs out of the whole sample show extra blackbody components in multiple time bins.

Acknowledgments

The GBM project is supported by NASA. Support for the German contribution to GBM was provided by the *Bundesministerium für Bildung und Forschung* via the *Deutsches Zentrum für Luft und Raumfahrt* under contract number 50 QV 0301. E.B. is supported by the Italian *Fondo di Sviluppo e Coesione 2007-2013 - APQ Ricerca Regione Puglia "Future In Research"*.

References

[1] P. N. Bhat et al., The Astrophysical Journal Supplement Series **223-2** 28 (2016)
[2] H.-F. Yu et al., Astronomy & Astrophysics **588** A135 (2016)
[3] V. Connaughton et al., The Astrophysical Journal Supplement Series **216-2**, 32 (2015)
[4] A. Hoover et al., Il Nuovo Cimento C **28**, 797-800 (2005)
[5] C. Fraley & A. E. Raftery, J. Am. Stat. Assoc. **97** 611 (2000)
[6] S. J. Gershman & D. M. Blei, Journal of Math. Psychology **56** 1 (2012)
[7] H.-F. Yu et al., Astronomy & Astrophysics **583** A129 (2015)
[8] M. Axelsson & L. Borgonovo, Monthly Notices of the Royal Astron. Society **447** 3150 (2015)

Fermi Bubbles: an elephant in the gamma-ray sky

Dmitry Malyshev[1,2,a]

[1] *Erlangen Centre for Astroparticle Physics, Erwin-Rommel-Str. 1, D-91058 Erlangen, Germany,*
[2] *on behalf of the Fermi LAT collaboration*

Abstract. The *Fermi* bubbles are one of the most remarkable features in the gamma-ray sky revealed by the *Fermi* Large Area Telescope (LAT). The nature of the gamma-ray emission and the origin of the bubbles are still open questions. In this note, we will review some basic features of leptonic and hadronic modes of gamma-ray production. At the moment, gamma rays are our best method to study the bubbles, but in order to resolve the origin of the bubbles multi-wavelength and multi-messenger observations will be crucial.

1 Introduction

The *Fermi* bubbles were originally discovered as a spherical gamma-ray haze emission [1] in a search for a gamma-ray counterpart of the microwave haze detected in the Wilkinson Microwave Anisotropy Probe (WMAP) data [2]. With more gamma-ray data, sharp boundaries of the bubbles were resolved at about 55° above and below the Galactic center [3]. The bubbles occupy a solid angle $\sim$ 1 sr, which is approximately equal to the area of a square of 3 meters on a side at a distance of 3 meters, i.e., about the solid angle of an elephant standing in a room. The spectrum of the bubbles is $\sim E^{-2}$ between 100 MeV and 100 GeV [3, 4] with a cutoff or significant softening above 100 GeV [4].

Possible origin of the bubbles includes cosmic ray (CR) acceleration by the supermassive black hole (SMBH) at the center of our Galaxy Sgr A* [3], a period of starburst activity, or an accumulation of CR for a long time from the regular star formation near the Galactic center (GC) [5]. In the latter case, a special arrangement of magnetic fields, e.g., magnetic draping, is necessary to keep the CR from escaping the Galaxy. Examples of bubble-like structures in other Galaxies include a pair of jets in Centaurus A, which is also discovered as an extended gamma-ray source [6], and a star-bust activity in the M82 galaxy. To understand the origin of the bubbles, it would be useful to know the gamma-ray production mechanism: whether the gamma-rays are produced in interactions of hadronic CR with interstellar gas or by inverse Compton (IC) scattering of high energy electrons with interstellar radiation fields.

2 Leptonic emission models

The spectrum of the bubbles can be relatively easily explained with IC gamma rays produced by scattering of interstellar radiation photons and high energy CR electrons. The spectrum of the electrons

[a] e-mail: dmitry.malyshev@fau.de

can be represented as a power law with an index −2.2 and an exponential cutoff at 1.25 GeV [4]. There is a tentative association of the *Fermi* bubbles emission with the microwave haze observed in the WMAP [2] and Planck [7] data. For a magnetic field $\approx 8.4\mu G$ [4], the microwave haze can be explained by the same population of electrons that produce the gamma-ray emission. Although the intensity of gamma-ray emission from the bubbles is approximately uniform up to $|b| \approx 55°$ while the microwave haze intensity decreases significantly above $|b| \approx 35°$, the difference in the morphology of the gamma-ray and the microwave emission can be explained by a decrease in the magnetic field further away from the Galactic plane [8].

The total gamma-ray luminosity of the *Fermi* bubbles $L_{\rm bbl} \approx 4\times10^{37}$erg/s. The power of injection of CR electrons should be $\gtrsim L_{\rm bbl}$. Upscattering of cosmic microwave background (CMB) photons to a few GeV energies requires electron energies ~ 1 TeV ($E_\gamma = 4/3h\nu\gamma^2$, where $h\nu \approx 2.4\times10^{-4}$ eV and $\gamma \sim 2\times10^6$). Since the cooling time of a 1 TeV electron is about 1 Myr = 3×10^{13} s, the total power contained in electrons above 1 TeV is ~ 10^{51} erg. Integrating the electron CR spectrum above 1 GeV gives $\approx 10^{52}$ erg [4]. Dividing by an approximate volume of the bubbles, one can derive the energy density of the CR electron population. Around 1 TeV, it turns out to be a factor of 2 to 3 larger than the local energy density of CR electrons [4], while around 100 GeV the energy density of electrons inside the bubbles is about the same as the local energy density.

The possibility to explain the *Fermi* bubbles together with the microwave haze by the same population of electrons is very appealing, one difficulty of the leptonic interpretation of the gamma-ray emission is the necessity to have electrons of 1 TeV energies around 10 kpc away from the Galactic plane (GP). If the electrons were produced near the GC, then, with the cooling time of $\lesssim 1$ Myr, the velocity of their transport to high latitudes should be at least 10 kpc/1 Myr $\approx 10^4$ km s^{-1}. Since this velocity is much larger than the speed of sound in the plasma around the GP, one would expect to see a shock front at the boundary of the bubbles, but no such shock front has been observed. Moreover, the observed velocities of gas outflow in the direction of the bubbles are $\lesssim 10^3$ km s^{-1} [9] . A possible solution is that the electrons are (re)accelerated via the 2nd order Fermi acceleration mechanism by the sound waves left behind a shock front that may have existed as the bubbles were forming [10] or from a series of shocks expanding from the Sgr A* [11].

Overall, leptonic model offers natural explanations for the gamma-ray spectrum of the bubbles and for the possibly associated microwave emission. The uniform spectrum of the bubbles as a function of latitude can be explained with re-acceleration of electrons by an ensemble of shock waves inside the bubbles.

3 Hadronic emission models

The second possible explanation of the gamma-ray emission from the *Fermi* bubbles is the interactions of high energy CR hadrons with the interstellar gas. The density of gas above a few hundred parsecs is at least a 100 times smaller than the density of gas in the GP, the energy density of CR required to explain the emission from the bubbles (which is about 10 - 30 times smaller than the gamma-ray emission from hadronic interactions in the GP above 10 GeV) is 3 - 10 times larger than the CR density in the GP above 100 GeV [4]. The spectrum of CR protons that can explain the gamma-ray spectrum of the bubbles is ~ E^{-2} with a cutoff around 10 TeV. There is a slight tension of the primary gamma-ray spectrum with the observed spectrum of the bubbles around 100 MeV due to the pion cutoff. This tension can be easily resolved if one takes into account secondary IC emission from the electrons and positrons produced in the hadronic interactions together with the primary gamma rays. However, the spectrum of the secondary leptons is too soft to explain the microwave haze [4]. Since IC gamma rays above 100 MeV are mostly produced by leptons with $E \gtrsim 100$ GeV up-scattering

CMB photons, while the microwave haze is produced by electrons with energies $\lesssim$ 30 GeV for a magnetic field $\gtrsim 5\,\mu$G, one needs a population of electrons below 100 GeV with a hard spectrum to explain the microwave haze in addition to the secondary leptons in hadronic interactions.

Integrating the energy density of CR hadrons over the volume of the bubbles, one gets the total energy contained in CR protons above 1 GeV of 3×10^{55} erg (for the gas density of 0.01 cm^{-3}). Assuming the scattering cross section of 30 mb, the characteristic interaction time is 3×10^9 yr, i.e., unless there is a mechanism to keep the protons inside the bubbles for several billions of years, the majority of protons will escape the bubbles without interacting. If there is a magnetic draping on the boundary of the bubbles, it can reflect the protons back into the volume of the bubbles and prevent the escape. In this case the energy density of the CR can be obtained by accumulating the output in CR from about 3×10^5 supernovae (SN) assuming that a characteristic kinetic energy of a SN shell is 10^{51} erg and efficiency of CR acceleration is 10% (see, e.g., discussion in [5]). This can be easily accumulated over $\sim$ billion of years assuming a SN rate in the inner part of the Galaxy ~ 1 kyr^{-1}.

If we divide, the total energy in CR to the volume of the bubbles (e.g., assuming that the bubbles consist of two spheres with 5 kpc radii), then we get the energy density in CR of ~ 1 eV cm^{-3}. This energy density can be compared to the kinetic energy density of plasma in the halo around the GP $\rho \approx 0.0035$ cm^{-3} and $kT \sim 340$ eV [12], which gives $W_{plasma} \sim 1.5$ eV cm^{-3}. The conclusion is that the kinetic energy of the CR is sufficient to push away the plasma and create a cavity that may be observed in X rays. The energy of CR electrons in leptonic scenario $\sim 10^{52}$ erg is at least three orders of magnitude smaller than the energy density of CR protons. As a result, one should not expect to see such cavity in leptonic models.

Although no clear cavity has been observed in ROSAT data, there are preliminary signs of emission measure change across the boundary of the bubbles with pointed observations by Suzaku [13]. The presence of the cavity may be confirmed or excluded by the future more sensitive all-sky X-ray survey with eROSITA.

4 Models of *Fermi* bubbles creation

The two most popular models of creation of the bubbles are an active galactic nucleus (AGN)-like activity of the SMBH Sgr A* at the center of our Galaxy and a starburst near the GC. Numerical simulations show that the bubbles can be inflated by two jets emitted from the SMBH [14–16], as well as by a spherical outflow, shaped into the bubbles by molecular clouds around the GC [17].

Since gamma-ray data have no jet-like structure associated with the bubbles, one has to rely on external data to distinguish AGN jets or an outflow from a starburst-driven wind as the main mechanism that has created the bubbles. Observations of external galaxies suggest [18] that AGN winds come together with significant photo-ionization, while starburst winds are associated with shock-ionization. Although the ultraviolet (UV) radiation during star formation has comparable luminosity to UV radiation from accretion onto the black hole, the UV radiation from young stars is largely absorbed by molecular clouds, where the star formation typically happens, in addition, there is a delay of several Myr or more between the star formation and SN explosions, which drive the wind [18]. Unfortunately no ionization has been detected near the *Fermi* bubbles that could help to distinguish the two scenarios. There is, however, an enhanced Hα emission in the Magellanic Stream towards the South Pole in Galactic coordinates, which is consistent with UV radiation produced in an AGN-like activity of Sgr A* around 0.5 – 3 Myr ago [19]. Although such level of ionization in the Magellanic Stream is improbable in the starburst scenario [19], it does not exclude the possibility that a starburst, e.g., star formation related to $\sim$ 6 Myr population of stars near the GC [20], produced the bubbles.

5 Conclusions

Although the *Fermi* bubbles are some of the brightest and most significant features in the gamma-ray sky, their origin and even the nature of the gamma-ray emission remains unresolved. Both leptonic IC scattering and hadronic production of gamma rays are viable options, while the formation of the bubbles can be either due to an AGN-like activity of Sgr A* or a starburst event near the GC.

One of the signatures of leptonic origin of the gamma-ray emission is the synchrotron emission from the electrons. There is a tentative association of the microwave haze emission and the *Fermi* bubbles which can be explained by the same population of electrons. A discovery of polarized emission with features correlated with the *Fermi* bubbles would strengthen this hypothesis.

For the hadronic model, the required CR energy density is comparable to the energy density of the halo plasma. As a result, the presence of the CR may create a cavity in the plasma, that can be detected with the future X-ray observations, most notably, eROSITA. There is also a possibility to detect an associated neutrino signal with new more sensitive neutrino detectors, such as PINGU in IceCube and KM3NeT.

Developing a model of the *Fermi* bubbles near the GC will play a significant role in disentangling their origin. If the bubbles were produced by a Sgr A* activity, then they should have a narrow base centered at the GC, while a starburst activity may result in a broader base of the bubbles not necessarily centered on the GC. Current and future Cherenkov telescopes can help to study the morphology of the *Fermi* bubbles near the GC at energies $\gtrsim$ 100 GeV, where the *Fermi* LAT loses sensitivity.

The *Fermi*-LAT Collaboration acknowledges support from NASA and DOE (United States), CEA/Irfu, IN2P3/CNRS, and CNES (France), ASI, INFN, and INAF (Italy), MEXT, KEK, and JAXA (Japan), and the K.A. Wallenberg Foundation, the Swedish Research Council, and the National Space Board (Sweden). This work was partially supported by NASA grant NNH13ZDA001N.

References

[1] G. Dobler, D.P. Finkbeiner, I. Cholis et al., ApJ, **717**, 825 (2010), 0910.4583
[2] D.P. Finkbeiner, ApJ, **614**, 186 (2004), arXiv:astro-ph/0311547
[3] M. Su, T.R. Slatyer, D.P. Finkbeiner, ApJ, **724**, 1044 (2010), 1005.5480
[4] M. Ackermann, A. Albert, W.B. Atwood et al., ApJ, **793**, 64 (2014), 1407.7905
[5] R.M. Crocker, F. Aharonian, Physical Review Letters **106**, 101102 (2011), 1008.2658
[6] A.A. Abdo, M. Ackermann, M. Ajello et al., Science **328**, 725 (2010), 1006.3986
[7] P.A.R. Ade, N. Aghanim, M. Arnaud et al., A&A, **554**, A139 (2013), 1208.5483
[8] G. Dobler, ApJ, **750**, 17 (2012), 1109.4418
[9] A.J. Fox, R. Bordoloi, B.D. Savage et al., ApJ, **799**, L7 (2015), 1412.1480
[10] P. Mertsch, S. Sarkar, Physical Review Letters **107**, 091101 (2011), 1104.3585
[11] K.S. Cheng, D.O. Chernyshov, V.A. Dogiel et al., ApJ, **731**, L17 (2011), 1103.1002
[12] S.L. Snowden, R. Egger, M.J. Freyberg et al., ApJ, **485**, 125 (1997)
[13] J. Kataoka, M. Tahara, T. Totani et al., ApJ, **779**, 57 (2013), 1310.3553
[14] F. Guo, W.G. Mathews, ApJ, **756**, 181 (2012), 1103.0055
[15] F. Guo, W.G. Mathews, G. Dobler et al., ApJ, **756**, 182 (2012), 1110.0834
[16] H.Y.K. Yang, M. Ruszkowski, P.M. Ricker et al., ApJ, **761**, 185 (2012), 1207.4185
[17] K. Zubovas, A.R. King, S. Nayakshin, MNRAS, **415**, L21 (2011), 1104.5443
[18] R.G. Sharp, J. Bland-Hawthorn, ApJ, **711**, 818 (2010), 1001.4315
[19] J. Bland-Hawthorn, P.R. Maloney, R.S. Sutherland et al., ApJ, **778**, 58 (2013), 1309.5455
[20] T. Paumard, R. Genzel, F. Martins et al., ApJ, **643**, 1011 (2006), arXiv:astro-ph/0601268

Supernova Remnants with Fermi Large Area Telescope

M. Caragiulo[1,2,a] and L. Di Venere[1,2]on behalf of *Fermi*-LAT Collaboration

[1]*Dipartimento Interateneo di Fisica "Michelangelo Merlin", Via Amendola 173, 70126 Bari, Italy*
[2]*Istituto Nazionale di Fisica Nucleare (INFN), Sezione INFN di Bari, Via Orabona 4, 70126 Bari, Italy*

Abstract. The Large Area Telescope (LAT), on-board the Fermi satellite, proved to be, after 8 years of data taking, an excellent instrument to detect and observe Supernova Remnants (SNRs) in a range of energies running from few hundred MeV up to few hundred GeV. It provides essential information on physical processes that occur at the source, involving both accelerated leptons and hadrons, in order to understand the mechanisms responsible for the primary Cosmic Ray (CR) acceleration. We show the latest results in the observation of Galactic SNRs by Fermi-LAT.

1 Introduction

The LAT, the primary instrument on-board the *Fermi Gamma-ray Space Telescope* (*Fermi*), is an imaging, wide-field-of-view, pair-conversion telescope, which detects photons from few tens of MeV to few hundreds of GeV. It contains a high-resolution converter/tracker (composed of 16 planes of tungsten to promote the conversion of γ-rays into pairs, interleaved with silicon strip detectors) for direction measurement of the incident γ-rays, a CsI(Tl) crystal electromagnetic calorimeter for energy measurement of the incoming photon, and an anti-coincidence detector to identify the background of charged particles, mostly due to charged CRs [1].

Recently, the *Fermi*-LAT collaboration has released the latest version of Instrument Response Functions (IRFs), along with the reprocessed data, obtained with the new event reconstruction (tracker, calorimeter and ACD reconstruction), known as PASS 8. All these modifications contribute to improve the effective area, the angular and energy resolutions. Combined with the large amount of data collected by the LAT since its launch, PASS 8 is a powerful tool to study the emission below 100 MeV and the extended sources, as it will be shown in section 3.2.

The *Fermi*-LAT is providing an unprecedented wealth of detections and observations of γ-ray sources. The detection of γ-rays is a clear signature that non-thermal processes are happening at the source, eventually accelerating particles. Since photons are not deflected by magnetic fields, they provide key information to study these phenomena.

2 Supernova Remnants as sources of Cosmic Rays

A Supernova Remnant (SNR) results from the interaction between the stellar material ejected by a Supernova (SN) explosion and the environmental gas. SNRs are considered the most probable

[a]e-mail: micaela.caragiulo@ba.infn.it

sources of Galactic CRs, with energy from about 1 GeV to about 10^8 GeV (*SNR paradigm*) [?], and the acceleration processes take place at the SNR shocks.

The acceleration mechanism, also known as *diffuse shock acceleration* (DSA) theory, naturally predicts that the accelerated particles distribution is well described by a power law $\propto E^{-q}$ with a universal slope, whose spectral index in case of strong SNR shocks is $q = 2$. A generalization of this theory, known as Non-Linear Diffusive Shock Acceleration (NLDSA) theory, takes into account the dynamical reaction of the accelerated particles on the shock, predicting a steeper particle distribution, compatible with the spectrum measured at Earth [2]. The CR flux observed at Earth $F(E)$ ($\propto E^{-2.75}$) is proportional to the injection spectrum at the source $Q_{inj}(E)$ ($\propto E^{-q}$) and the Galactic permanence time $\tau_{esc}(E)$ ($\propto E^{-\delta}$), due to the propagation of CRs through the Galaxy:

$$F(E) = Q_{inj}(E)\tau_{esc}(E) \propto E^{-(q+\delta)}. \tag{1}$$

The model for CR propagation, based on the CR isotope and secondary-to-primary composition, provides a value of $\delta \sim 0.3 - 0.6$, suggesting a injection spectrum index $q = 2.1 - 2.4$, compatible with the NLDSA predictions.

SNRs would be able to provide the required CR energetics, indeed the acceleration mechanism is efficient enough to account for the observed CR energy density $\rho_{CR} \sim 1eV/cm^3$. ρ_{CR} can be expressed as:

$$\rho_{CR} = R_{SN}E_{SN}\tau_{esc}(E)\epsilon, \tag{2}$$

where $R_{SN} \sim 3$ SN/century is the SN explosion rate in our Galaxy, $E_{SN} \sim 10^{51}$ erg is the typical explosion energy of a SN, $\tau_{esc}(E) \sim 10^7$ years is the Galactic CR confinement time in the Galaxy and ϵ is the required acceleration efficiency, which has to be close to $\sim$10%. Some models have already shown that NLDSA can provide such acceleration efficiencies [3].

An important evidence of the NLDSA is given by the observation of narrow filaments of non-thermal X-ray radiation along the shock front in young SNRs [4]. They are due to the synchrotron emission of accelerated electrons in a strong amplified magnetic field, which might be originated from the interaction of the charged accelerated particles with the shock.

3 Gamma-ray observations

The accelerated CRs interact with the SNR environment producing photons of energy range from radio to TeV. The predominant processes in MeV-GeV energy range are:

- Inverse Compton (IC) scattering of accelerated electrons on local photons from Cosmic Microwave Background and infrared radiation due to dust emission,
- bremsstrahlung radiation of the same electron population deflected in the Coulomb field, generated by charged particles of the gas surrounding the remnant,
- hadronic interaction between accelerated particles and this gas, which produces many hadrons, especially neutral pions, which then decay in photons.

The γ-ray spectrum due to the π^0-decay, with its peculiar shape due to the kinematic cut-off around 100 MeV of the π^0 decay process, represents the hint of the so-called *SNR paradigm* for CRs since its production involves accelerated hadrons. For this reason, the observation of this kind of γ-ray emission from SNRs is essential to infer about the accelerations mechanisms.

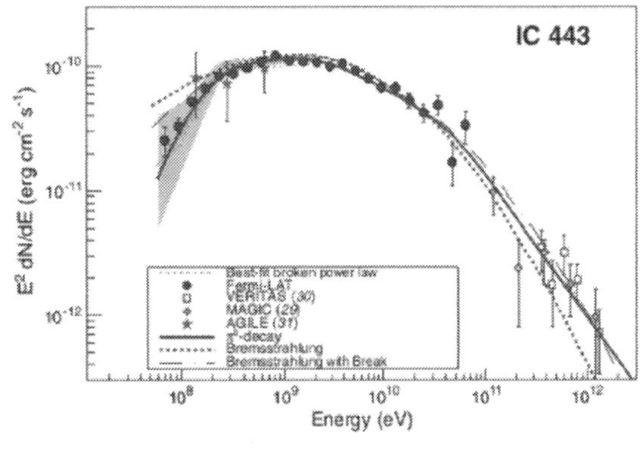

(a)

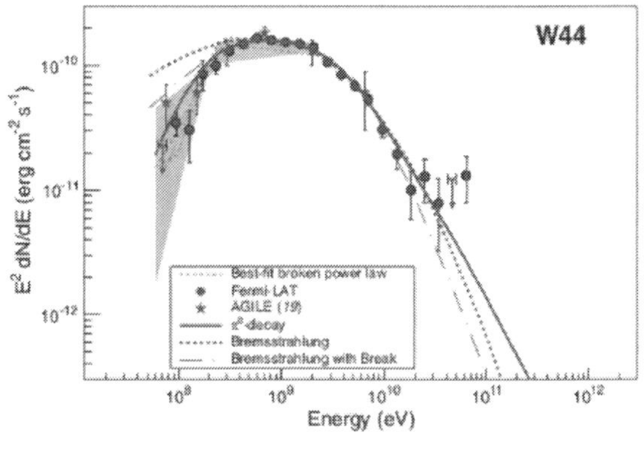

(b)

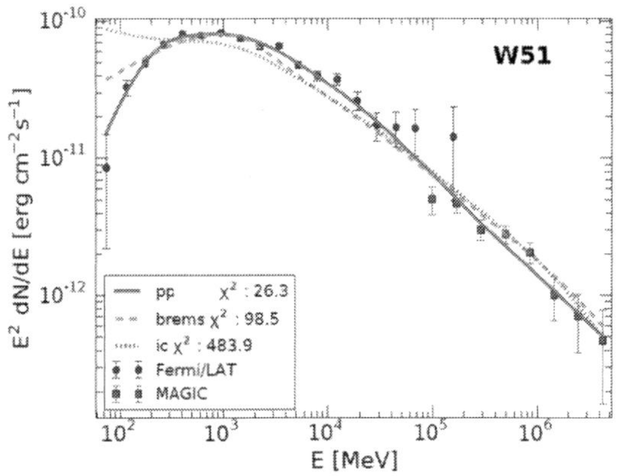

(c)

Figure 1: Spectral Energy Distribution of three SNRs, overlaid with theoretical models based either on a hadronic or leptonic interpretation of γ-ray emission. The shape of the observed γ-ray spectrum favors the hadronic interpretation. (a): IC443 [6]. (b): W44 [6]. (c): W51 [7].

3.1 *Fermi*-LAT detection of hadronic interaction

SNRs interacting with molecular clouds (MCs) are old objects in which the acceleration process is not very efficient anymore, as suggested by the steep γ-ray spectrum at high energies, which is probably due to the escaping of high energy CRs. Indeed, their strong γ-ray emission is not due to a high flux of relativistic hadrons, but to a very dense cloud target in the vicinity of the remnant, which interacts with the latter via nuclear collisions. Therefore, they do not give the best view of the CR acceleration but they are useful to investigate the CR propagation around sources and the escape from them.

The first detection of γ-ray spectrum due to the hadronic interaction has been found in the observation of tree middle aged SNRs interacting with MCs, few years ago W 44, IC 443 [5] [6] and recently W51 [7]. Figure 1 shows the Spectral Energy Distributions (SEDs) of these three SNRs. The data points at energies below 200 MeV strongly support the hadronic interpretation.

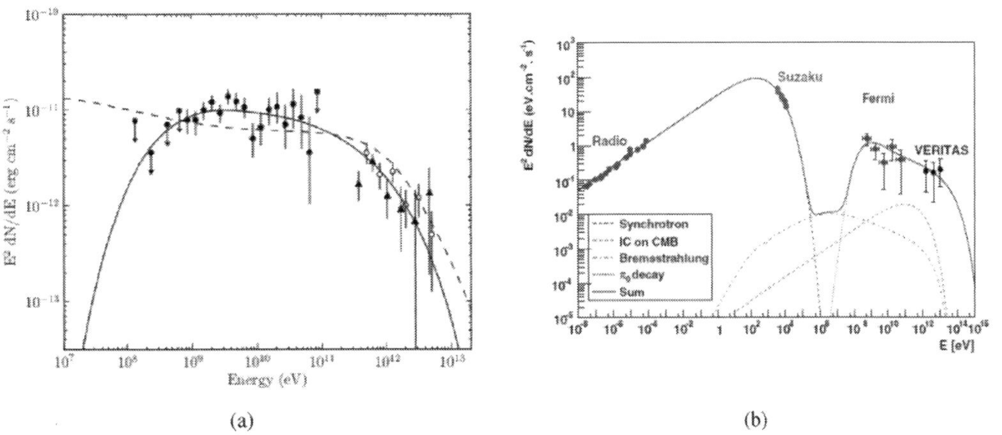

(a) (b)

Figure 2: Spectral Energy Distribution of two young SNRs, overlaid with theoretical models. (a): Cassiopeia A. The lines represent two different interpretation models based on a hadronic (solid) or leptonic (dashed) origin of γ-ray emission [9]. (b): Tycho. The lines represent the contribution of different emission mechanisms. The total γ-ray emission (solid line) is dominated by the π^0-decay spectrum [8].

Young SNRs ($\sim$ few thousand years old), instead, are at the initial stage of their evolution, so they are evolving in cleaner (and in most cases low-density) environments. For this reason, a detection of their γ-ray emission could give clear information about the acceleration processes taking place in these SNRs.

The best candidates as CR accelerators seem to be Tycho [8] and Cassiopeia A (Cas A) [9]. Their γ-ray spectrum detected by *Fermi*-LAT seems to be compatible with the hadronic model. In particular, the γ-ray flux of Cas A is characterized by a peculiar shape of the spectrum at energies below 100 MeV, where the π^0-decay spectrum presents a break due to the threshold energy of the pion production. On the other hand, other young SNRs, such as RX J1713.7-3946 [10] and RCW 86 [11], show different spectrum shapes (with a very hard spectral index in the MeV–GeV energy range), which support the leptonic model, based on the production of γ-ray via IC scattering. The leptonic interpretation is also favored when considering the H.E.S.S. results [12] [13]. For these two SNRs, the absence of γ-rays from π^0-decay does not exclude the possibility of an efficient accelerated of CRs, but might be due to a low gas density around the source, which reduces the hadronic contribution to the photon flux. Figures 2 and 3 show the SEDs of these young SNRs with the best-fit interpretation models overlaid.

3.2 Morphology study with PASS 8

An important aspect of the study of γ-ray sources is the detection of their spatial shape, since it is an important characteristic for correctly associating γ-ray emitting sources with their counterparts at other wavelengths. This problem is complicated by the large PSF of the instrument. However, the improvement in PSF and effective area introduced by the PASS 8 reconstruction method has allowed the study of the spatial extension of some sources.

RCW 86 is one of the SNRs which have been recently detected as extended with PASS 8. Figure 4 shows a test statistic map of this source observed by the LAT, compared with the contours of the same sources observed by the H.E.S.S. experiment [13]. In order to determine the best morphology of RCW

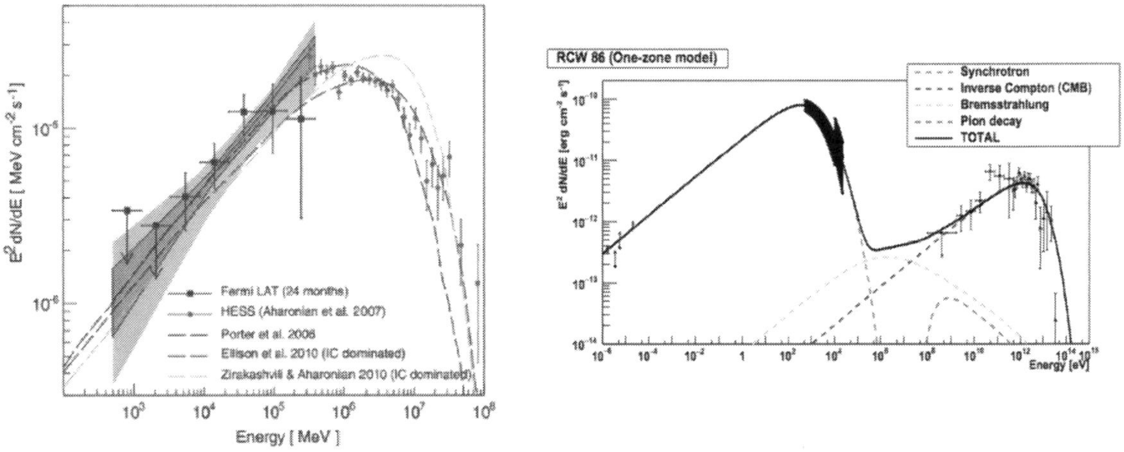

(a) (b)

Figure 3: Spectral Energy Distribution of two young SNRs, overlaid with theoretical models based on a leptonic interpretation of the γ-ray emission. (a): RX J1713.7-3946. The lines represent different interpretation models in which the γ-ray emission (solid line) is dominated by the IC scattering [10]. (b): RCW 86. The lines represent the contribution of different emission mechanisms. The total γ-ray emission (solid line) is dominated by the IC scattering [11].

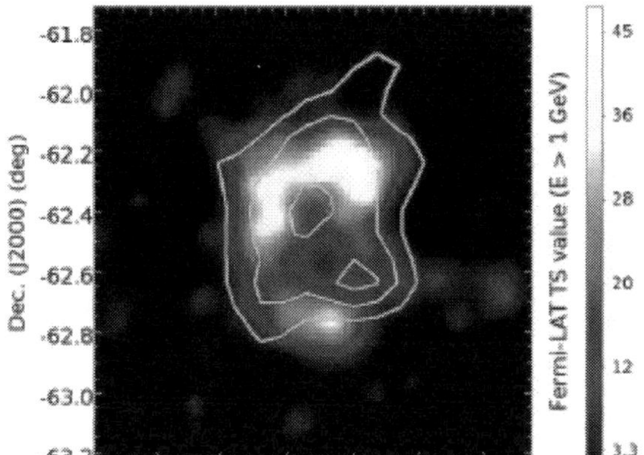

Figure 4: Test statistic maps of RCW 86 SNR seen by *Fermi*-LAT. The map is overlaid to the contours of the same source observed by H.E.S.S. experiment [11].

 DOI: 10.1051/epjconf/201713603012

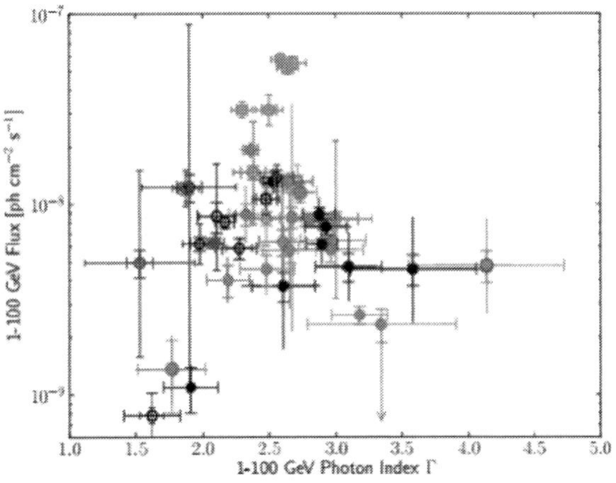

Figure 5: Scatter plot between γ-ray flux and spectral index of the SNRs detected in the first SNR catalog. Black: classified SNRs, grey: marginal candidates. Red: SNRs interacting with a molecular cloud, blue: young SNRs. Full circles: point-like sources, empty circles: extended sources [14].

86, the data have been fitted with different spatial models, simple or derived from measurements in other energy ranges [11]. The analysis revealed that the template provided by H.E.S.S., which detects γ-rays with energy above $\sim$ 1 TeV, is the one that better describes the LAT data. This supports the hypothesis that the γ-ray emission observed by the two experiments are due to the same particle population and the same emission mechanism. In this case, the whole SNR modeling favors the leptonic model, based on the IC scattering of accelerated electrons on the CMB photons, because of the very hard spectral index in the MeV-GeV energy range.

3.3 The 1st Supernova Remnant Catalog

A systematic study of an entire population of SNRs has been recently completed by the *Fermi*-LAT collaboration, producing the 1st Supernova Remnant catalog [14]. Three years of data between 1 GeV and 100 GeV have been analyzed to study the SNRs already detected in other energy ranges (274 SNRs included in the catalog of Green [15], plus five additional SNRs identified after its publication), resulting in the observation of 36 SNRs, with the detection of 14 new sources. A detailed analysis of the localization and extension of the sources has been performed, as well as of the systematic errors due to the uncertainty on the instrument response and on the diffuse background modeling, which significantly affects the estimation of the source characteristics. A multi-wavelength study has also been done when possible, searching for example for possible correlations between γ-ray and radio or X-ray spectral index. Figure 5 summarizes the results of the catalog, showing the scatter plot between the γ-ray flux and spectral index of the detected sources.

4 Conclusions

Fermi-LAT has an excellent capability to investigate the γ-ray spectrum of the astrophysical objects in the Galactic plane, such as SNRs. It is essential to understand the origin of these γ-ray emission, giving also information on non-thermal processes involving accelerated particles. Many SNRs have been detected and studied, with different characteristics, but a more systematic study is necessary to have a definite proof of CR acceleration in SNRs. For this purpose, the 1st SNR catalog has been produced and future updates with more data and the new PASS 8 performances will provide more information.

Furthermore, the improved spatial resolution introduced with PASS 8 is allowing the identification and also more precise studies of the morphology of the brightest extended sources, giving the possibility to compare them with the templates observed at other wavelengths. New results have to be expected from Fermi-LAT in the next years.

Acknowledgments

The *Fermi*-LAT Collaboration acknowledges support for LAT development, operation and data analysis from NASA and DOE (United States), CEA/Irfu and IN2P3/CNRS (France), ASI and INFN (Italy), MEXT, KEK, and JAXA (Japan), and the K.A. Wallenberg Foundation, the Swedish Research Council and the National Space Board (Sweden). Science analysis support in the operations phase from INAF (Italy) and CNES (France) is also gratefully acknowledged.

References

[1] W.B. Atwood et al., Astrophys. J. **697**, 1071 (2009), 0902.1089
[2] P. Blasi, Astron. Astrophys. Rev. **21**, 70 (2013), 1311.7346
[3] G. Morlino, D. Caprioli, Astron. Astrophys. **538**, A81 (2012), 1105.6342
[4] S. Katsuda et al., Astrophys. J. **709**, 1387 (2010)
[5] A. Giuliani et al., Astrophys. J. Lett. **742**, L30 (2011)
[6] M. Ackermann et al., Science **339**, 807 (2013), 1302.3307
[7] T. Jogler, S. Funk, Astrophys. J. **816**, 100 (2016)
[8] F. Giordano et al., Astrophys. J. Lett. **744**, L2 (2012), 1108.0265
[9] Y. Yuan et al., Astrophys. J. **779**, 117 (2013), 1310.8287
[10] A.A. Abdo et al., Astrophys. J. **734**, 28 (2011), 1103.5727
[11] M. Ajello et al., Astrophys. J. **819**, 98 (2016)
[12] F. Aharonian et al., Astron. Astrophys. **464**, 235 (2007), astro-ph/0611813
[13] A. Abramowski et al., arXiv:1601.04461 (2016), 1601.04461
[14] F. Acero et al., Astrophys. J. Suppl. Series **224**, 8 (2016), 1511.06778
[15] D.A. Green, Bulletin of the Astronomical Society of India **37**, 45 (2009), 0905.3699

EPJ Web of Conferences **136**, 03013 (2017) DOI: 10.1051/epjconf/201713603013
RICAP16

LATTES: a new gamma-ray detector concept for South America

P. Assis[1], U. Barres de Almeida[2], A. Blanco[3], R. Conceição[1,a], B. D'Ettoree Piazzoli[4], A. De Angelis[5,6,1], M. Doro[7,5], P. Fonte[3], L. Lopes[3], G. Matthiae[8], M. Pimenta[1], R. Shellard[2], and B. Tomé[1]

[1]*LIP Lisboa and IST Lisboa, Portugal*
[2]*CBPF, Rio de Janeiro, Brazil*
[3]*LIP Coimbra and University of Coimbra, Portugal*
[4]*Università di Napoli "Federico II" and INFN Roma Tor Vergata, Italy*
[5]*INFN Padova, Italy*
[6]*Università di Udine, ItalyUniversità di Padova, Italy*
[7]*INFN and Università di Roma Tor Vergata, Roma, Italy*

Abstract. Currently the detection of Very High Energy gamma-rays for astrophysics rely on the measurement of the Extensive Air Showers (EAS) either using Cherenkov detectors or EAS arrays with larger field of views but also larger energy thresholds. In this talk we present a novel hybrid detector concept for a EAS array with an improved sensitivity in the lower energies ($\sim$ 100 GeV). We discuss its main features, capabilities and present preliminary results on its expected perfomances and sensitivities.This wide field of view experiment is planned to be installed at high altitude in South America making it a complementary project to the planned Cherenkov telescope experiments and a powerful tool to trigger further observations of variable sources and to detect transients phenomena.

1 Introduction

The study of high-energy and very-high-energy gamma-rays is very important to probe extreme phenomena that takes place in the Universe. Moreover, being neutral, this radiation can pin-point to their emission source, as they are not deflect by the surrounding magnetic fields.

The detection of gamma-rays at lower energies (below $\sim$ 100 GeV) can be done using instruments placed in artificial satellites, for instance Fermi. However, as the gamma-ray energy increases, its flux at Earth becomes increasingly smaller, to the point where the available collection areas aboard satellites are not enough to study them. Fortunately, the interaction of gamma-rays with the Earth atmosphere produces Extensive Air Shower (EAS) whose secondaries can be sampled by detector arrays, or one could collect the Cherenkov light produced by the secondaries using Imaging Array Cherenkov Telescopes (IACTs). These ground based detectors attain different advantages/disadvantages with respect to each other: IACTs have a lower energy threshold, and have better angular and energy resolution, as they can image the shower development; on the other hand EAS array have significantly wider field-of-views.

[a]e-mail: ruben@lip.pt(presenter)

In figure 1 it is shown the sensitivities of current and future gamma-ray experiments with wide field-of-views. Two things become evident: there is no wide field-of-view experiment covering the Southern hemisphere sky; there is a gap in the 100 GeV region. Such wide FoV experiment with a low energy threshold and a large duty cycle would be fully complementary to the powerful narrow-FoV Cherenkov Telescope Array (CTA) as it would be able not only to issue alerts of transient phenomena but would also enable long term observations of variable sources and help on the search for emissions from extended regions, such as the Fermi bubbles or dark matter annihilations from the centre of our galaxy. Hence, we propose a novel hybrid detector to be installed at ~ 5200 m a.s.l which ensures an improved sensitivity to the 100 GeV energy region.

This manuscript is organised as follows: in section 2 we describe the detector and the layout of the experiment. In section 3 we discussed the performance of such detector and in section 4 we present the achieved sensitivities. We end with final remarks,

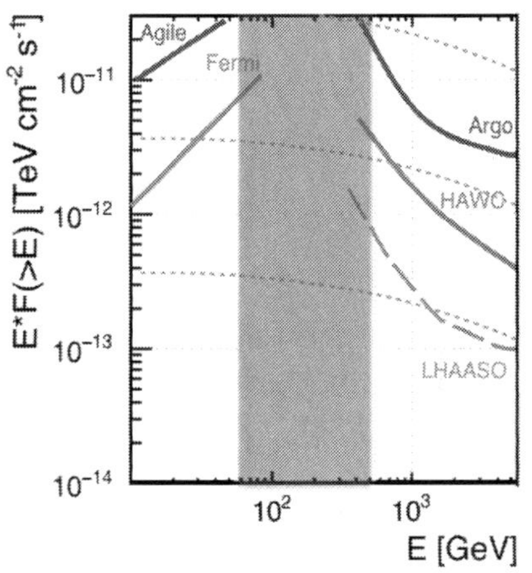

Figure 1. Integrated sensitivity of several gamma-ray experiments with a wide field-of-view as a function of the primary energy. (Taken from [1].)

2 Detector description

In order to surpass the limitations of the previous EAS arrays experiments, and be able to lower the energy threshold while maintaining a reasonable energy and angular resolution, we propose to build a dense array with an area of $10\,000\,m^2$ constituted by modular hybrid detectors. Each station is composed by two low-cost Resistive Plate Chambers (RPC) on top of a Water Cherenkov Detector (WCD), as shown in figure 2. Each RPC has 16 charge collecting pads covering a total area of $1.5 \times .1.5\,m^2$. The WCD has a rectangular structure with dimensions $3 \times 1.5 \times 0.5\,m^3$. The signals are read by two photomultipliers (PMTs) at both ends of the smallest vertical face of the WCD. On the top of the RPCs there is a thin lead plate (5.6 mm) to convert secondary photons. The conversion of the photons is important to improve the geometrical reconstruction as photons have a stronger correlation with the primary direction with respect to secondary shower electrons. The success of such hybrid detector concept lies on the fact that the RPCs contribute with its high segmentation and time resolution while the WCD provides a calorimetric measure of the shower secondary particles allowing to lower the energy threshold. Moreover, with this detector concept, it is possible to trigger in the WCD which allows the RPCs to operate at a low threshold while minimising several sources of noise (detector, electronics, environment).

Figure 2. Basic detector station with 1 WCD covered with RPCs and a thin lead plate. The green lines show the tracks of Cherenkov photons produced by charged particles in the WCD. The blue semi-spheres inside the WCD are the two PMTs.

3 Experiment performance

The performance of this detector has been assessed using an end-to-end realistic Monte Carlo simulation. The EAS have been simulated using CORSIKA (COsmic Ray Simulations for KAscade) and the detector response was treated by a Geant4 dedicated simulation.

We generated 10000 CORSIKA simulations for gammas and protons between 10 GeV and 5 TeV. To save computational times, the simulations were generated using a power law differential energy spectrum with an index −1, and afterwards were weighted for the corresponding particle fluxes. The zenith angle for gammas was fixed to 10°, while for protons the range was between 5 and 15 degrees. The detector was assumed to be placed at an altitude of 5200 m a.s.l.

To evaluate the effective area at the trigger level, we required that at least 3 stations have detected a signal. The trigger condition for the station requires at least 5 photonelectrons in each PMT. The effective area for this array, has been computed using simulations. We have found that for gamma primaries with an energy of 100 GeV we still have around 10^3 m^2, even considering selection quality cuts, which will be described below. The energy estimation has been done using the total signal (S_{tot}) recorded in all WCD stations for each event. A calibration curve was derived and used to get the reconstructed energy (E_{rec}) for each event from S_{tot}. From this curve, it is possible to evaluate the energy resolution that one could achieve with such detector as a function of E_{rec}. This is shown in figure 3 (left) where it is possible to see that the energy resolution improves as the shower energy increases and, at the lowest energies, one still has a reasonable energy resolution of 100%, being this mostly dominated by shower-to-shower fluctuations.

The geometric reconstruction was done taking advantage of the RPC segmentation and fast timing (it was used in the simulations a time resolution of 1 ns). The position and time of the recorded hits in the RPC were fitted to a shower front plane model in order to the reconstruct the primary direction. The quality of the reconstruction can be improved applying a cut on the number of active RPCs' pads in the event: it was required that the event has at least 10 hits. The reconstructed angle was compared to the simulated one, and we calculate the 68% containment angle, $\sigma_{\theta,68}$. The result is shown in figure 3 (right) where it can be seen that, at energies around 100 GeV, a reasonable resolution can be achieved, better that 2°.

4 Sensitivity

In order to compute this detector sensitivity to steady sources, one needs to know, apart from the efective area, the energy and angular resolution, the discrimination capabilities between gamma and hadrons. Although we strongly believe that this hybrid detector could combine strategies explored in previous experiments such, as HAWC [2] and ARGO[3], the complexity of such required study is out of the scope of this manuscript. Therefore, conservatively, we assumed no background rejection below 300 GeV. As in this manuscript we aim to focus on the lowest energies, above 300 GeV we took HAWC gamma/hadron capabilities as an ansatz for the highest energies [2]. This should, of course, be carefully studied in a future work.

In figure 4 it is shown the differential sensitivity of this detector to study sources. We compute the sensitivity as the flux of a source giving $N_{excess}/\sqrt{N_{bkg}} = 5$ after 1 year of effective observation. It was

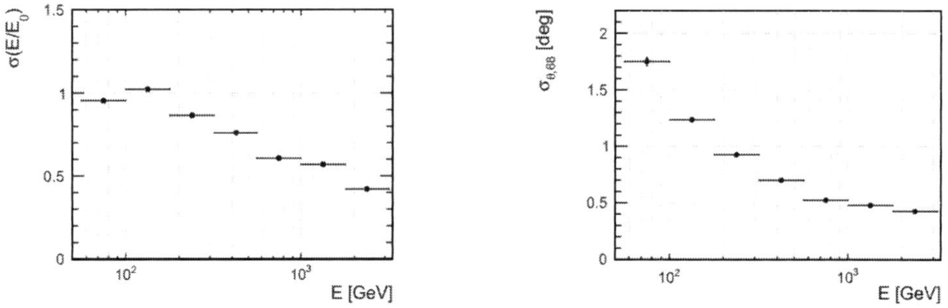

Figure 3. (Left) Reconstructed energy resolution as a function of the reconstructed energy for proton-initiated showers. (Right) Angular resolution for gamma-ray primaries with a zenith angle of 10° as a function of the reconstruction energy.

assumed that the source is visible one fourth of the time, which is roughly the time that the galactic centre is visible in the Southern tropic. The obtained results are compared with the 1 year sensitivities of FERMI and HAWC. One can clearly see that this detector would be able to cover the gap between the two of most sensitive experiments in this energy range.

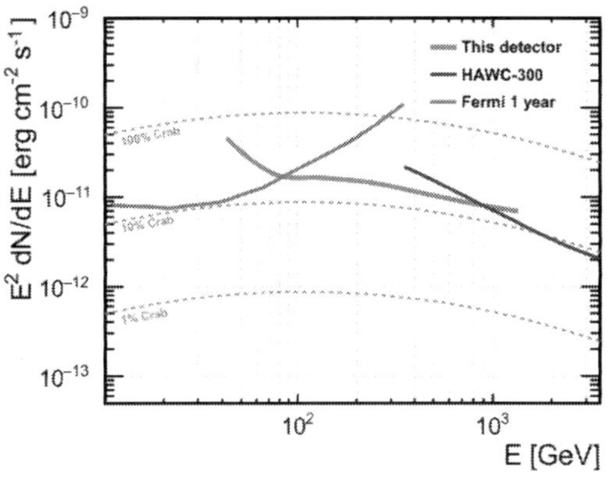

Figure 4. Differential sensitivity for steady sources after one year of effective time. See legend for details.

5 Final remarks

We have presented a novel hybrid detector able to extend of previous experiments down to the region of 100 GeV (more information can be found in [4]). This modular compact and low cost detector has given encouraging results, but its capabilities are far from explored, in particular, in what respects gamma-hadron discrimination. With the advent of the Cherenkov Telescope Array, this experiment would be a complementary project as it could provide not only triggers to transient phenomena as it could do long term observations of variable sources.

Acknowledgments R. Conceição acknowledges the financial support given by FCT-Portugal SFRH/BPD/73270/2010.

References

[1] A. De Angelis, M. Pimenta, *Introduction to particle and astroparticle physics* (Springer, 2015)
[2] A.U. Abeysekara, et al. (HAWC), Astropart. Phys. **50-52**, 26 (2013)
[3] M. Iacovacci et al. (ARGO-YBJ), Nucl.Phys.Proc.Suppl. **239-240**, 250 (2013)
[4] P. Assis et al. (2016), 1607.03051

Anomalies in the gamma-ray diffuse emission of the Galaxy and implications for the interpretation of IceCube results

D. Grasso[1,a], D. Gaggero[2], A. Marinelli[1], M. Taoso[3], A. Urbano[4], and M. Valli[5]

[1] *INFN and Dipartimento di Fisica "E. Fermi", Università di Pisa, Largo B. Pontecorvo 3, I-56127 Pisa, Italy*
[2] *GRAPPA Institute, University of Amsterdam, Science Park 904, 1090 GL Amsterdam, The Netherlands*
[3] *CERN, Theory division, CH-1211 Genève 23, Switzerland*
[4] *Instituto de Fsica Terica (IFT) UAM/CSIC, calle N. Cabrera 13-15, 28049 Cantoblanco, Madrid, Spain*
[5] *INFN and Dipartimento di Fisica, Università di Roma "la Sapienza", Largo A. Moro, Roma, Italy*

Abstract. Several independent analyzes of Fermi-LAT results found evidences of an excess of γ-ray diffuse emission along the inner Galactic plane and of a related spatial dependence of the cosmic ray (CR) proton spectral index. These features are not accounted for by conventional models of CR transport. We show that a phenomenological model accounting for those results in terms of spatial dependent CR transport also reproduces the γ-ray excess found by Milagro at 15 TeV in the inner Galactic plane and by H.E.S.S. in the Galactic center. We then use that model to compute the neutrino emission along the Galactic plane finding that is significantly larger than expected on the basis of conventional models. This emission is compatible with ANTARES upper limits and may soon be detected by IceCube or, more likely, by Km3NeT.

1 Introduction

The γ-ray diffuse emission of the Galaxy is produced by the interaction of the cosmic ray (CR) sea with the interstellar medium (ISM) gas. This emission is modeled integrating along the line of sight the product of the (energy dependent) scattering cross-section, the gas density distribution and the CR spectral density. The latter quantity is probed only locally so that its large scale distribution has to be inferred solving the CR transport equation for a given source distribution using suitable analytical or numerical tools. Until recently this was done under the hypothesis that the propagation properties are the same in the whole Galaxy. This assumption, which is not supported by any compelling theoretical argument, has been recently questioned by several independent analysis of Fermi-LAT measurements which found evidences of a dependence of the CR proton spectral index on the Galactocentric radius [1–3]. It was shown that this behaviour is at the origin of the excess of γ-ray diffuse emission found by Fermi-LAT along the inner Galactic plane [4] growing with energy above few GeV.

A phenomenological model (KRA$_\gamma$) has been proposed in [1] which accounts for those findings in terms of a radial dependent diffusion coefficient. The model also assumes that the CR spectral hardening found by PAMELA [5] and AMS-02 [6] at $\sim$ 250 GeV, as required to consistently match CREAM data [7], is a feature present in the whole Galaxy.

[a] e-mail: dario.grasso@pi.infn.it

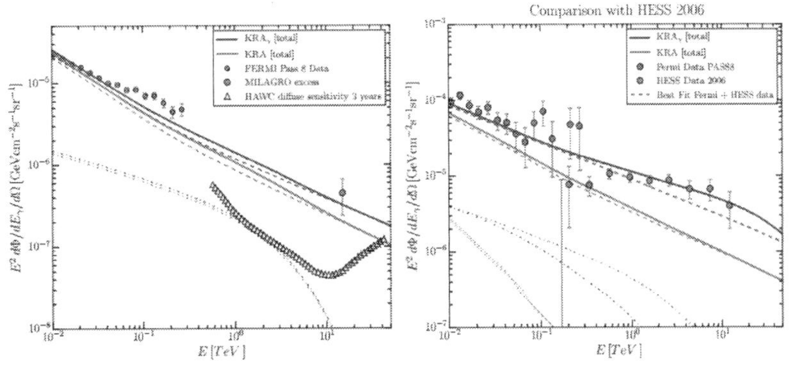

Figure 1. *Left panel*: The diffuse emission γ-ray spectrum measured by Fermi-LAT (PASS8) and Milagro in the inner Galactic plane ($|b| < 2°$, $30° < l < 65°$) is compared with the KRA (conventional) and KRA$_\gamma$ model predictions. The expected sensitivity of HAWC is reported. The main spectral components (π^0 decay: dashed; Inverse Compton: dot-dashed; bremsstrahlung: dotted) are also shown. *Right panel*: The same models are compared with H.E.S.S. and Fermi-LAT (PASS8) data for the diffuse emission in the Galactic ridge region: $|l| < 0.8°$, $|b| < 0.3°$. Point sources from the 3FGL Fermi catalogue are subtracted from those data. A single power-law fit of H.E.S.S. + Fermi-LAT data is also reported.

In Sec. 3 we will compare the KRA$_\gamma$ predictions with γ-ray data taken by Milagro [8] and H.E.S.S. [9] at energies above the TeV in the inner Galactic plane (GP) and Galactic center (GC) regions respectively. Since in the inner GP the γ-ray diffuse emission is mainly of hadronic origin, our results have direct implications for neutrino physics which will be discussed in Sec.4.

2 The model

The model proposed in [1] assumes that the exponent δ determining the rigidity dependence of the CR diffusion coefficient has the following Galactocentric radial dependence: $\delta(R) = AR + B$ where $A = 0.035$ kpc^{-1} and $B = 0.21$ so that $\delta(R_\odot) = 0.5$. The model also adopts a convective wind for $R < 6.5$ kpc with velocity $V_C(z)\hat{z}$ (z is the distance from the GP) vanishing at $z = 0$ and growing as $dV_c/dz = 100$ km s^{-1} kpc^{-1} as motivated by the X-ray ROSAT observations.

The observed γ-ray spectra at both low and mid Galactic latitudes, including the Galactic center, are reproduced by this model without spoiling local CR observables: proton, antiproton and Helium spectra, B/C and ^{10}Be/^{9}Be ratios. We implement the setup with `DRAGON`, a numerical code designed to compute the propagation of all CR species [10] in the general framework of position-dependent diffusion. We consider only proton and Helium CR nuclei since heavier species give a negligible contribution to the γ-ray emission. For their primary spectra here we assume a broken power law with index $\Gamma = 2.35/2.48$ below/above ~ 250 GeV/n and an exponential cutoff at $E_{\rm cut} = 50$ PeV such to reproduce PAMELA, CREAM and KASCADE-Grande [11] data.

3 Comparison with γ-ray data

Milagro water Cherenkov observatory measured the γ-ray flux in the sky window with $|b| < 2°$ and $30° < l < 65°$ at a median energy of 15 TeV. This was found to be 4σ above the predictions of a conventional models tuned on CR data available in 2008 [8]. Almost forgotten until recently, *the Milagro anomaly* holds however also considering updated conventional models based on Femi data.

DOI: 10.1051/epjconf/201713603014

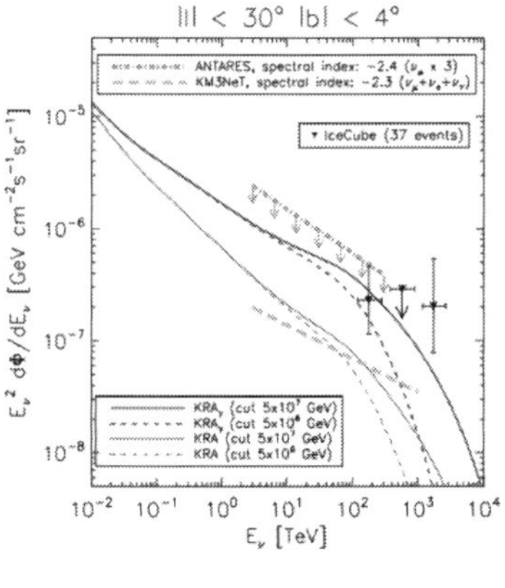

Figure 2. Total neutrino spectra in the inner Galactic plane computed for the conventional KRA and the novel KRA_γ models for two different cutoff of CR primaries. We also show the maximal flux, estimated considering 3 years of IceCube HESE events, the constraint from ANTARES experiment as well as the deduced sensitivity of the future Mediterranean observatory KM3NeT after 4 years of lifetime.

This is visible in Fig. 1 (left panel) where a reference conventional (KRA) model tuned to reproduce local CR data as well as the large scale γ-ray diffuse emission measured by Fermi-LAT falls short of Milagro data point by several sigmas. The same discrepancy holds for other conventional models. Also the KRA model used here adopts the same CR hardening at ~ 250 GeV/n as the KRA_γ one. It is evident from Fig.1 that the presence of that feature in the whole Galaxy is not sufficient to explain *the Milagro anomaly*. From the same figure the reader can see as the KRA_γ model is more successful as it matches the Milagro point and it is in better agreement with PASS8 Fermi-LAT data at lower energies.

We now check the KRA_γ model against H.E.S.S. and Fermi-LAT data in the Galactic ridge region: $|l| < 0.8°$, $|b| < 0.3°$. We show here, for the first time, as the PASS8 reconstruction algorithm allows to bridge Fermi-LAT data with H.E.S.S. so to cover the 10 GeV – 10 TeV energy interval. The combined spectrum can be fit with a single power-law with index 2.49 ± 0.3. Again, this is significantly harder than expected for conventional models (see red line in Fig. 1, right panel).

Rather, we see from the same figure as the KRA_γ model, predicting a harder CR spectrum in the GC, is in good agreement with the data. Noticeably, not only the slope but also the spectrum normalization are correctly reproduced by this model. This finding can hardly be interpreted as a coincidence and provides a further evidence in favor of the scenario proposed in [1]. We notice that respect to that work we used here the more detailed gas density distribution model in the central molecular zone described in [12].

4 Implications for Neutrino Astronomy

The hadronic component of the diffuse γ-ray emission discussed in the previous section is accompanied by a neutrino emission of similar intensity. We computed its spectrum as described in [13, 14] accounting for neutrino oscillations the effect of which is to redistribute the flux among the three flavors almost equally.

Here we consider the sky window $|l| < 30°$ and $|b| < 4°$ where the Galactic neutrino emission is expected to be dominant. For this region the ANTARES collaboration provided an upper limit on the muon neutrino flux based on the result of an unblinding analysis regarding the events collected between 2007 and 2013 in the energy range [3 ÷ 300] TeV [15].

In Fig. 2 we compare the ν_μ flux computed with the KRA (conventional model) and KRA_γ setups with that experimental constraint. We notice the large enhancement (almost a factor of 5 at 100 TeV) obtained with the KRA_γ model respect to the conventional scenario. Indeed, while – in agreement with previous results – we find that the flux corresponding to the KRA model may require long time of observation even by the KM3NeT observatory, our prediction for the KRA_γ model is instead well above the sensitivity reachable by that experiment in 4 years and it is almost within the ANTARES observation capabilities. Interestingly, our result is in good agreement with the maximal flux which we inferred from the fraction of IceCube HESE events compatible with that region. A good agreement with IceCube results was also found on the whole Galactic plane (see e.g. Fig. 1 in [16]).

On the whole sky, the diffuse Galactic emission computed with the KRA_γ model can account up to ~ 15% (to be compared to ~ 8% obtained for the conventional set-up [17]) of the flux measured by IceCube. Clearly an extra-Galactic (EG) contribution must be invoked to account for all IceCube event as well as for their almost isotropic distribution.

5 Conclusions

We showed that a Galactic CR model adopting a proper radial dependence for the diffusion coefficient so to reproduce Fermi-LAT results [1], also matches Milagro in the inner Galactic plane and H.E.S.S. measurements in the Galactic ridge. We showed that the Galactic neutrino emission computed in the same model is significantly larger than the predictions of conventional CR propagation models. Our results are in agreement with ANTARES and IceCube upper limits and will be testable by Km3NeT.

References

[1] D. Gaggero, A. Urbano, M. Valli, P. Ullio, Phys. Rev. **D91**, 083012 (2015), 1411.7623
[2] F. Acero et al. (Fermi-LAT), Astrophys. J. Suppl. **223**, 26 (2016), 1602.07246
[3] R. Yang, F. Aharonian, C. Evoli, Phys. Rev. **D93**, 123007 (2016), 1602.04710
[4] M. Ackermann et al. (Fermi-LAT), Astrophys. J. **750**, 3 (2012), 1202.4039
[5] O. Adriani et al. (PAMELA), Science **332**, 69 (2011), 1103.4055
[6] M. Aguilar et al. (AMS), Phys. Rev. Lett. **114**, 171103 (2015)
[7] H.S. Ahn et al., Astrophys. J. **714**, L89 (2010), 1004.1123
[8] A.A. Abdo et al., Astrophys. J. **688**, 1078 (2008), 0805.0417
[9] F. Aharonian et al. (H.E.S.S.), Nature **439**, 695 (2006), astro-ph/0603021
[10] C. Evoli, D. Gaggero, D. Grasso, L. Maccione, JCAP **0810**, 018 (2008)
[11] W.D. Apel et al., Adv. Space Res. **48**, 1295 (2011), 1303.7068
[12] K. Ferriere, W. Gillard, P. Jean, Astron. Astrophys. **467**, 611 (2007), astro-ph/0702532
[13] V. Cavasinni, D. Grasso, L. Maccione, Astropart. Phys. **26**, 41 (2006), astro-ph/0604004
[14] *et al.*. D. Gaggero, Astrophys. J. **815**, L25 (2015), 1504.00227
[15] S. Adrian-Martinez et al. (ANTARES), Phys. Lett. **B760**, 143 (2016), 1602.03036
[16] A. Marinelli, D. Gaggero, D. Grasso, A. Urbano, M. Valli, EPJ Web Conf. **116**, 04009 (2016), 1604.05776
[17] M. Ahlers, Y. Bai, V. Barger, R. Lu, Phys. Rev. **D93**, 013009 (2016), 1505.03156

EPJ Web of Conferences **136**, 03015 (2017) DOI: 10.1051/epjconf/201713603015
RICAP16

Gamma ray astronomy above 30 TeV and the IceCube results

Silvia Vernetto[1,a] and Paolo Lipari[2,b]

[1] *OATO-INAF and INFN Sezione di Torino, Torino, Italy.*
[2] *INFN, Sezione di Roma, Roma, Italy.*

Abstract. The study of the diffuse Galactic gamma ray emission is of fundamental importance to understand the properties of cosmic ray propagation in the Milky Way, and extending the measurements to $E \gtrsim 30$ TeV is of great interest. In the same energy range the IceCube detector has also recently observed a flux of astrophysical neutrinos, and it is important to test experimentally if the neutrino production is accompanied by a comparable emission of high energy photons. For $E \gtrsim 30$ TeV, the absorption effects due to e^+e^- pair production when the high energy photons interact with radiation fields present in space are not negligible and must be taken into account. Gamma rays, in good approximation, are completely absorbed if they have an extragalactic origin, but the absorption is significant also for Galactic photons. In this case the size and angular dependence of the absorption depends on the space distribution of the emission. In this work we estimate the absorption for different models of the space distribution of the gamma ray emission, and discuss the potential of future detectors.

1 Introduction

The observation of the diffuse Galactic gamma ray emission allows to study the space and energy distribution of cosmic rays (CR) in the Milky Way. This information is essential to determine the properties of Galactic CR propagation. The extension of the measurements of the diffuse emission to higher energy ($E_\gamma \approx 10$–10^3 TeV) is very important to shed light on this problem.

An additional motivation for the study of the diffuse emission at high energy comes from the recent results of the IceCube detector [1, 2] that has observed a signal of astrophysical neutrinos that emerges above the atmospheric foreground in the energy range $E_\nu \simeq 30$–2000 TeV. The IceCube signal is consistent with an isotropic flux, suggesting an extragalactic origin of the neutrinos, however it is also possible that a significant fraction of the events are generated in our Galaxy. At present, the low statistics and the large error boxes of most event directions prevent a definite conclusion.

The creation of high energy neutrinos in astrophysical sources is accompanied by a gamma ray emission with a similar spectral shape and intensity, since both neutrinos and gamma rays originate in the decay of pions produced in the hadronic interactions of relativistic protons and nuclei. The detection of an associate gamma ray flux could be of great importance for the identification of the neutrino sources.

[a] e-mail: vernetto@to.infn.it
[b] e-mail: paolo.lipari@roma1.infn.it

DOI: 10.1051/epjconf/201713603015

The new generation gamma ray telescopes, as LHAASO [3], HISCORE [4] and CTA [5], will have a much larger sensitivity at energies above 30 TeV with respect to the current instruments. Furthermore, the large field of view of the future air shower arrays, as LHAASO, will allow the measurement of diffuse fluxes, as those produced by cosmic ray interactions in the interstellar medium or originating by other processes inside our Galaxy. However, while neutrinos cross cosmological distances suffering a negligible absorption, gamma rays undergo pair production by interacting with the low energy radiation fields. The absorption is significant for photons travelling for intergalactic distances, making impossible to detect gamma rays of energy above 30 TeV unless the sources are very close to our Galaxy ($z < 0.01$). At these energy, however, the absorption is also not negligible for Galactic gamma rays, and modifies the spectrum with an absorption pattern that depends on the source spatial distribution.

In this paper we describe the effects of the absorption on a Galactic gamma ray flux, assuming different distributions of the sources, according to some models proposed in the literature on the possible origin of the IceCube neutrinos.

2 Attenuation of gamma rays

The gamma ray energy threshold for pair production is $E_\gamma^{th} = 0.52/[\epsilon(1 - cos\theta)]$ TeV where ϵ is the target photon energy in electronvolts. Since the maximum absorption occurs when $E_\gamma = 1.97\ E_\gamma^{th}$, the shape of the spectrum of the target photons produces well defined absorption features in the spectrum of gamma rays. To calculate the survival probability of gamma rays with a given energy, it necessary to know the number density, the spectrum and the angular distribution of the target photons in any point of the line connecting the source to the Sun.

The radiation field in our Galaxy is the sum of four components: two of extragalactic origin, permeating the Universe with a uniform and isotropic flux, and two originating in our Galaxy, highly anisotropic, with a larger flux from the direction of the Galactic center. The most intense extragalactic photon field is the Cosmic Microwave Background Radiation (CMBR), a pure blackbody radiation of temperature 2.725°K, that affects mostly gamma rays in the PeV range. The intensity of the CMBR is known with high accuracy, hence it is possible to evaluate precisely the relative absorption. A much weaker component is the Extragalactic Background Light (EBL), produced by the contribution of stars and dust of all galaxies during the history of the universe. The EBL absorption effects can be dramatic for gamma rays travelling in the extragalactic space, but are almost negligible in our Galaxy.

The most important Galactic component is the infrared light emitted by the dust heated by stars. The emission peaks around 0.01 eV and mostly affects gamma rays with energy of order 100 TeV. Since this is the energy most suitable to study the sources of neutrinos, an accurate evaluation of its intensity is necessary. The last component is the starlight, that peaks around 1 eV. The starlight mostly interacts with gamma rays of energy of order 1 TeV, but the absorption effect is almost negligible due to the small photon density. The flux of the radiation emitted by stars and dust have been measured locally, but a model is necessary to evaluate its spectrum and angular distribution in other locations in the Galaxy. Starting from a parametrization of the infrared radiation made by Misiriotis et al. [10] we developed a simplified Galactic emission model [7], whose results are in fair agreement with the available data, and with previous estimates [6]. Fig.1 shows the number density of photons for the four radiation components, according to the model. The infrared flux measured by COBE-FIRAS [8] and IRAS [9] are shown in the same figure. For the EBL spectrum, we used the parametrization by Franceschini et al. [11].

Our model allows the calculation of the absorption for any gamma ray trajectory in the Galaxy. As an example, the resulting gamma ray survival probability for three source positions in the Galactic

DOI: 10.1051/epjconf/201713603015

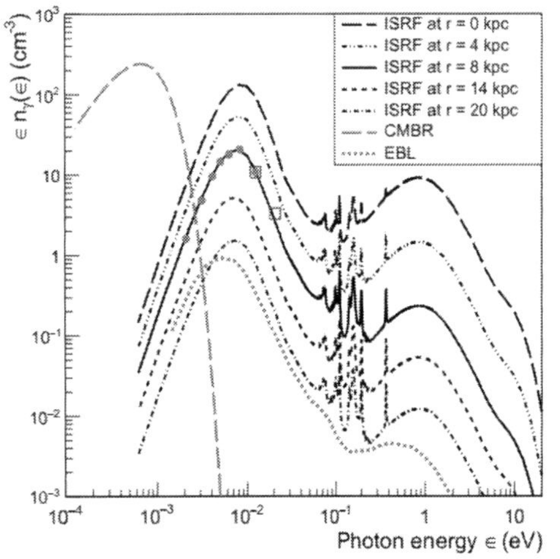

Figure 1. Photon number density (times energy) for the CMBR, the EBL and the Interstellar Radiation field (ISRF, i.e. the sum of star and dust emission). The ISRF is given for points on the Galactic plane at different distances from the Galactic center. Red dots (blue squares) indicate the COBE-FIRAS (IRAS) data.

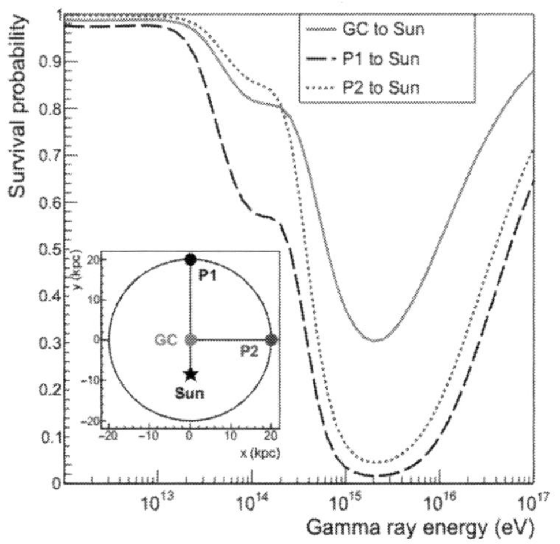

Figure 2. Survival probability of gamma rays for three different trajectories in the Galactic plane, as a function of the gamma ray energy. The inset shows a top view of the three sources positions in the Galaxy.

plane are shown in Fig.2. Up to ~20 TeV the flux attenuation is less than a few percent for every source position. Above ~20-30 TeV the absorption increases due to the interaction with the infrared radiation and reaches its maximum at ~150 TeV. Above ~200 TeV the CMBR becomes the major source of absorption, and practically only depends on the source distance. From these results one can conclude that the absorption is not a severe obstacle to measurements up to ~200-300 TeV, while at PeV energies the fluxes can be seriously affected when the source distance is larger than a few kiloparsecs.

3 Diffuse Galactic gamma ray flux

Cosmic rays confined in the Galaxy by magnetic fields generate gamma rays and neutrinos when they interact with the gas and radiation fields present in interstellar space. The gamma ray emission generates a diffuse Galactic flux that, in the region from 0.1 to 100 GeV, has been accurately measured over the entire sky by the FERMI telescope [12]. The dominant contribution to the diffuse flux is the production and decay of neutral pions in the interactions of protons and nuclei. Additional contributions are generated by Inverse Compton scattering and bremsstrahlung of CR electrons. The spectrum of the diffuse gamma rays measured by FERMI for $E_\gamma \gtrsim 10$ GeV has in good approximation a power law form with a spectral index of order $\alpha \simeq 2.65 \pm 0.05$ that reflects the shape of the spectra of CR protons and nuclei in interstellar space. The diffuse flux is is concentrated in a narrow region in Galactic latitude around the equator (approximately 50% of the emission is contained in the region $|b| \lesssim 5.6°$). The flux is also larger towards the Galactic center, with the flux from the direction of the center approximately four times larger than the flux from the anticenter.

Measurements of the diffuse Galactic gamma ray flux at TeV energies for some limited angular region of the Galactic disk have been obtained by the air shower arrays ARGO-YBJ [13] and Milagro [14, 15]. These measurements are consistent with a smooth extrapolation of the FERMI observations. No detection exist above 30 TeV, where the best upper limits have been obtained by the CASA-MIA air shower array in the energy rage 0.14–3 PeV [16]. The measurements and upper limits of the Galactic diffuse emission are shown in Fig. 3.

To extrapolate the FERMI observations to higher energy we have made the following assumptions: (i) π° decay is the dominant mechanism of the emission; and (ii) the spectral shapes of CR protons and nuclei in different points of the Galaxy are approximately equal to what is observed at the Earth. From these assumptions one can deduce that the energy and angular distribution of the diffuse Galactic flux calculated at the position of the Sun *neglecting absorption* factorize into an angle independent spectral shape, and an energy independent angular distribution (approximately equal to the one observed at $E_\gamma \simeq 10$ GeV). The predicted energy distribution of the emission reflects the spectral shape of the interacting CR and therefore follows the power law behaviour measured at $E_\gamma \simeq 10$ GeV, with a softening at 100 TeV due to the presence of the "knee" in CR spectra. The effects of the knee have been calculated numerically using a simple model of π° production in hadronic interactions and the model of the CR spectra given in [17].

To calculate the effects of gamma ray absorption one needs to estimate the space distribution of the Galactic diffuse emission rate. With the assumptions discussed above the emission rate has an energy independent form and is proportional to the product of the density of CR and of gas in the Galaxy: $q(\vec{r}) \propto n_{\rm ism}(\vec{x}) \times n_{\rm CR}(\vec{x})$. The form of this distribution can be estimated from the angular distribution of the diffuse flux when absorption is negligible. The main features of the diffuse flux angular distribution can be reasonably well reproduced with a simple axysimmetric, exponential form for the source density $q(r, z) \propto \exp(-r/r_0 - z/z_0)$ with r and z cylindrical coordinates. The parameters r_0 and z_0 can be estimated by fitting the observed angular distribution of the diffuse emission for $E_\gamma \simeq 10$ GeV with the result $r_0 \simeq 3.9$ kpc and $z_0 \simeq 0.27$ kpc.

It is now straightforward to compute the angular and energy distribution of the observable flux by summing the contributions of all points in the Galaxy, taking into account the absorption for each trajectory. The result is shown in Fig. 3, where the expected spectrum in the same galactic plane region explored by ARGO-YBJ is represented by a shaded band. The band width is determined by the uncertainties of the Fermi spectrum slope and normalization. It is interesting to note that at energies above 100 TeV the estimated spectrum is just below the CASA-MIA upper limits.

Future detectors have the potential to observe the extrapolation of the diffuse Galactic spectrum discussed above. Among the current projects, LHAASO has the highest sensitivity at ~100 TeV. A rough evaluation of its sensitivity to a diffuse flux can be obtained by multiplying the LHAASO point source sensitivity given in [18] by the factor $f_c = (\Omega_{PSF}\Omega_{GP})^{-1/2}$, where Ω_{PSF} is the solid angle of the observational window used for point sources and Ω_{GP} in the solid angle of the Galactic plane region to be studied. According to this calculation the minimum flux detectable at 5 sigma by LHAASO in one year at 100 TeV in the same region observed by ARGO-YBJ is $F_{min} \sim 7\times10^{-16}$ photons cm^{-2} s^{-1} TeV^{-1} sr^{-1}, a factor five smaller than the CASA-MIA upper limits at a median energy of 140 TeV, and below the expected flux produced by cosmic ray interactions in the same energy region.

4 The IceCube neutrinos and the Galactic gamma ray flux

The IceCube collaboration finds that the excess of neutrino events is consistent with an isotropic flux generated by the ensemble of all extragalactic sources. In this case the associated gamma ray emission is in good approximation entirely absorbed during propagation and is not observable. On

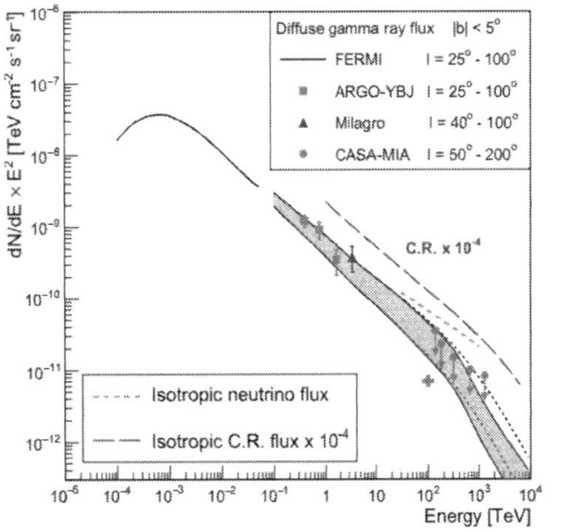

Figure 3. Galactic gamma ray diffuse flux measurements by different experiments. The shaded band shows the predicted flux in the energy range 100 GeV–1 PeV for the angular region of the ARGO-YBJ measurements. The dotted lines delimit the unabsorbed spectrum. The expected LHAASO sensitivity is indicated by a cross. The CR spectrum (rescaled by a factor 10^{-4}) and the isotropic neutrino flux are also shown.

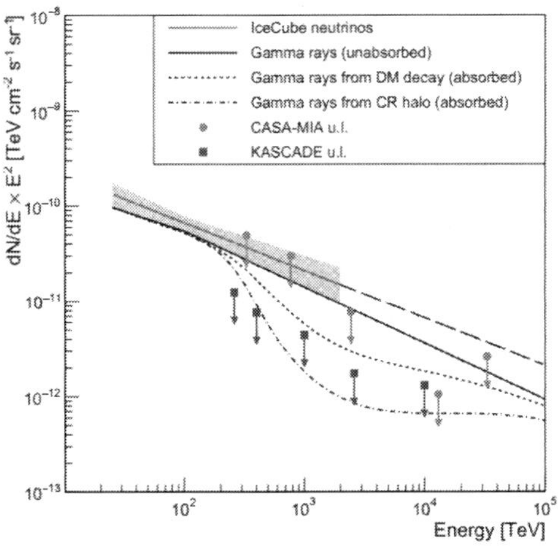

Figure 4. Effect of absorption on gamma rays emitted with a spatial distribution according to the DM model and the extended halo model. The shaded band indicates the IceCube neutrino spectrum with one sigma error. The arrows represent the 90% C.L. upper limits for the isotropic gamma ray flux measured by CASA-MIA and KASCADE.

the other hand several authors have discussed the possibility that all (or a not negligible fraction of) the astrophysical neutrinos are of Galactic origin. In this case the gamma rays associated to the neutrino emission are only partially absorbed and these Galactic models can be tested experimentally with gamma ray observations. In fact the existing limits of the diffuse gamma ray fluxes discussed above are already significant constraints. These studies require a detailed calculation of the absorption effects that can significantly reduce the observable gamma ray fluxes.

Two models that discuss an entirely Galactic origin for the IceCube neutrino signal are those of Esmaili and Serpico [19], and of Taylor, Gabici and Aharonian [20]. Esmaili and Serpico consider a model where the high energy neutrinos are generated in the decay of a very massive, unstable Dark Matter (DM) particle. In this case the space distribution of the emission corresponds to the mass density of the Galactic DM, and can be modeled with the (spherically symmetric) Navarro-Frenk-White [21] form $\rho_{DM}(r) = \rho_0/(r/r_c(1 + r/r_c)^2)$ with $r_c \simeq 20$ kpc. Taylor, Gabici and Aharonian hypothesize that neutrinos are produced by cosmic rays interacting in a large halo of gas extending at distances of order 100 kpc (in the following we will describe this halo as a simple Gaussian function $\rho(r) \propto \exp(-r^2/2r_0^2)$ with $r_0 = 57$ kpc, so that $\sqrt{\langle r^2 \rangle} = 100$ kpc). In both cases the associated gamma ray flux is not limited to the Galactic plane. Crossing regions with a lower infrared radiation, gamma rays of ~100 TeV suffer a smaller absorption than travelling inside the disk. On the other hand the average larger source distance increases the CMBR absorption in the PeV energy range.

Fig.4 shows the absorption effects (averaged over all solid angle) calculated for the two models discussed above. The unabsorbed gamma ray flux has been assumed equal to the lower limit of the neutrino spectrum. Note that in the case of the DM model the relation between the ν and γ emission depends of the branching ratios into different decay channels, and therefore on the assumed properties of the DM particle. The figure also shows the existing upper limits on the isotropic flux obtained by KASCADE [22] and CASA-MIA [23].

The KASCADE upper limits (only published on conference proceedings) are in tension with the DM model. It has to be noted however that both sets of upper limits has been obtained with observations from the northern hemisphere. Since the Sun has an offset of ~8 kpc from the Galactic center, a halo distribution will produce an anisotropic flux, higher from the direction of the Galactic center, that is not visible from the locations of KASCADE and CASA-MIA. Since the absorption too will be larger for gamma rays from the direction of the Galactic center, the anisotropy will be in some measure reduced. All these effects must be accurately taken into account to make a correct comparison between models and data.

Since the cosmic ray flux, the major background source for gamma ray observations, is about a factor 10^4 larger than the neutrino flux (see Fig.3), a background rejection better than 10^{-4} is necessary for future air shower arrays to detect the possible gamma ray flux associated to the IceCube neutrinos. Observation made from locations in both hemispheres will be advantageus, allowing the study of the global signal anisotropy, a fundamental ingredient to investigate the spatial distribution and the nature of the neutrino sources.

References

[1] M.G.Aartsen et al., PRL **113**, 101101 (2014)
[2] M.G.Aartsen et al., ApJ **809**, 98 (2015)
[3] S.W.Cui et al., Astroparticle Physics **54**, 86 (2014)
[4] M.Tluczykont et al., Astroparticle Physics **56**, 42 (2014)
[5] B.S.Acharya et al., Astroparticle Physics **43**, 3 (2013)
[6] V.Moskalenko, T.A.Porter and A.W.Straw, ApJL **640**, L155 (2006)
[7] S.Vernetto and P.Lipari, Phys.Rev.D, **94**, 063009 (2016)
[8] D.Finkbeiner et al., ApJ **524**, 867 (1999)
[9] M.A.Miville-Dechenes and G.Lagache, ApJS **157**, 302 (2005)
[10] A.Misiriotis et al., A&A **459**, 113 (2006)
[11] A.Franceschini et al., A&A **487**, 837 (2008)
[12] F.Acero et al., ApJS, **223**, 26 (2016)
[13] B.Bartoli et al., ApJ **806**, 20, (2015)
[14] R.Atkins et al., PRL **95**, 251103 (2005)
[15] A.A.Abdo et al., ApJ **688**, 1078 (2008)
[16] A.Borione et al., ApJ **493**, 175 (1998)
[17] T.K.Gaisser, T.Stanev and S.Tilav, Front.Phys.(Beijing) **8**, 748 (2013)
[18] S.Vernetto et al., Journal of Physics, Conf. Series, **718** (2016)
[19] A.Esmaili & P.D.Serpico, JCAP **10**, 014 (2015)
[20] A.M.Taylor, S.Gabici and F.Aharonian, Phys.Rev.D **89**, 103003 (2014)
[21] J.F.Navarro, C.S.Frenk and S.D.M.White, ApJ **462**, 563 (1996)
[22] G.Schatz et al., Proc. 28th ICRC (2003)
[23] M.C.Chantell et al., Phys.Rev.Lett. **79**, 1805 (1997)

Modeling the Galactic center emission from GeV to PeV

A. Marinelli[1,a], D. Gaggero[2], D. Grasso[1], M. Taoso[3], A. Urbano[4], and M. Valli[5]

[1] *INFN and Dipartimento di Fisica "E. Fermi", Università di Pisa, Largo B. Pontecorvo 3, I-56127 Pisa, Italy*
[2] *GRAPPA Institute, University of Amsterdam, Science Park 904, 1090 GL Amsterdam, The Netherlands*
[3] *CERN, Theory division, CH-1211 Genève 23, Switzerland*
[4] *Instituto de Fisica Teorica (IFT) UAM/CSIC, calle N. Cabrera 13-15, 28049 Cantoblanco, Madrid, Spain*
[5] *INFN and Dipartimento di Fisica, Università di Roma "la Sapienza", Largo A. Moro, Roma, Italy*

Abstract. The H.E.S.S. collaboration recently reported a new analysis of the gamma-ray diffuse emission from a "semi-annulus" region of 1.4×10^{-4} steradiants around Sagittarius A*. The gamma-ray spectral energy distribution measured from this region suggests the presence of interacting cosmic rays at PeV energies. This analysis adds a important piece to the previous measurements obtained with Fermi-LAT, H.E.S.S. and VERITAS telescopes in this particular region of the Central Molecular Zone. Here we describe this diffuse gamma-ray emission observed around the galactic center within a comprensive model implying a cosmic-ray population with a harder spectrum in the central part of the galaxy respect to the standard scenarios. With this phenomenological model we obtain the expected diffuse gamma-ray and neutrino components for this target rich region considering the energy range from GeV to PeV. While for the gamma-ray expectations we compare the results with the data of the mentioned experiments, for the expected neutrino spectra we consider the possibility to observe the signal with Cherenkov telescopes.

1 Introduction

In 2006 the H.E.S.S. telescope reported the measurement of diffuse gamma-ray emission from a central ridge [1] of our galaxy after the subtraction of point-like contributions. The diffuse gamma-ray spectral energy distribution (SED) obtained from the ridge cannot be explained with a standard scenario where the transport of cosmic rays (CR) doesn't change moving from the periphery of the Galaxy to the central molecular zone (CMZ). A more recent publication of H.E.S.S. collaboration reported the results of a longer campaign [2], around ~ 250 hours, focused on a "semi-annulus" region of 1.4×10^{-4} steradiants around Sagittarius A*. The SED obtained with H.E.S.S. observations, from this region around Sagittarius A*, can be fitted with a single power law up to tens of TeV without needing a cut-off and representing the first evidence of PeV cosmic-ray population in our Galaxy. In the past Sagittarius A* had periods of high activities in x-ray [3] and a outflow was also observed [4]; however his contribution to Galactic cosmic rays is still unknown. The H.E.S.S. collaboration argues that the measured spectrum from the "semi-annulus" region can be linked to the past emission [5] of Sagittarius A* [2]. However, in the H.E.S.S. analysis, the gamma-ray contribution from the interactions of cosmic-ray sea with the interstellar gas around Sagittarius A* was not taken into account. The

[a] e-mail: antoino.marinelli@pi.infn.it

expected diffuse gamma-ray and neutrino fluxes produced by these processes are strongly dependent on the considered CR transport scenario and the assumed gas distribution. Here we present the expected diffuse gamma-ray and neutrino fluxes produced in the "semi-annulus" region, considering a recently introduced scenario (KRA_γ) with a radially-dependent diffusion coefficient [6, 7]. To extend the gamma-ray analysis to lower energies we use the Fermi-LAT data extracted from the PASS8 catalog. The PASS8 gamma-ray events were recently obtained with a new selection algorithm released by the Fermi-LAT collaboration [8]. The new selection criteria increased the effective area, the energy estimation and the angular resolution of the detector, especially at high energy. After subtracting the point-like contribution from the Fermi-LAT data in the diffuse region around Sagittarius A* and adding the H.E.S.S. measurements we build a new SED lasting from 10 GeV to 40 TeV. We compare the whole spectrum with the expectations from the cited KRA_γ scenario and the standard KRA scenario where the CR propagation is homogeneous for the whole galaxy. For both cases we constrain the possible diffuse emission directly linked to Sagittarius A* activity. Finally we compute the KRA and KRA_γ diffuse neutrino expectation for the "semi-annulus" region introducing also a possible emission from RXJ 1745-290 [9].

2 Comparison between data and models

The first analysis we show in this work represents the comparison between the expected diffuse gamma-ray flux, obtained by the interaction of CR sea and the interstellar medium, and the measurements of H.E.S.S. and Fermi-LAT experiment in the defined "semi-annulus" [2] around sagittarius A*. Two different diffuse gamma-ray expectations are computed considering, the standard scenario with a homogeneous CR propagation for the whole galaxy (KRA) and the case with a radially-dependent CR diffusion properties inducing a hardening of CR spectrum toward the Galactic center regions (KRA_γ). As a target of CR interaction when producing gamma rays and neutrinos we use the 3-dimensional semi-analytic gas distribution presented in [10]. Where the molecular H_2 Hydrogen is obtained from the CO emission line with a conversion parameter $X_{CO}(r \sim 0) \simeq 0.5 \times 10^{20}\ cm^{-2}\ K^{-1}\ km^{-1}$ with a factor 2 uncertainty, while a number density of $0.11\ (n_{HI} + 2n_{H_2})$ is assumed for the atomic HI gas. The computed gamma-ray expectations are then compared with a wide energy range data, the low-energy ones extracted from Fermi-LAT PASS8 sample while the high-energy ones obtained by the new campaign published in 2016 by the H.E.S.S. collaboration, covering the window from 10 GeV to 40 TeV. It is interesting to notice that the entire studied SED can be fitted with a single power law, as shown in the left plot of fig. 1, suggesting the common origin for the gamma rays emitted in this energy range. The best fit for the two data samples considering a simple power law without cutoff $F = F_0 E^{-\alpha}$ is obtained for $F_0 = 1.9 \times 10^{-12}\ TeV cm^{-2} s^{-1}$ and $\alpha = 2.39$. On the right plot of fig. 1 it is shown also the difference between the best-fit analysis and the two computed scenarios (KRA and KRA_γ). The sum of hadronic and leptonic components of KRA_γ, represented on the right plot of fig. 1 by the blu solid line, results very close to the gamma-ray best fit over the entire SED leaving a small slot for possible diffuse component directly related to the central engine Sagittarius A*. In other words the gamma-ray component produced by the CR sea in the "semi-annulus" region account for the quasi-total measured flux, when the KRA_γ model is considered, and even in the standard CR scenario the diffuse sea component seems not negligible. The second analysis we present here is the expected neutrino emission from the "semi-annulus" region considering the two scenarios, KRA and KRA_γ for the diffuse galactic CR production. Taking into account the optimistic angular resolution of 1° for neutrino telescopes at TeV energy ranges, when reconstructing track-like events, it is opportune to add the possibility of having neutrino contribution from central point source $RXJ1745 - 290$, consistent with the Sagittarius A* position within 5″. The same studies done through gamma-ray

DOI: 10.1051/epjconf/201713603016

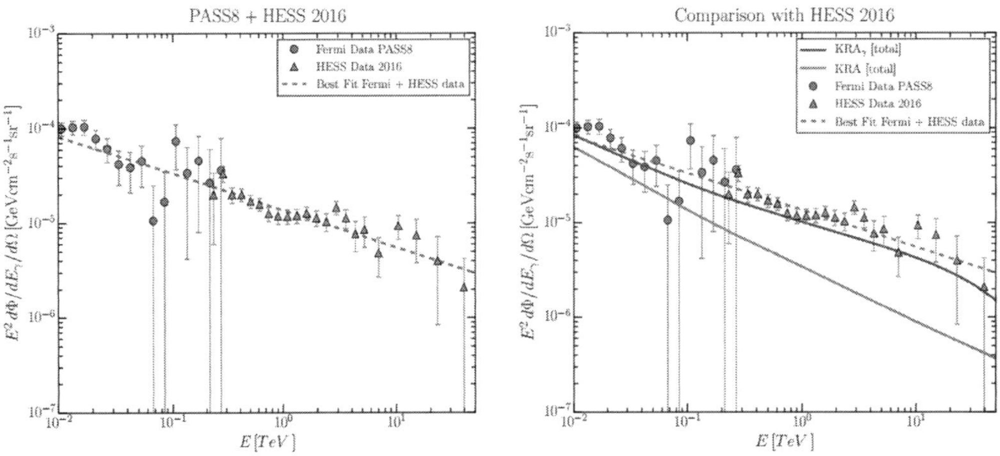

Figure 1. On left side are reported the differential flux measured by H.E.S.S. collaboration from the "semi-annulus" around Sagittarius A* and the differential flux obtained from the Fermi-LAT PASS8 sample for the same region. With the green dashed line the best-fit of the entire SED from 10 GeV to 40 TeV. On the right side are reported the same data compared with two different scenarios; the KRA_γ representing the case with a variable diffusion coefficient for the galactic CR in blu (the sum of leptonic and hadronic components is considered) and the *KRA* representing the standard case with a homogeneous CR transport for the whole Galaxy in red (also here we consider the sum of leptonic and hadronic components)

detectors for the "semi-annulus" region are not feasible with Cherenkov neutrino telescopes for both statistics of events and angular resolution. However the standard *KRA* and the new KRA_γ scenarios give us the possible neutrino expectations from this region of the CMZ. This region is not preferential for the IceCube detectors being in southern hemisphere while represents a good target for the incoming KM3NeT detector [11]. However considering the sensitivies showed in [12] more than 5 years of data can be needed to observe the total SED indicated with the black solid line in fig. 2.

3 Conclusions

In this contribution we present the study of the emission from the "semi-annulus" region around Sagittarius A* indentified by the H.E.S.S. collaboration as a possible Petaelectronvolt CR engine. We show the best-fit analysis of the gamma-ray SED in the region between 10 GeV and 40 TeV collecting togheter the PASS8 data, from Fermi-LAT, and the H.E.S.S. data after the subtraction of point-like contribution. Then we compare the best-fit analysis with the expected gamma-ray diffuse emission produced by CR sea from this region of the sky considering the KRA_γ scenario with a radial dependent CR propagation and the standard *KRA* scenario with the same CR propagation properties for the whole galactic plane. The entire gamma-ray SED considered, well fitted with a single power law without cutoffs, is compatible with the KRA_γ SED (expected from the interaction of CR sea with the gas) leaving a small room for possible diffuse emission directly linked to the central source Sagittarius A*. We show also the study of the expected neutrino emission from "semi-annulus" region considering the *KRA* and KRA_γ scenarios and introducing a possible additional contribution from RXJ 1275-290. The expected total neutrino signal can be a challenge for the incoming KM3NeT telescope while can be hardly detected by the actual IceCube detector. With all mentioned results we show that the KRA_γ

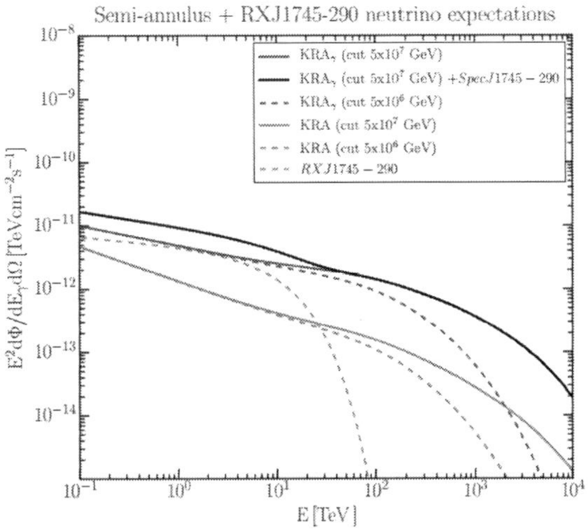

Figure 2. Neutrino expectation from the "semi-annulus" region considering also a possible additional component from RXJ 1745-290. The neutrino SED from KRA_γ scenario is reported with the blu solid line for the protons cutoff at 50 PeV and with the blu dashed line for the protons cutoff at 5 PeV. The neutrino SED from KRA scenario is reported with the red solid line for the protons cutoff at 50 PeV and with the red dashed line for the protons cutoff at 5 PeV. The grey dashed line is the possible RXJ 1745-290 contribution assuming the entire spectrum measured by H.E.S.S. [2], $\Phi_0 = (2.55 \pm 0.04_{stat} \pm 0.37_{syst}) \times 10^{-12} TeV^{-1} cm^{-2} s^{-1}$ with $\Gamma_0 = 2.14 \pm 0.02_{stat} \pm 0.10_{syst}$ and an energy cutoff of $E_{cut} = 10.7 \pm 2.0_{stat} \pm 2.1_{syst}$, of hadronic origin. The black solid line represent the sum of the KRA_γ (with protons cutoff at 50 PeV) expectations and the possible $RXJ1745-290$ neutrino contribution.

model, tuned on larger galactic scale, gives a more consistent picture, respect to the standard scenario (KRA), also for this small region of the Galactic center.

References

[1] F. Aharonian et al. (H.E.S.S.), Nature **439**, 695 (2006), astro-ph/0603021
[2] A. Abramowski et al. (H.E.S.S.), Nature **531**, 476 (2016), 1603.07730
[3] M. Clavel, R. Terrier, A. Goldwurm, M.R. Morris, G. Ponti, S. Soldi, G. Trap, A&A**558**, A32 (2013), 1307.3954
[4] M. Su, T.R. Slatyer, D.P. Finkbeiner, ApJ**724**, 1044 (2010), 1005.5480
[5] G. Bélanger, A. Goldwurm, M. Renaud, R. Terrier, F. Melia, N. Lund, J. Paul, G. Skinner, F. Yusef-Zadeh, ApJ**636**, 275 (2006), astro-ph/0508128
[6] D. Gaggero, A. Urbano, M. Valli, P. Ullio, Phys. Rev. **D91**, 083012 (2015), 1411.7623
[7] D. Gaggero, D. Grasso, A. Marinelli, A. Urbano, M. Valli, Astrophys. J. **815**, L25 (2015), 1504.00227
[8] W. Atwood, A. Albert, L. Baldini, M. Tinivella, J. Bregeon, M. Pesce-Rollins, C. Sgrò, P. Bruel, E. Charles, A. Drlica-Wagner et al., ArXiv e-prints (2013), 1303.3514
[9] F. Aharonian, A.G. Akhperjanian, K.M. Aye, A.R. Bazer-Bachi, M. Beilicke, W. Benbow, D. Berge, P. Berghaus, K. Bernlöhr, O. Bolz et al., A&A**425**, L13 (2004), astro-ph/0406658
[10] K. Ferriere, W. Gillard, P. Jean, Astron. Astrophys. **467**, 611 (2007), astro-ph/0702532

[11] D. Gaggero, D. Grasso, A. Marinelli, A. Urbano, M. Valli, ArXiv e-prints (2015), 1508.03681
[12] S. Adrián-Martínez, M. Ageron, F. Aharonian, S. Aiello, A. Albert, F. Ameli, E. Anassontzis, M. Andre, G. Androulakis, M. Anghinolfi et al., Journal of Physics G Nuclear Physics **43**, 084001 (2016), 1601.07459

The inner 300 parsecs of the Milky Way seen by H.E.S.S.: a Pevatron in the Galactic Centre

Emmanuel Moulin[1,a] for the H.E.S.S. collaboration

[1] *DRF/Irfu, Service de Physique des Particules, CEA Saclay, F-91191 Gif-Sur-Yvette Cedex, France*

Abstract. The Galactic Centre region has been observed by the High Energy Stereoscopic System (H.E.S.S.) array of ground-based Cherenkov telescopes since 2004 leading to the detection of the very-high-energy γ-ray source HESS J1745-290 spatially coincident with the supermassive black hole Sagittarius A*. Diffuse TeV gamma-ray emission has been detected along the Galactic plane, most likely due to hadronic cosmic-ray interactions with the dense gas of the Central Molecular Zone. The rich 2004-2013 dataset permits detailed spectral and morphological studies of the diffuse emission in the inner 300 pc of the Galactic Centre region. The new results provide an important statement regarding the location and origin of the accelerator of PeV protons. The H.E.S.S. observations of the Pevatron are discussed in the context of the origin of Galactic cosmic rays.

1 Introduction

The Galactic center (GC) region has been longtime observed from radio to X-ray wavelengths with increasing accuracy. Since more than a decade with the advent of the LAT instrument onboard the Fermi satellite and ground-based Cherenkov telescopes, observations of the inner GC region provide a efficient probe of the most violent phenomena in the universe. The very-high-energy (VHE, $E_\gamma > 100$ GeV) γ-ray observations with arrays of imaging atmospheric Cherenkov telescopes (IACT) provide crucial insights to study cosmic-ray (CR) particle acceleration and propagation up to PeV energies.

While totally obscured at optical wavelengths, the GC is visible in a broad-band spectrum of electromagnetic wavelengths. The infrared observations probe star formation regions where the UV/optical light from massive young stars is converted by dust to infrared light. Non-thermal radio, X-ray and γ-ray radiations probe charged particles accelerated to high energies in cosmic accelerators such a supernova remnants, pulsar wind nebulae, or massive star forming regions. Radio and X-ray emissions result from energetic electron interaction with magnetic fields through synchrotron radiation while GeV/TeV emissions are produced by bremsstrahlung and inverse Compton of electrons scattering off gas or ambient radiation fields. Another mechanism of high-energy (HE, 10 MeV - 100 GeV) and VHE γ-ray emissions proceeds through π_0 decays produced by inelastic collision of energetic protons and nuclei with the ambient gas.

During the last decade, detailed studies of the HE/VHE observations of the GC shed light to acceleration mechanisms taking place in this region. Observations with the Fermi-LAT instrument

[a] e-mail: emmanuel.moulin@cea.fr

and IACTs such as H.E.S.S., MAGIC and VERITAS, led to the detection of numerous sources in the HE/VHE regimes such as the pulsar wind nebula G0.9+0.1, supernova remnants, a central emission coincident in position with the supermassive black hole Sagittarius (Sgr) A*, as well as a diffuse GeV/TeV emission.

2 A strong TeV emission from the Galactic Center

The GC has been long time recognized as a multi-TeV particle accelerator and then been subject to IACT observations of utmost importance. A central VHE signal has been detected in 2004 by H.E.S.S. [1], hereafter referred as to HESS J1745-290, and further observations with H.E.S.S. led to the detection an energy cut-off in the 10 TeV energy range [2]. With 10 years of observations, H.E.S.S. provides the most detailed view so far at VHE (see Fig. 1). The position of the HESS

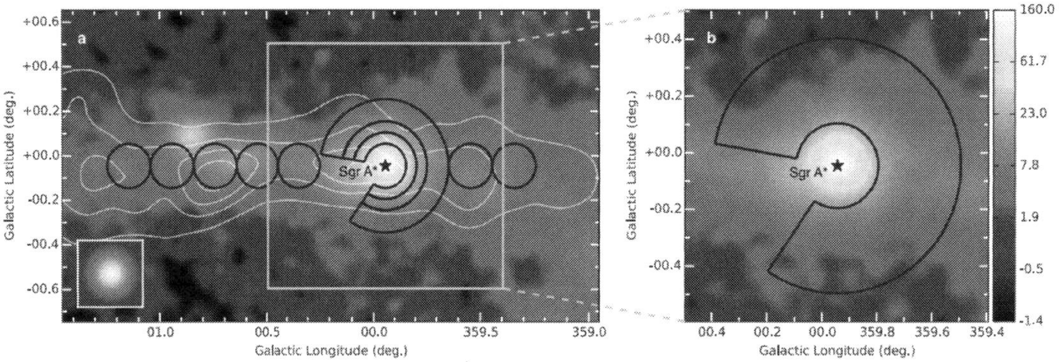

Figure 1. VHE γ-ray image of the Galactic Centre region. Left panel: The black lines outline the regions used to calculate the cosmic-ray energy density throughout the central molecular zone. A section of 66° is excluded from the annuli to avoid contamination from a newly detected source. White contour lines indicate the density distribution of molecular gas, as traced by its CS line emission. The location of Sgr A* is given by the black star. The inset shows the simulation of a point-like source. Right panel: Zoomed view of the inner 70 pc and the contour of the region used to extract the spectrum of the diffuse emission.

J1745-290 centroid is coincident with Sgr A* and the pulsar wind nebula G359.95-0.04 positions within 13″ [3]. The energy spectrum of HESS J1745-290 derived from 10 year observations with H.E.S.S. is shown in the left panel of Fig. 2. Despite the rich statistics accumulated so far towards the GC, the nature of HESS J1745-290 is still unknown and compelling production mechanisms for this VHE emission possibly include: *(i)* a CR source accelerating high energy protons in the vicinity of Sgr A* which produces VHE γ-rays from π_0 decays originating from hadronic CR penetrating the interstellar medium gas; *(ii)* the pulsar wind nebula G359.95-0.04; and *(iii)* a spike of annihilating dark matter particles. With the detection of the 2FHL J1745.7-2900 and 3FGL J1745.6-2859c sources with Fermi-LAT, the GeV emission from the pulsar wind nebula model significantly underestimates by far the luminosity of these sources, given the pronounced peak-like structure exhibited by the inverse Compton emission. Assuming the GeV source 3FGL J1745.6-2859c and the TeV source HESS J1745-290 originate from the same production mechanism, any emission model should at least meet the following requirements: *(i)* a power-law spectrum from 100 MeV to 20 TeV with an energy cut-off at ~10 TeV, *(ii)* no hint for variability on timescale from minutes to years, *(iii)* the emission region of the TeV source is point-like and coincident with the position of Sgr A*, and its intrinsic size is less than 1′; and *(iv)* a possible small extension of the GeV source.

 DOI: 10.1051/epjconf/201713603017

3 Diffusion emission at TeV energies

Follow-up VHE observations provided a more detailed view of the overall Central Molecular Zone (CMZ) and led to the detection of a diffuse emission in the central 300 pc of the GC [4] extended along the Galactic plane. The spatial correlation of the TeV emission with the giant molecular clouds of the CMZ first hinted for acceleration of hadronic CRs in this region [4], where the γ-rays result from decays of π_0 produced by the interactions of relativistic protons with the ambient gas.

The large photon statistics accumulated over 10 year observations together with improvements in the data analysis method, provide now a refined picture of the spectral and spatial properties of the diffuse emission of the CMZ [5]. As shown in Fig. 1, the sky map in VHE γ-rays shows a strong, although not linear, correlation between the brightness distribution of γ-rays and the locations of massive gas-rich complexes. The γ-ray morphology of the CMZ is both determined by the location and the particle injection rate history of the CR accelerator(s) responsible for the relativistic protons, and by the gas density distribution. The spatial morphology seen now in VHE γ-rays is a unique proxy to probe the CR distribution within the CMZ.

If the γ-ray emission is completely due to the decay of π_0 produced in pp interactions, the γ-ray luminosity L_γ above energy E is related to the total energy of CR protons W_p by $L_\gamma(>E) \simeq \eta_N W_p(>10\,E)/t_{pp\rightarrow\pi_0}$, where $t_{pp\rightarrow\pi_0} = 1.6 \times 10^8\,\mathrm{yr}\,(1\mathrm{cm}^{-3}/n_\mathrm{H})$ is the proton energy loss timescale due to π_0 production in an environment of hydrogen gas of density n_H, and $\eta_\mathrm{N} \simeq 1.5$ accounts for the presence of nuclei heavier than hydrogen in both CRs and interstellar matter. With the measurements of the

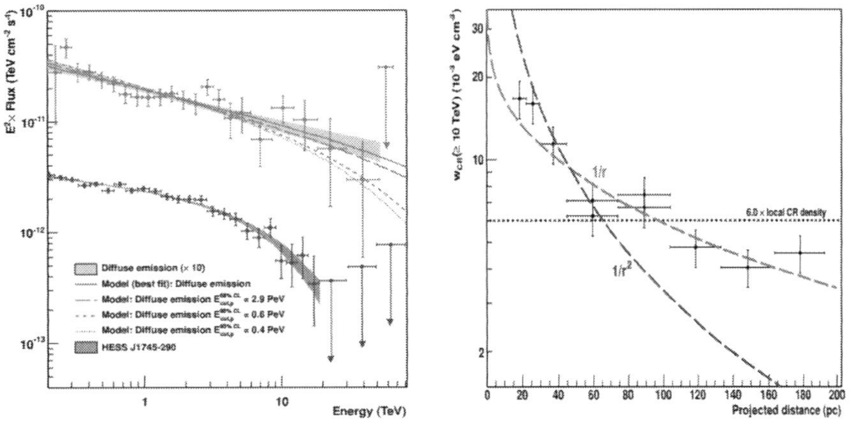

Figure 2. Left panel: VHE γ-ray spectra of the diffuse emission and HESS J1745-290. Arrows represent 2σ flux upper limits. The 1σ confidence bands of the best-fit spectra of the diffuse and HESS J1745-290 are shown in red and blue shaded areas, respectively. The red lines show the numerical computations assuming that γ-rays result from the decay of neutral pions produced by pp interactions. Right panel: Spatial distribution of the CR energy density versus projected distance from Sgr A*. The vertical and horizontal error bars show the 1σ statistical plus systematic errors and the bin size, respectively. Fits to the data of a $1/r$ (red line), a $1/r^2$ (blue line) and a homogeneous (black line) CR density radial profile integrated along the line of sight are shown. The best fit of a $1/r^\alpha$ profile to the data is found for $\alpha = 1.10 \pm 0.12\ (1\sigma)$.

target masses which can be inferred using tracer molecules such as CS, $^{12}C^{16}O$ and HCN [5], the CR energy density w_CR averaged along the line of sight, can be obtained. The right panel of Fig. 2 shows the radial profile of w_CR for E>10 TeV CRs up to a radius of 200 pc for a GC distance of 8.5 kpc, determined from the γ-ray luminosity and the amount of target gas. This high energy density in the CMZ is found to be an order of magnitude larger than that of the CR "sea" that universally fills the Galaxy, while the energy density of low energy (GeV) CRs in this region has a level comparable to it [5]. This requires the presence of one or more accelerators of multi-TeV particles operating in the CMZ. The measurement of the CR density profile clearly supports the $1/r$ dependence over the entire CMZ. The $1/r^2$ and constant profiles, the former being expected if CRs are advected in a wind, and the latter in the case of a single burst-like event of CR injection, are significantly disfavored.

The CR radial profile points towards an accelerator located in the inner 10 pc of the GC. The $1/r$ behavior of the CR density up to 200 pc indicates a quasi-continuous injection of protons that diffuse into the CMZ from a centrally-located accelerator within the 10 pc of Sgr A*. The average injection rate of particles is found to be $Q_p(> 10\,\mathrm{TeV}) \simeq 4 \times 10^{37}(D/10^{30}\mathrm{cm}^2\mathrm{s}^{-1})$ erg s^{-1}. The diffusion coefficient D depends on the power spectrum of the turbulent magnetic field, which is unknown in the CMZ region. Yet, the diffusive nature of the propagation is constrained by the condition $R^2/6D \ll R/c$. For a radius R of the CMZ region of 200 pc, this implies $Q_p \ll 1.2 \times 10^{38}$erg s^{-1}. The supermassive black hole Sgr A* at the GC is the most plausible supplier of ultra-relativistic protons and nuclei where these particles could have been accelerated either in the accretion flow (*i.e.* in the immediate vicinity of the black hole) or somewhat further away - for example, at the site of termination of an outflow [5].

4 The discovery of a Pevatron

Given the available statistics, the energy spectrum of the diffuse VHE γ-ray emission has been extracted from an annulus centred at Sgr A* (see right panel of Fig. 1). The best fit to the data is found for a spectrum following a power law extending with a photon index of ~2.3 to energies up to tens of TeV without an energy cut-off or break, as shown in the left panel of Fig. 2. This is the first time that such a γ-ray spectrum, arising from hadronic interactions, has been detected. Since these γ-rays result from decays of π_0 produced by pp interactions, the derivation of such a hard power-law spectrum implies that the spectrum of the parent protons should extend to energies close to 1 PeV. Assuming a cut-off in the parent proton spectrum, the corresponding secondary γ-ray spectrum deviates from the H.E.S.S. data at 68%, 90% and 95% confidence levels for cut-offs at 2.9 PeV, 0.6 PeV and 0.4 PeV, respectively. This establishes the first robust detection of a VHE cosmic hadronic accelerator which operates as a source of PeV particles, *i.e.* a Pevatron.

If Sgr A* is indeed the particles' source, the required acceleration rate would exceed by two or three orders of magnitude its current bolometric luminosity, and would constitute at least 1% of the current power produced by accretion onto the supermassive black hole. Given the relatively-modest current accretion rate of Sgr A* and that at certain epochs this supermassive black hole could have operated at a much higher accretion rate, this higher rate could also have facilitated greater CR production rates. An average acceleration rate of 10^{39} erg s^{-1} of E>10 TeV protons over the last 10^6-10^7 years would be sufficient to explain the flux of CRs around the "knee", an energy spectrum feature around 1 PeV. If so, Sgr A* should be considered as a viable alternative to supernova remnants as a source of PeV Galactic cosmic rays.

References

[1] F. Aharonian et al. (H.E.S.S. Collaboration), Astron. Astrophys. **425**, L13 (2004)
[2] F. Aharonian et al. (H.E.S.S. Collaboration), Astron. Astrophys. **503**, 817 (2009)
[3] F. Acero et al. (H.E.S.S. Collaboration), Mon. Not. Roy. Ast. Soc. **402**, 1877 (2010)
[4] F. Aharonian et al. (H.E.S.S. Collaboration), Nature **439**, 695 (2006)
[5] A. Abramowski et al. (H.E.S.S. Collaboration), Nature **531**, 476 (2016)

Limits on Lorentz invariance violation at the Planck energy scale from H.E.S.S. spectral analysis of the blazar Mrk 501

Matthias Lorentz[1,a] and Pierre Brun[1,b], for the H.E.S.S. collaboration

[1] *Irfu, CEA Saclay, 91191 Gif-sur-Yvette France*

Abstract. Some extensions to the Standard Model lead to the introduction of Lorentz symmetry breaking terms, expected to induce deviations from Lorentz symmetry around the Planck scale. A parameterization of effects due to Lorentz invariance violation (LIV) can be introduced by adding an effective term to the photon dispersion relation. This affects the kinematics of electron-positron pair creation by TeV γ rays on the extragalactic background light (EBL) and translates into modifications of the standard EBL opacity for the TeV photon spectra of extragalactic sources. Exclusion limits are presented, obtained with the spectral analysis of H.E.S.S. observations taken on the blazar Mrk 501 during the exceptional 2014 flare. The energy spectrum, extending very significantly above 10 TeV, allows to place strong limits on LIV in the photon sector at the level of the Planck energy scale for linear perturbations in the photon dispersion relation, and provides the strongest constraints presently for the case of quadratic perturbations.

1 Introduction

Special relativity is a pillar of modern physics and Lorentz symmetry has been established to be an exact symmetry of Nature up to the precision of current experiments. It has been suggested however that this symmetry could only be approximate and that deviations from Lorentz invariance could appear at an energy scale beyond our current grasp. A generic approach to introduce such effects consists of adding effective terms in the dispersion relation of particles, *i.e.* for photons

$$E_\gamma^2 = p_\gamma^2 \pm E_\gamma^2 \left(\frac{E_\gamma}{E_{\rm LIV}} \right)^n , \tag{1}$$

where $E_{\rm LIV}$ is the hypothetical energy scale at which Lorentz symmetry could stop being exact, and n the order of the leading correction. In some approaches to quantum gravity $E_{\rm LIV}$ is expected to be of the order of Planck energy $E_{\rm Planck} = \sqrt{\hbar c^5/G} \simeq 1.22 \times 10^{28}$ eV [3]. As such deviations are only expected for photons at the highest energies, astrophysical γ-rays can be used to probe potential LIV effects. The most widely-used approach is to look for energy-dependent time delays for photons produced by distant γ-ray bursts (GRB) or during TeV flares of active galaxy nuclei (AGN) , see *e.g.* [21]. An attractive alternative possibility takes advantage of the fact that the modified dispersion relation for photons that could be induced by LIV would affect the kinematics for the e^+e^- pair production of

[a] e-mail: matthias.lorentz@cea.fr
[b] e-mail: pierre.brun@cea.fr

DOI: 10.1051/epjconf/201713603018

TeV γ rays coming from AGNs on the EBL resulting in a modified opacity to extragalactic γ rays, see *e.g.* [11]. In the following we consider LIV affecting only photons (like in [12, 13]), not electrons as the constraints on LIV for electrons are very stringent due to observations of synchrotron radiation from the Crab Nebula [18].

2 Modified EBL opacity in the presence of LIV

The EBL is the background photon field originating from the integrated starlight and its re-processing by the interstellar medium over cosmic history. Its spectral energy distribution has two main components, an optical ($\sim$ 1 eV) and an infrared ($\sim 10^{-2}$ eV) component. Extragalactic very high-energy (VHE, E > 100 GeV) γ rays can be used as an independent way to probe this background radiation, as such γ rays interact with EBL photons via e^+e^- pair production [1], resulting in an attenuated observed flux (for a review see [2]). The optical depth for a VHE photon of energy E_γ traveling through a medium with EBL physical density $n(\epsilon, z)$ from a source at z_s is:

$$\tau(E_\gamma, z_s) = c \int_0^{z_s} dz \frac{dt}{dz} \int_0^2 d\mu \frac{\mu}{2} \int_{\epsilon_{thr}}^{\infty} d\epsilon \frac{dn(\epsilon, z)}{d\epsilon} \sigma_{\gamma\gamma} \left(E_\gamma(1+z), \epsilon, \mu\right), \tag{2}$$

where $dt/dz = \left(H_0(1+z)\sqrt{\Omega_M(1+z)^3 + \Omega_\Lambda}\right)^{-1}$ [1], $\mu = 1 - \cos(\theta)$, $\epsilon_{thr}(E_\gamma, z) = \frac{2m_e^2c^4}{E_\gamma\mu(1+z)}$ and $\sigma_{\gamma\gamma}$ is the Bethe-Heitler cross section for pair production. The absorption effect on the intrinsic spectrum of an extragalactic source is expressed as $\Phi_{\rm obs}(E_\gamma) = \Phi_{\rm int}(E_\gamma)\, e^{-\tau(E_\gamma, z_s)}$. EBL absorption then leaves a typical redshift and energy-dependent imprint on the observed spectrum of extragalactic sources. Knowledge of the EBL spectral energy distribution has greatly improved over the last decade, constraints from VHE γ rays (see *e.g.* [6–8]), predictions from models (see *e.g.* [4, 5]), and results from an empirical determination [9] agree in between lower and upper limits.
The effective dispersion relation in the presence of LIV Eq.1 propagates into the optical depth given in Eq.2, the invariant center-of-mass energy squared s and threshold energy $\epsilon_{\rm thr}$ become :

$$s \rightarrow s \pm \frac{E_\gamma^{n+2}}{E_{\rm LIV}^n}, \text{ and } \epsilon_{\rm thr} \rightarrow \epsilon_{\rm thr} \mp \frac{1}{4}\frac{E_\gamma^{n+1}}{E_{\rm LIV}^n} \tag{3}$$

We assume, as in [12], that the modified expression of s can still be considered as an invariant quantity in the LIV framework (for a discussion see Appendix A. in [13]). We only consider the subluminal case (minus sign in Eq. 1) : if non negligible, the effective term will induce lower values for s suppressing pair creation on the EBL, causing an excess of transparency for γ rays [2], see Fig. 1.

3 H.E.S.S. observations of Mrk 501 during the 2014 flare

3.1 H.E.S.S. experiment

The High Energy Stereoscopic System (H.E.S.S.) is an array of five imaging atmospheric Cherenkov telescopes located 1800 m above see level in the Khomas Highland, Namibia, detecting γ-rays ranging from $\sim$ 100 GeV to a few tens of TeV.

[1] We assume a flat ΛCDM cosmology with $\Omega_M = 0.3$, $\Omega_\Lambda = 0.7$ and $H_0 = 70$ km s^{-1} Mpc^{-1}.

[2] Alternatively, in the superluminal case the threshold energy would be lowered implying an enhanced pair production for γ rays that could more easily interact with the cosmic microwave background. This would result in a strong cut-off in the observed energy spectra of AGNs. This LIV scenario is unlikely with respect to current observations and also theoretically disfavored.

3.2 Mrk 501

Mrk 501 is a well known AGN at redshift $z = 0.034$ which belongs to the class of blazars, *i.e.* with its relativistic jet closely aligned to our line of sight. It is known to be highly variable from radio to VHE γ rays and is referred to as a high-frequency-peaked blazar with a flux-dependent spectral hardening observed during flaring states. Its spectral characteristics and its relatively low redshift allow for the detection of among the most energetic extragalactic γ rays, making this source ideal to investigate LIV through spectral studies as it has already been done [7, 10, 13] with the historically highest VHE flux recorded in 1997 by the HEGRA [17] and CAT [16] telescopes.

3.3 Flare data set

The 2014 H.E.S.S. observations of Mrk 501 were triggered following high fluxes reported by the FACT collaboration. Observations taken during the night of June 23-24 2014 (MJD 56831-56832) revealed an exceptional flare with highest fluxes of Mrk 501 ever recorded by H.E.S.S. [14]. These observations were performed with full array of all five telescopes, however for this study requiring optimal sensitivity at highest energies, data from the central large telescope are not essential. The mean zenith angle of observations was $\sim 63°$. The Model analysis with loose cuts [19] was performed leading to an excess of more than 1200 photons with a $\sim 67\sigma$ significance for the 2 hours of observations taken that night. Spectral analysis was performed using the forward folding technique described in [20]. The spectrum, extending significantly up to $\sim$ 20 TeV, is well fitted (χ^2/n.d.f = 8.5/8) by a simple EBL-absorbed power law using the EBL model of [4]. There is no evidence for intrinsic curvature nor cut-off. The fitted intrinsic index is 2.15 ± 0.06 [15].

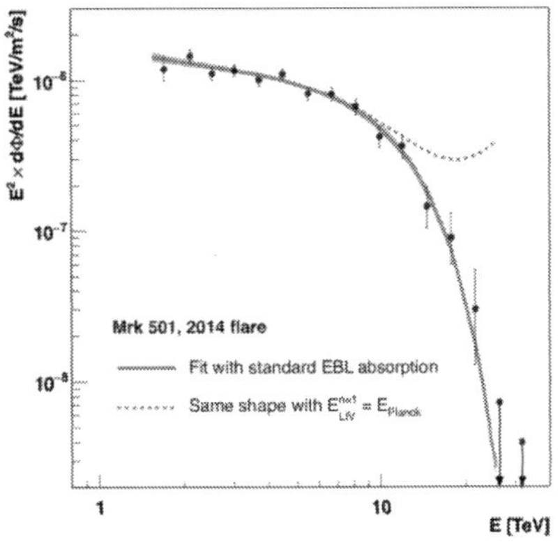

Figure 1. Energy spectrum of Mrk 501 obtained from the H.E.S.S. phase-I analysis of the 2014 flare data. The fitted EBL-absorbed power law for the standard case is showed by the solid line, as well with the corresponding $1-\sigma$ confidence band. For comparison the same intrinsic power law with modified EBL absorption due to linear Planck scale perturbations is represented by the dashed line.

4 Results and discussion

The maximum likelihood forward folding method for spectrum determination is performed assuming an intrinsic power law absorbed with the EBL model of [4]. The optical depths are computed considering modifications due to LIV as explained in Sec. 2. Values of $E_{\rm LIV}$ are scanned logarithmically in the range of interest for linear (n=1) and quadratic (n=2) scenarios. As the data show no evidence for a high-energy upturn, the fit prefers LIV-free optical depth values. Indeed log-likelihood values reach plateaus corresponding to the standard case with no deviations from Lorentz symmetry in both cases. This allows to compute exclusion limits on E_{LIV}, as summarized in Tab. 1.

Table 1. Exclusion limits obtained on E_{LIV} at different confidence levels.

	$2\,\sigma$	$3\,\sigma$	$5\,\sigma$
n=1	2.8×10^{28} eV ($2.29 \times \mathrm{E_{Planck}}$)	1.9×10^{28} eV ($1.6 \times \mathrm{E_{Planck}}$)	1.04×10^{28} eV ($0.86 \times \mathrm{E_{Planck}}$)
n=2	7.5×10^{20} eV	6.4×10^{20} eV	4.7×10^{20} eV

These strong constraints naturally come from the exceptional spectrum of the 2014 flare data-set where the power law intrinsic emission extends up to 20 TeV. A cross-check analysis using independent calibration and reconstruction chains lead to a compatible spectrum. The same LIV analysis using the EBL model of [5] leads to very similar exclusion limits.

5 Conclusions

The non observation of deviations from standard EBL absorption in the multi-TeV spectrum of Mrk 501 observed by H.E.S.S. during the 2014 flare allows us to derive strong limits on E_{LIV} in the photon sector, currently the best limits obtained with an AGN. This confirms the result obtained with GRB 090510 [22] that standard photon dispersion relation holds up to the Planck energy scale in the case of linear perturbations, and pushes higher the current limit in the case of quadratic perturbations.

Acknowledgments

Please see standard acknowledgements in H.E.S.S. papers, not reproduced here due to lack of space.

References

[1] Gould, R. J., & Schréder, G. P. 1967, Physical Review, 155, 1404
[2] Dwek, E., & Krennrich, F. 2013, Astroparticle Physics, 43, 112
[3] G. Amelino-Camelia, Living Rev. Rel. **16** (2013) 5 [arXiv:0806.0339 [gr-qc]].
[4] Franceschini, A., Rodighiero, G., & Vaccari, M. 2008, A&A, 487, 837
[5] Dominguez, A., Primack, J. R., Rosario, D. J., et al. 2011, MNRAS, 410, 2556
[6] H.E.S.S. Collaboration, Abramowski, A., Acero, F., et al. 2013, A&A, 550, A4
[7] J. Biteau and D. A. Williams, Astrophys. J. **812** (2015) no.1, 60 [arXiv:1502.04166 [astro-ph.CO]].
[8] M. Lorentz *et al.* [HESS Collaboration], arXiv:1509.03477 [astro-ph.HE].
[9] F. W. Stecker, S. T. Scully and M. A. Malkan, arXiv:1605.01382 [astro-ph.HE].
[10] F. W. Stecker and S. L. Glashow, Astropart. Phys. **16** (2001) 97 [astro-ph/0102226].
[11] U. Jacob and T. Piran, Phys. Rev. D **78** (2008) 124010 [arXiv:0810.1318 [astro-ph]].
[12] M. Fairbairn, A. Nilsson, J. Ellis, J. Hinton and R. White, JCAP **1406** (2014) 005 [arXiv:1401.8178 [astro-ph.HE]].
[13] Tavecchio, F., & Bonnoli, G. 2016, A&A, 585, A25
[14] N. Chakraborty *et al.* [HESS Collaboration], arXiv:1509.04893 [astro-ph.HE].
[15] G. Cologna *et al.* [HESS Collaboration], arXiv:1509.04458 [astro-ph.HE].
[16] D A. Djannati-Atai *et al.*, Astron. Astrophys. **350** (1999) 17 [astro-ph/9906060].
[17] F. Aharonian [HEGRA Collaboration], Astron. Astrophys. **349** (1999) 11 [astro-ph/9903386].
[18] T. Jacobson, S. Liberati and D. Mattingly, Nature **424** (2003) 1019 [astro-ph/0212190].
[19] de Naurois, M., & Rolland, L. 2009, Astroparticle Physics, 32, 231
[20] F. Piron *et al.*, Astron. Astrophys. **374** (2001) 895 [astro-ph/0106196].
[21] A. Abramowski *et al.* [HESS Collaboration], Astropart. Phys. **34** (2011) 738 [arXiv:1101.3650 [astro-ph.HE]].
[22] V. Vasileiou *et al.*, Phys. Rev. D **87** (2013) no.12, 122001 [arXiv:1305.3463 [astro-ph.HE]].

Prospects for Gamma-Ray Burst detection by the Cherenkov Telescope Array

E. Bissaldi[1,2,a], T. Di Girolamo[3,4], F. Di Pierro[5], T. Gasparetto[6], F. Longo[6,7], P. Vallania[5,8], and C. Vigorito[8,9] for the CTA Consortium[10]

[1] *Politecnico di Bari, Bari, Italy*
[2] *INFN – Sezione di Bari, Bari, Italy*
[3] *Università degli Studi di Napoli "Federico II", Napoli, Italy*
[4] *INFN – Sezione di Napoli, Napoli, Italy*
[5] *INAF – OAT, Torino, Italy*
[6] *Università degli Studi di Trieste, Trieste, Italy*
[7] *INFN – Sezione di Trieste-Udine, Trieste, Italy*
[8] *INFN – Sezione di Torino, Torino, Italy*
[9] *Università degli Studi di Torino, Torino, Italy*
[10] *See* www.cta-observatory.org *for full author & affiliation list*

Abstract. The Large Area Telescope (LAT) on the Fermi satellite is expected to publish a catalogue with more than 100 Gamma-Ray Bursts (GRBs) detected above 100 MeV thanks to a new detection algorithm and a new event reconstruction. This work aims at revising the prospects for GRB alerts with the Cherenkov Telescope Array (CTA) based on the new LAT results. We start considering the simulation of the observations with the full CTA of two extremely bright events, the long GRB 130427A and the short GRB 090510, then we investigate how these GRBs would be observed by a particular configuration of the array with the telescopes pointing to different directions in what is called the "coupled divergent mode".

1 The CTA observatory

The Cherenkov Telescope Array (CTA) is a worldwide project aiming at building and operating the next generation of Imaging Atmospheric Cherenkov Telescopes (IACTs). In its baseline design, two huge arrays are foreseen, one in the Southern (Armazones, Chile) and one in the Northern (La Palma, Spain) hemisphere. In its current design, more than 100 telescopes will extend the energy range of currently operating IACTs especially at higher energies and will improve the sensitivity by about one order of magnitude with better angular and energy resolutions. In order to achieve these goals, a combination of different types of telescopes is necessary. Large Size Telescopes (LSTs), with a diameter of 23 m, will observe the energy region between ~20 and ~200 GeV, having a compact placement in both Northern and Southern sites. Near the low-energy threshold, the number of source photons is relatively high but the Cherenkov image is poor, therefore a few huge telescopes are used to collect

[a] *speaker*, e-mail: elisabetta.bissaldi@ba.infn.it

faint showers. Medium Size Telescopes (MSTs), with a diameter of 12 m, will be sensitive in the energy range between ~100 GeV and ~10 TeV, arranged on about a km^2 area. At these energies, the Cherenkov signal starts to increase while the source flux rapidly fades, so an intermediate choice for number and dimensions of the telescopes is optimal. Small Size Telescopes (SSTs), with a diameter of 4 m, will operate in the energy range between ~5 and ~300 TeV, and will cover a 3–4 km^2 area. The current CTA baseline array layout consists of 4 LSTs and 25 MSTs in the Southern Site and 4 LSTs and 15 MSTs in the Northern Site. 70 SSTs will be built only in the Southern site, where most of the Galactic plane is visible.

2 High-energy GRB observations

Gamma-Ray Bursts (GRBs) represent a very interesting case study in astrophysics, mainly due to their multi-disciplinary nature. At present time, a GRB can trigger one or more of the dedicated instruments based on several satellites orbiting around the Earth, such as Swift, Fermi, MAXI or INTEGRAL. The observed keV-MeV prompt emission may be accompanied by an X-ray, optical or radio afterglow. Rapid follow-up of the prompt keV-MeV emission is possible thanks to communication through the Gamma-ray Coordinates Network (GCN), where the GRB position is spread out in real time to all other observatories. This includes all currently operative IACTs like MAGIC, H.E.S.S., and VERITAS. Unfortunately, none of them ever succeeded in capturing a high-energy signal from a GRB, but several upper limits from a single or from a sample of bursts were published by each collaboration over the last years.

Current prospects for very high-energy GRB observations by CTA [1, 2] are based on extrapolations taken either from the GRB spectral parameters published in the catalogs of the Burst and Transient Source Experiment (BATSE, 20 keV–2 MeV) and of the Swift Burst Alert Telescope (BAT, 15–150 keV), or from some very energetic GRBs detected by the Fermi instruments before 2012.

Our aim is to expand these samples with the newest results obtained by both instruments on-board Fermi, namely the Gamma-Ray Burst Monitor (GBM, 8 keV–40 MeV) and the Large Area Telescope (LAT, ~30 MeV–300 GeV). We are currently preparing a library of GRBs observed at different post-trigger epochs, which includes ~1000 GRBs from the second GBM spectral catalog [3], 35 GRBs from the first LAT GRB catalog [4], and ~130 new GRBs from the second LAT GRB catalog (in preparation, [5]). A subsample of these GRBs also comes with redshift information.

For the highest energy bursts, we extrapolate the flux to CTA energies by assuming that the TeV flux follows the time evolution of the GeV flux. Here we present the analysis results for two particular bursts detected by LAT, namely the long GRB 130427A and the short GRB 090510.

3 Simulation of GRB observations

To estimate the detectability of a burst, we make use of the `ctools`, a software package specifically developed for the scientific analysis of CTA data (v0.9.0)[1].

Our first test case is GRB 130427A, which represents the current record holder of the highest energy photon ever observed from a GRB (95 GeV). Moreover, it was detected at a low redshift of 0.34 and therefore followed–up by a very large number of telescopes. Observations by VERITAS began almost 20 hours post trigger and led to an upper limit (UL) of 3.3×10^{-12} erg $cm^{-2}s^{-1}$. This UL was used to cross-check our extrapolation to the CTA energies of the spectrum measured by LAT, which is described by a spectral index γ = -2.2 [6], almost constant from 400 s up to 70 ks post trigger,

[1] http://cta.irap.omp.eu/ctools/index.html

DOI: 10.1051/epjconf/201713603019

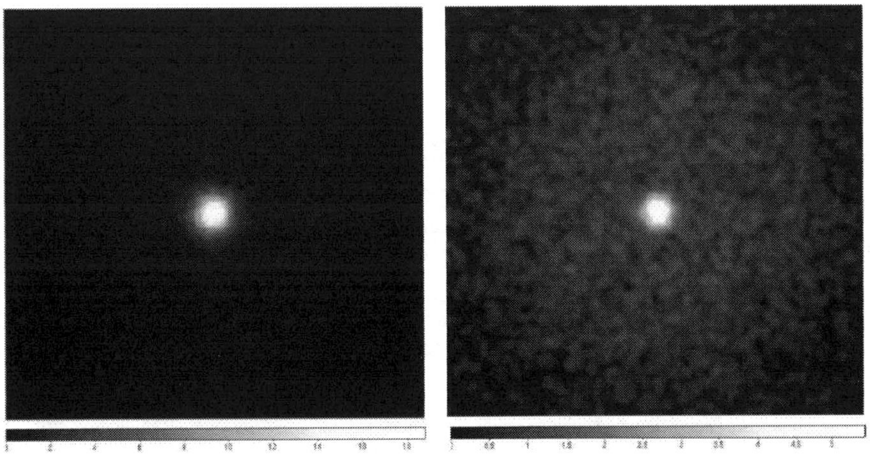

Figure 1. Simulations of GRB 130427A with `ctools`. The camera images are divided into 200×200 bins of 0.02°. The color scale gives the counts/bin after Gaussian smoothing. Left panel: Count map as seen by CTA North in 10 min starting from 1 ks after the trigger. Right panel: Count map as seen by CTA North in 30 min starting from 10 ks after the trigger.

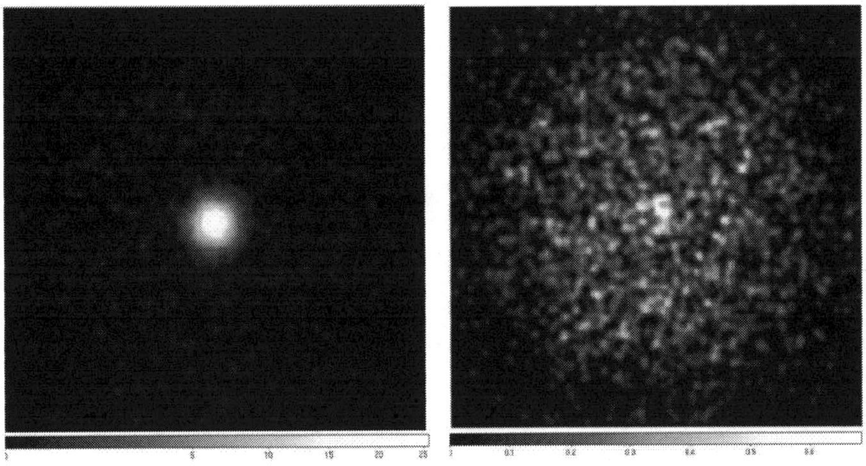

Figure 2. Simulations of GRB 090510 with `ctools`. Binning and color code as in Fig. 1. Left panel: Count map as seen by CTA North in 100 s starting from 100 s after the trigger. Right panel: Count map as seen by CTA North in 500 s starting from 1 ks after the trigger.

and by a power-law decay with a temporal index $\tau = -1.35$ (valid for t > 380 s). For each GRB, an intrinsic spectrum of $(dN/dE) \propto E^{\gamma}$ is assumed.

The second test case is the short burst GRB 090510, also detected by LAT and located at a relatively low redshift of 0.9. In this case, the highest-energy photon detected is a 31 GeV one. The spectral parameters adopted in the simulation (taken from [7]) are $\gamma = -1.6$ for $t < 200$ s and $\gamma = -2.5$ at later times, while the temporal parameter (taken from [8]) is $\tau = -1.38$.

We assume that the bursts are observed on axis with respect to the CTA array and at a zenith angle $\theta = 20°$, a value provided in the latest CTA large-scale Monte Carlo simulations (*Prod-2*) [9].

In each case, we simulated two possible observations by CTA, namely (A) a small time-window at early post-trigger times, covering the end of the GRB prompt emission period, and (B) a large time-window at late post-trigger times, covering the GRB afterglow emission period. For GRB 130427A, we simulated observation (A) lasting 10 min at t = 1 ks post trigger and observation (B) lasting half an hour at t = 10 ks post trigger. For GRB 090510, observation (A) starts at t = 100 s post trigger and lasts 100 s while observation (B) starts at t = 1 ks post trigger and lasts 500 s. In the case of GRB 130427A, we also included the effect of the EBL absorption, using the EBL model of [10]. For simplicity, in our simulations we extended the spectrum only up to 1 TeV, the maximum energy after which the source is assumed to be totally absorbed. Therefore, the energy range used in our GRB 130427A simulations is 50 GeV – 1 TeV. In the case of GRB 090510 we neglected the effect of EBL absorption and limited the energy range to 50–100 GeV, since the burst occured at a higher redshift than GRB 130427A.

The count maps of the simulated observations of GRB 130427A and GRB 090510 are displayed in Figures 1 and 2, respectively. In each case, left panels refer to observation (A) and right panels to observation (B). Count maps were obtained using the `ctools` functions `ctobssim` and `ctbin` and adopting the official CTA instrument response functions (IRFs)[2]. We made use of the *North_0.5h* IRF for observation (A) and of the *North_5h* IRF for observation (B). The decision of using a more conservative IRF for longer observations was driven by the need of reducing the increased background contribution.
A preliminary `ctlike` analysis was performed on all simulated observations, getting a significant detection in cases (A) and (B) of GRB 130427A and in case (A) of GRB 090510. Further analysis of other bright Fermi GRBs with known redshift is in progress.

In the future, we plan to use different pointing configurations of the array and to study the performance of the "coupled divergent mode" for the CTA observatory, in which pairs of telescopes are pointing to slightly different positions on the sky. Moreover, we intend to explore deeper the contribution of the MSTs in GRB searches.

Acknowledgments

We gratefully acknowledge support from the agencies and organizations listed under Funding Agencies at this website: http://www.cta-observatory.org/. E.B. is supported by the Italian *Fondo di Sviluppo e Coesione 2007-2013 - APQ Ricerca Regione Puglia "Future In Research"*.

References

[1] S. Inoue et al., Astroparticle Physics **43** 252 (2012)
[2] R. Gilmore et al., Experimental Astronomy **35** 413 (2013)
[3] D. Gruber et al., The Astrophysical Journal Supplement Series **211** 12 (2014)
[4] M. Ackermann et al., The Astrophysical Journal Supplement Series **209** 11 (2013)
[5] G. Vianello et al., AAS Meeting Proceedings **227** id. 416.01 (2016)
[6] M. Ackermann et al., Science **343** 42 (2014)
[7] M. Ackermann et al., The Astrophysical Journal **716** 1178 (2010)
[8] M. De Pasquale et al., The Astrophysical Journal Letters **709** 146 (2010)
[9] T. Hassan et al., Proceedings of Science (ICRC2015) **971** (2015)
[10] A. Franceschini et al., Astronomy & Astrophysics **487** 837 (2008)

[2]https://portal.cta-observatory.org/Pages/CTA-Performance.aspx

The Fermi GBM and LAT follow-up of GW150914

E. Bissaldi[1,2,a], V. Connaughton[3], N. Omodei[4], E. Burns[5], A. Goldstein[6], and G. Vianello[4]

[1] *Politecnico di Bari (Italy)*
[2] *INFN – Sezione di Bari (Italy)*
[3] *Universities Space Research Association, AL (USA)*
[4] *W. W. Hansen Experimental Physics Laboratory, Kavli Institute for Particle Astrophysics and Cosmology, Department of Physics and SLAC National Accelerator Laboratory, Stanford University, CA (USA)*
[5] *Physics Department, University of Alabama in Huntsville, AL (USA)*
[6] *NASA/Marshall Space Flight Center, AL (USA)*

Abstract. As the first detection of Gravitation Wave (GW) event arising from the coalescence of two stellar-mass Black Holes (BH) was announced by LIGO, a new era for astronomy began. Searches for electromagnetic (EM) counterparts of GW events is of fundamental importance, as they increase the confidence in the GW detection and help characterize the parameters of the merger. The *Fermi* gamma-ray space telescope has the best sensitivity to simultaneously observe a large fraction of the sky from 10 keV to more than 300 GeV, providing the unique capability of rapidly covering the entire probability region from a LIGO candidate.

Here we present observations by the *Fermi* Gamma-Ray Burst Monitor (GBM) [1] and by the Large Area Telescope (LAT) [2] of the LIGO Gravitational Wave event GW150914, which has been associated to the merger of two stellar-mass BHs. We report the presence of a weak transient event in GBM data, close in time to the LIGO one. We discuss the characteristics of this GBM transient, which are consistent with a weak short GRB arriving at a large angle to the direction in which *Fermi* was pointing. Furthermore, we report LAT upper limits (ULs) for GW150914, and we present the strategy for follow-up observations of GW events with the LAT.

1 GBM observations of GW150914

On September 16th, 2015 the LIGO and Virgo collaborations reported that a candidate event had been identified in data recorded on September 14th [3]. The candidate was subsequently characterized as being consistent with a signal from the merger of a stellar-mass BH binary system and was denominated GW150914 [4]. Although there are no predictions or well established mechanisms for detectable EM emission from these kind of mergers to guide a search for counterparts in the GBM data, we carried out a methodical search around the time and sky location of GW150914.

GBM has an instantaneous sky coverage of about 70%, with the remainder blocked by the Earth. On September 14th 2015, GBM was recording data continuously from nearly 2 hr before to over 7 hr after the GW event, without interruptions due to passages in the South Atlantic Anomaly (SAA).

[a] e-mail: elisabetta.bissaldi@ba.infn.it

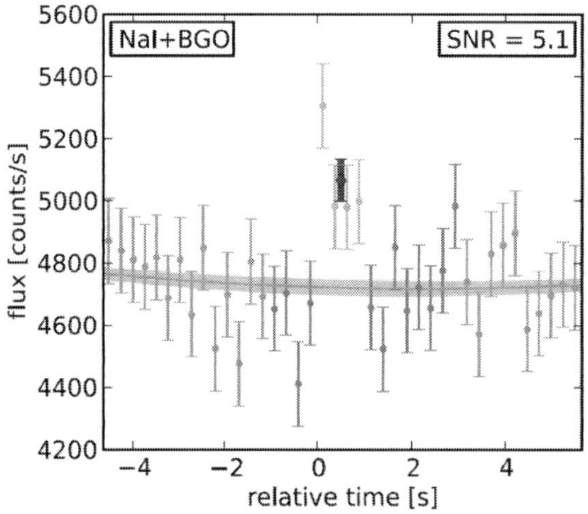

Figure 1. Model-dependent count rates detected as a function of time relative to the start of GW150914-GBM. The raw count rates are weighted and summed to maximize the signal to noise for a modeled source. Green data points are used in the background fit, gold points are the counts in the time period that shows significant emission, gray points are outside this time period, and the blue point shows an average over the gold points. For a single spectrum and sky location, detector counts for each energy channel are weighted according to the modeled rate and inverse noise variance due to background. The weighted counts from all NaI and BGO detectors are then summed to obtain a signal-to-noise optimized light curve for that model. A likelihood is assigned to each model by the targeted search based on the foreground counts (in the region of time spanned by the gold points), and this is used to marginalize the light curve over the unknown source location and spectrum.

GBM did not record any on board trigger around the time of the GW detection, at 09:50:45.391 UT. The triggers closest in time were from two events on the same day which are consistent with particle precipitation in or near the spacecraft, at 04:09:23 UT on entering the SAA and at 14:21:34 UT when *Fermi* was at high geomagnetic latitude, nearly 6 hr before and 4.5 hr after the GW event, respectively.

There are two offline search pipelines that check for impulsive events too weak to trigger on board *Fermi*, or from a sky position unfavorable to the two-detector on board triggering requirement. The first undirected search is based on analysis of the GBM Continuous Time-tagged Event (CTTE) data over four energy bands and 10 timescales, and it reported no candidates above the detection threshold on the day of the GW event. The second directed search is a targeted one [5]. By scanning both the GW and GBM data it aims to strengthen the significance of a sub-threshold signal in one data set by the detection of a signal in the other, provided that the false positive rate of the joint search is characterized and the detection levels in both instruments are selected accordingly. The directed search of the GBM data is seeded with the time and (optionally) the sky location of any LIGO/Virgo candidate event. A coherent search over all GBM detectors (NaI and BGO) using the full instrument response at each sky position is performed over a user-specified time window, assuming one of three template source spectra, revealing short-duration candidates typically between 0.256 to 8 s in duration. The candidates are ranked by a Bayesian likelihood statistic.

We searched 30 s of GBM data before and after the LIGO coalescence time for a plausible counterpart and found two candidate events. The first transient, occurring 11 s after GW150914, lasted 2 s, was visible only below 50 keV and its position was computed to be near the galactic center, well separated from and incompatible with the LIGO localization region. The second, hard transient occurred

0.4 s after GW150914, lasted 1 s and was reported in [6] as "GW150914-GBM". Figure 1 shows the model-dependent light curve of GW150914-GBM, where the detector data have been summed using weights that maximize the signal to noise for a given source model, and the unknown source model itself is weighted according to its likelihood in the data. Using 220 ks of GBM data from September 2015, we calculated the FAR for GW150914-GBM to be 1.6×10^{-4} Hz. After considering both the frequency of occurrence and the proximity to the GW trigger time, we finally derived a post-trial false-alarm probability (FAP) of 0.0022 (2.9σ). In the following section we explore whether the GBM data for GW150914-GBM suggest an astrophysical origin and, if so, whether the source is consistent with GW150914 or can be attributed to other causes.

1.1 Characteristics of GW150914-GBM

Localization - GBM is capable of localizing an event from an arrival direction beneath the spacecraft, from which nearly equal count rates are expected in most of the NaI detectors. The rough localization of GW150914-GBM spans a region covering 3000 square degrees (68% confidence level) and places the source direction underneath the spacecraft, at an angle of 163° to the spacecraft pointing direction. This explains the unusual detector pattern of GW150914-GBM, with all of the individual detector count rates being slightly above background, simultaneously.

Spectrum - The data for GW150914-GBM imply a weak but significant hard X-ray source with a spectrum that extends into the MeV range. The count spectrum from the NaI detectors (summed) is consistent with the count spectrum from the BGO detectors (summed), indicating a reasonable physical spectrum that peaks in the BGO energy range. In order to compute the GBM instrument response functions we sample a range of 11 arrival directions along the observed LIGO location arc, using the data and associated responses for the detectors at each location that are most favorably oriented to the arrival direction. We find that for all considered directions, a power- law fit to the data from GW150914-GBM can be constrained. Weighting the sampling along the arc according to the LIGO localization probability contained near each point on the arc, we obtain a best-fit power-law index $-1.40^{+0.18}_{-0.24}$ and amplitude $0.002^{+0.002}_{-0.001}$ photons s^{-1} cm^{-2} keV^{-1} yielding a fluence between 10 and 1000 keV of $2.4^{+1.7}_{-1.0} \times 10^{-7}$ erg cm^{-2}. These are typical values for weak, short GRBs. If GW150914-GBM is part of the short GRB population, then its fluence is not atypical but its unfortunate arrival direction yields only a weak signal in GBM. At the time of the GW event, *Fermi* was at low geomagnetic latitude and was not near the SAA. While we cannot exclude a magnetospheric origin for GW150914-GBM, the observing conditions were not conducive to such an event, nor is the light curve typical of magnetospheric activity, which is usually manifested as longer and smoother (tens of seconds) bumps above background. Moreover, using various search techniques, we found (i) no evidence for long-term steady emission from the direction of GW150914-GBM, (ii) no evidence for contamination by known sources of hard X-ray emission of any search for emission related to GW150914-GBM, and (iii) no evidence for non-impulsive emission related to the GW event in the days surrounding the event. A search for longer-term emission in the months before and after the GW event also revealed no signal.

2 LAT observations of GW150914

Fermi was operating in normal survey mode at the time of the LIGO trigger. Hence, the LAT autonomously observed the entire LIGO localization region within ~70 min of the GW trigger, independently of any notification from LIGO, in the high-energy gamma-ray band. The LAT Collaboration reported a preliminary search throughout the LIGO localization area that did not reveal any new gamma-ray sources [7]. Here we describe LAT observations of the localization area of GW150914 around the time of the trigger. We performed two complementary sets of searches for transient high-energy gamma-ray emission: automated searches that are performed routinely on all LAT data and targeted searches in the LIGO localization region on short and long time baselines that exploit the full sensitivity of the standard LAT analysis chain.

Automated searches to the counterpart for GW150914 were based on three main LAT pipelines: (i) The Burst Advocate (BA) Tool and the LAT Transient Factory (LTF); (ii) The Automated Science Processing (ASP); and (iii) the *Fermi* All-sky Variability Analysis (FAVA). None of the LAT pipelines found a possible counterpart to GW150914. During the 6 hr interval containing the LIGO trigger GW150914, ASP detected ($> 3\sigma$) 12 known gamma-ray sources and 3 low-significance ($> 1\sigma$) unidentified transients, none of which were consistent with the LIGO event localization.

For what concerns targeted searches, we checked different time windows by carrying out two customized analyses of the LAT data, which are based on the standard maximum likelihood analysis technique used for LAT data. In all of our searches we included in the likelihood model all sources (point-like and extended) from the LAT source catalog "3FGL" [8] as well as the Galactic and isotropic diffuse templates provided by the *Fermi*-LAT Collaboration[1]. We used the Pass 8 `P8_TRANSIENTR010E_V6` event class and the corresponding instrument response functions.

2.1 Short-baseline search

This search focuses on the hours immediately after the GW trigger t_{GW} and it is the most likely to find a counterpart to GW150914 if it is similar to a short GRB (sGRB). We consider a point in the sky observable by LAT if it is within the 65° radius FoV and has an angle with respect to the local zenith smaller than 100°. The LAT coverage represents the integral of the probability densities of all points in the LIGO localization probability map observable by LAT at a given time. It was between 50% and 90% in the hours before the trigger, while at t_{GW} the LAT was unfortunately viewing the opposite side of the sky from the LIGO localization region. Moreover, the coverage was zero until $t_{GW} \sim 4200$. The

[1] http://fermi.gsfc.nasa.gov/ssc/data/access/lat/BackgroundModels.html

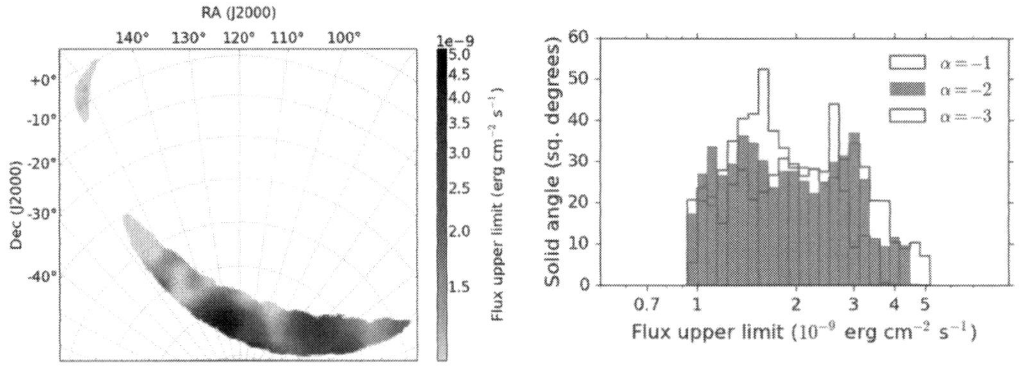

Figure 2. Flux upper limits (95% c.l.) in the energy range 100 MeV–1 GeV for GW150914 during the interval T_1 (4442–4867 s from t_{GW}). *Left panel*: the upper limits map covering the 90% region of the LIGO probability map. *Right panel*: the histogram of the upper limits in the map. We assumed a power-law spectrum for the source, with a photon index α of -2 (typical of afterglows of GRBs; green histogram), -1 (blue histogram), and -3 (red histogram). While the distributions are slightly different for the three cases, the ranges spanned by the upper limits are largely independent of the photon index.

time interval $t_{GW} + 4442 - 4867$ s (T_1) had a coverage > 90%, while during $t_{GW} + 4867 - 10,000$ s the coverage varied between 50% and 98%, decreasing back to zero at around t_{GW} + 10 ks. We searched for a transient source in the time interval having more than 90% coverage (T_1) and we did not find any significant excess.

We then derived upper limits for the gamma-ray flux of GW150914. Because the sensitivity of the LAT depends strongly on the angle from the source to the boresight of the instrument, the continuous variation of the LAT viewing direction in survey mode makes any flux limit for a particular source position time-dependent. Flux limits are also sensitive to astrophysical backgrounds, particularly in the Galactic plane, so that positions along the LIGO arc will have different flux limits even for the same observing conditions. These effects mean that flux limits vary according to both the time of observation and the position in the localization region. We show a map of the derived upper limits (95% c.l.) for the gamma-ray flux of GW150914 in the band 100 MeV–1 GeV in the left-hand panel of figure 2 and a histogram of the upper limits in the right-hand panel, both for interval T_1. Assuming a power-law spectrum for the source with a photon index of $\alpha = -2$, which is typical for GRB afterglows at LAT energies, the upper limits we find have a median of 1.7×10^{-9} erg cm^{-2}s^{-1}, and 5% and 95% percentiles of 0.9×10^{-9} and 3.7×10^{-9} erg cm^{-2}s^{-1}, respectively. These upper limits are only weakly dependent on the choice of α as shown in the right-hand panel of figure 2.

2.2 Long-baseline search

In this second search we considered data gathered during a two-month interval centered on t_{GW}. To increase the number of gamma-rays, we included all photons with energies between 60 MeV and 100 GeV. Because the PSF at 60 MeV is broad, we applied a zenith cut of 95° to further limit Earth limb contamination. We looked both for a long-duration signal of the order of one day as well as for a short-duration signal, but not necessarily in strict temporal coincidence with the LIGO trigger. To this end, we covered the entire 90% probability region provided by LIGO with a set of nine partially overlapping ROIs, each with a radius of 10°. For the first analysis of the second search, we divided the data in 10 ks time bins. For each time bin and for each ROI, we calculated a TS map and determined the location of the grid position with the maximum TS (TS_{max}). We considered the position of TS_{max} as the location of a possible counterpart and we ran an unbinned likelihood analysis adding a point source at the position of TS_{max}. This gave a value of TS_{src} (which is normally similar to TS_{max}). In these maps derived from low-statistics data, single high-energy gamma-rays can cause a high value of TS_{max}. To reduce the number of false positives from random coincidences, we required that the number of photons N_γ that have a probability larger than 0.9 to be associated with the candidate counterpart to be greater than 2. No excesses met this requirement. We repeated the same analysis considering time bins of one day and again did not find any significant excess. We also considered the possibility of excesses over shorter timescales (< 1 hr), similar to the typical duration of high-energy emission from GRBs but not in temporal coincidence with the GW trigger. We calculated the entry and exit times for each ROI in the FoV of the LAT (a "FoV passage"), requiring that the distance between the LAT boresight and the center of the ROI be <60°. In standard survey mode the duration of a FoV passage varies from a few hundred seconds to nearly one hour. Because we do not know if an EM signal would be in temporal coincidence with the GW signal, we searched for possible excesses in every passage, corresponding to a total of 6615 passages for each ROI. We did not detect any significant excess in any of the passages before or after t_{GW} for any ROI.

To validate our interpretations of TS values we performed a detailed Monte Carlo simulation of two months of data (the same interval used in our analysis). The actual pointing history of the satellite was used; therefore, the correct exposure of the sky was automatically taken into account. All of the sources from the 3FGL catalog were kept fixed at their 3FGL catalog fluxes. As a result, the

DOI: 10.1051/epjconf/201713603020

simulation is suitable for computing the distribution of TS under the null hypothesis that no transient signal is present. With the simulated data we repeated exactly the same analysis used on real data previously described. The Monte Carlo distributions proved to be a good match to the distributions of the TS_{src} values obtained from the flight data and the good absolute agreement is consistent with no statistically significant transient counterpart being present in the flight data. Also, given the large number of trials, relatively high values of TS can be obtained in Monte Carlo simulations even if no transient signal was added.

3 Conclusions

Future joint observations of GW events by LIGO/Virgo and *Fermi*-GBM could reveal whether the weak transient GW150914-GBM is a plausible counterpart to GW150914 or just a chance coincidence. As advanced LIGO begins operations, we eagerly anticipate the detection of gravitational waves in coincidence with a gamma-ray signal from GBM and LAT, likely from a short GRB arising from the merger of two neutron stars. Offline searches for weak GRBs that fail to trigger on-board *Fermi* indicate that additional short GRBs can be detected in the GBM data and dedicated analysis of LAT data can results in sub-threshold detections that can greatly improve our knowledge of the source of GW events and affect follow-up strategies for counterpart searches by other observers.

Acknowledgements

The *Fermi*-LAT Collaboration acknowledges support for LAT development, operation and data analysis from NASA and DOE (United States), CEA/Irfu and IN2P3/CNRS (France), ASI and INFN (Italy), MEXT, KEK, and JAXA (Japan), and the K. A. Wallenberg Foundation, the Swedish Research Council and the National Space Board (Sweden). Science analysis support in the operations phase from INAF (Italy) and CNES (France) is also gratefully acknowledged. E.B. is supported by the Italian *Fondo di Sviluppo e Coesione 2007-2013 - APQ Ricerca Regione Puglia "Future In Research"*.

References

[1] V. Connaughton et al., The Astrophysical Journal Letters **826** L6 (2016)
[2] M. Ackermann et al., The Astrophysical Journal Letters **823** L2 (2016)
[3] The LIGO Scientific Collaboration & Virgo, GCN **18330** (2015)
[4] B. P. Abbott et al., Physical Review Letters **116** 061102 (2016)
[5] L. Blackburn et al., The Astrophysical Journal Supplement Series **217** 8 (2015)
[6] L. Blackburn et al., GCN **18339** (2015)
[7] N. Omodei et al., GCN **18709** (2015)
[8] F. Acero et al., The Astrophysical Journal Supplement Series **218** 23 (2015)

Measuring space-time fuzziness with high energy γ-ray detectors

Paolo Walter Cattaneo[1,a] and Andrea Rappoldi[1,b]

[1] *INFN Pavia, Via Bassi 6, Pavia, I-27100, Italy*

Abstract. There are several suggestions to probe space-time fuzziness (also known as space-time foam) due to the quantum mechanics nature of space-time. These effects are predicted to be very small, being related to the Planck length, so that the only hope to experimentally detect them is to look at particles propagating along cosmological distances. Some phenomenological approaches suggest that photons originating from point-like sources at cosmological distance experience path length fluctuation that could be detected. Also the direction of flight of such photons may be subject to a dispersion such that the image of a point-like source is blurred and detected as a disk. An experimentally accessible signature may be images of point-like sources larger that the size due to the Point Spread Function of the instrument. This additional broadening should increase with distance and photon energy. Some concrete examples that can be studied with the AGILE and FERMI-LAT γ-ray satellite experiments are discussed.

1 Theoretical framework

The theoretical framework used as reference is presented in [1] and references therein [2–6].
A highly simplified view is the following: a unified point of view of general relativity and quantum mechanics implies that at the the Planck scale, $\ell_P = \sqrt{\frac{\hbar G}{c^3}}$, space-time cannot be treated as smooth and structureless. It would rather appear like *quantum foam*, whose features are unknown both at the theoretical and experimental levels.
Even if a quantitative experimental prediction of the influence of this quantum foam is beyond the existing theoretical framework, some phenomenological approaches are attempting qualitative predictions that are potentially subject to experimental measurements.
Following [1, 7], we assume that the accuracy with which a length ℓ can be measured due to fluctuation of the space-time is

$$\delta\ell \sim N\ell^{(1-\alpha)}\ell_P^{\alpha} \tag{1}$$

where $N \sim 1$ and $\alpha \leq 1$ is a parameter defining different space-time models. The models discussed in [1] suggest values $\alpha \in (\frac{1}{2}, \frac{2}{3})$ depending on the assumptions.
The goal would be to study variation of a fraction of wave length over astronomical scale. That seems

[a]Corresponding author. e-mail: paolo.cattaneo@pv.infn.it
[b]e-mail: andrea.rappoldi@pv.infn.it

an impossible task at first sight; it is so if we interpret Eq.1 in term of variation of time of arrival of the different photons

$$\delta t \sim \delta \ell / c \sim N t^{(1-\alpha)} t_P^{\alpha} \tag{2}$$

where $t_P = \sqrt{\frac{\hbar G}{c^5}} \sim 5.4 \times 10^{-44}$ s is the Planck time.
The best result in this line of thought is due to FERMI-LAT [8] that sets limits on the spread of time arrival due to quantum effect $\delta t \leq 1$ s for a propagation time $t \sim 7 \times 10^9$ y. This result sets a limit $\alpha > 0.3$ that is not very significant.
A much more powerful approach relies on the fact that different parts of the photon wave front traverse different parts of the space-time foam and experience different path length fluctuations. That results in different phases on the spherical front of the wave that, at the time of interaction with the detector, results in the direction perpendicular to the front (the photon direction) fluctuating randomly. That translates in an angular uncertainty

$$\delta\psi \sim N(\ell/\lambda)^{(1-\alpha)}(\ell_P/\lambda)^{\alpha} \tag{3}$$

This angular uncertainty can be measured at wave lengths from optical to radio and, at some extent, in the soft X-ray, using interference techniques (see [1] for discussion). Figure 1 shows the angular resolutions of several telescopes versus the wave length.
A complementary approach relies on making use of high energy γ-rays, that permit the measurement of the photon direction on an event by event basis. Existing detectors, based on conversion in $e^+ - e^-$ pairs provide resolutions $O(1°)$ or less in the energy range $E_\gamma \geq 100$ MeV (see [9] for AGILE, [10] for FERMI-LAT).
Figure 1 reports approximate estimations of the angular resolutions for such experiments. It demonstrates that high energy γ-ray detectors offer the opportunity of exploring a region of the parameter space otherwise inaccessible.

2 Experimental search

The search of space-time foam could make use of AGILE and FERMI-LAT data looking at extra-galactic sources far away such that their transverse size is small in comparison to the detector resolution. Being the γ-ray detector resolutions much worse than those of optical telescopes, that is easier. The images from these sources should be carefully reconstructed and compared with the PSF of the detectors. The detector resolutions must therefore be carefully measured and simulated on ground and in flight.
If the measured sizes exceed the expected angular resolutions and if this trend increases (linearly) with the distance and with the energy, it would be a strong sign of quantum mechanical effects on the propagation of photons through space-time foam.
The dependency on the distance can be studied measuring the spot sizes for different extra-galactic sources at different distances while the dependency on the energy can be studied separately for each source as long as the spot sizes can be determined with sufficient statistics.
In Fig. 2 the angular resolution of the FERMI-LAT and AGILE detectors are shown versus the energy compared with the blurring due to space-time foam for $\alpha = 0.75$ and z = 1. The effect should be detectable above a few GeVs.

3 Conclusions

The quantum mechanics nature of space-time is possibly subject to experimental investigation studying the path length fluctuations due to the space-time foam. The smallness of these effects require

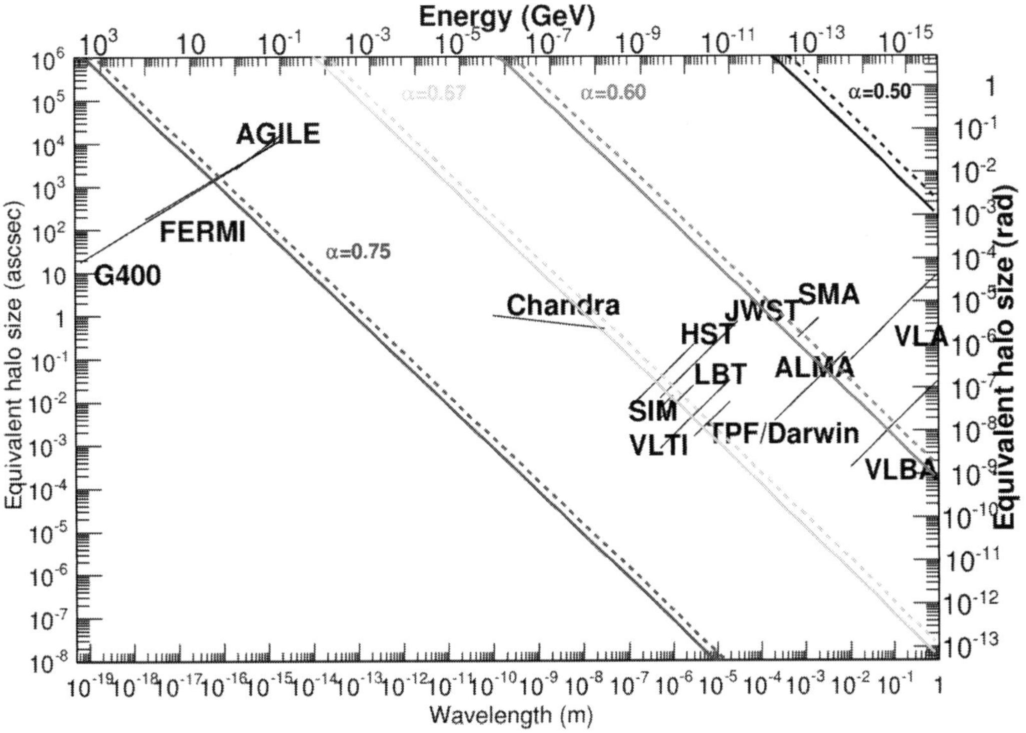

Figure 1. The detectability of various models of foamy space-time with existing and planned telescope/detectors. Diagonal tracks are shown for four models of foamy space-time, namely α= 0.5, 0.6, 2/3, 0.75 for z = 1 and 4 (respectively the lower and upper tracks for each model) and N = 1.9. Also shown are the observing ranges and resolution limits (i.e., PSF size) for a wide variety of telescopes/detectors, both current and planned.

measurements on astronomical scale. The size of a point-like source is expected to increase with distance and energy. That makes attractive high energy γ-ray detectors with good angular resolution to study so far unexplored regions of the parameter spaces. An approximated estimation of the capability of investigating space-time foam with AGILE and FERMI-LAT detectors is provided.

References

[1] W.A. Christiansen, Y.J. Ng, D.J. Floyd, E.S. Perlman, Phys.Rev. **D83**, 084003 (2011), 0912.0535

[2] W. Christiansen, Y.J. Ng, H. van Dam, Phys.Rev.Lett. **96**, 051301 (2006), gr-qc/0508121

[3] E.S. Perlman, Y.J. Ng, D.J. Floyd, W.A. Christiansen, Astron. & Astrophys. **535**, L9 (2011), 1110.4986

[4] G. Amelino-Camelia, V. Astuti, G. Rosati, Phys. Rev. D **87**, 084023 (2013), 1304.7630

[5] Y.J. Ng, H. van Dam, W. Christiansen, Astrophys.J. **591**, L87 (2003), astro-ph/0302372

[6] V. Vasileiou, J. Granot, T. Piran, G. Amelino-Camelia, Nat Phys **11**, 344 (2015)

[7] Y.J. Ng, Mod.Phys.Lett. **A18**, 1073 (2003), gr-qc/0305019

[8] A.A. Abdo et al., Nature **462**, 331 (2009)

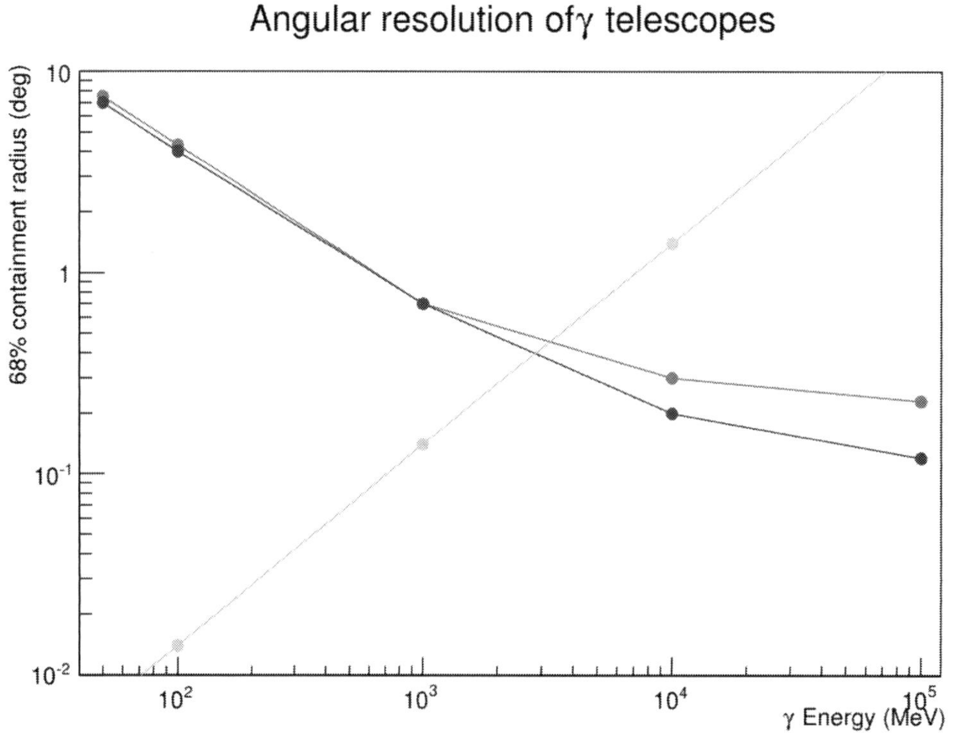

Figure 2. Angular resolution of the FERMI-LAT detector (blue) and of AGILE-GRID detector (red) and the blurring of the source due to the foam for $\alpha = 0.75$ and z = 1.

[9] A.W. Chen, A. Argan, A. Bulgarelli, P. Cattaneo, T. Contessi et al., Astron. & Astrophys. **558**, A37 (2013), 1310.1594

[10] A.A. Abdo et al. (Fermi LAT Collaboration), Astropart.Phys. **32**, 193 (2009), 0904.2226

Towards the development of a SiPM-based camera for the Cherenkov Telescope Array

G. Ambrosi[1], E. Bissaldi[2,3,a], L. Di Venere[3,4], E. Fiandrini[1], N. Giglietto[2,3], F. Giordano[3,4], M. Ionica[1], R. Paoletti[5,6], D. Simone[3], and V. Vagelli[1] for the CTA Consortium[7]

[1] *INFN – Sezione di Perugia, Perugia, Italy*
[2] *Politecnico di Bari, Bari, Italy*
[3] *INFN – Sezione di Bari, Bari, Italy*
[4] *Università di Bari, Bari, Italy*
[5] *INFN – Sezione di Pisa, Pisa, Italy*
[6] *Università di Siena, Siena, Italy*
[7] *See* `www.cta-observatory.org` *for full author & affiliation list*

Abstract. The Italian National Institute for Nuclear Physics (INFN) is involved in the development of a prototype for a camera based on Silicon Photomultipliers (SiPMs) for the Cherenkov Telescope Array (CTA), a new generation of telescopes for ground-based gamma-ray astronomy. In this framework, an R&D program within the 'Progetto Premiale TElescopi CHErenkov made in Italy (TECHE.it)' for the development of SiPMs suitable for Cherenkov light detection in the Near-Ultraviolet (NUV) has been carried out. The developed device is a NUV High-Density (NUV-HD) SiPM based on a micro cell of 30 μm × 30 μm and an area of 6 mm × 6 mm, produced by Fondazione Bruno Kessler (FBK). A full characterization of the single NUV-HD SiPM will be presented. A matrix of 8 × 8 single NUV-HD SiPMs will be part of the focal plane of the Schwarzschild-Couder Telescope prototype (pSCT) for CTA. An update on recent tests on the detectors arranged in this matrix configuration and on the front-end electronics will be given.

1 Characterization of the NUV-HD SiPM

A first extensive test campaign was conducted at the end of 2015 at the INFN laboratory of the Physics Department of Bari and it was devoted to the study of the NUV-HD SiPM produced by FBK [1] with pixel size of $30 \times 30\,\mu\text{m}^2$ and with area of 6×6 mm^2, in single configuration. All tests were performed in a dark box at room temperature (i.e., at 25 °C). In order to characterize the SiPM, we illuminated it with laser LEDs at 300 nm, 345 nm, 380 nm and 460 nm. The current signal was converted to a voltage signal by the op-amp based trans-impedance amplifier AdvanSiD [2], which includes a pole-zero compensation network. We obtained the distribution of signal events by analyzing the aquired waveforms in terms of amplitude and in terms of charge. Two distributions for 8 V of overvoltage (OV) are shown in figure 1. The left panel displays the histogram of the maximum amplitude of each waveform, while the right panel shows the histogram obtained by integrating the waveform from the onset of the signal up to a certain integration time t_{int}. Here we chose t_{int} = 75 ns. We then performed

[a] *speaker*, e-mail: elisabetta.bissaldi@ba.infn.it

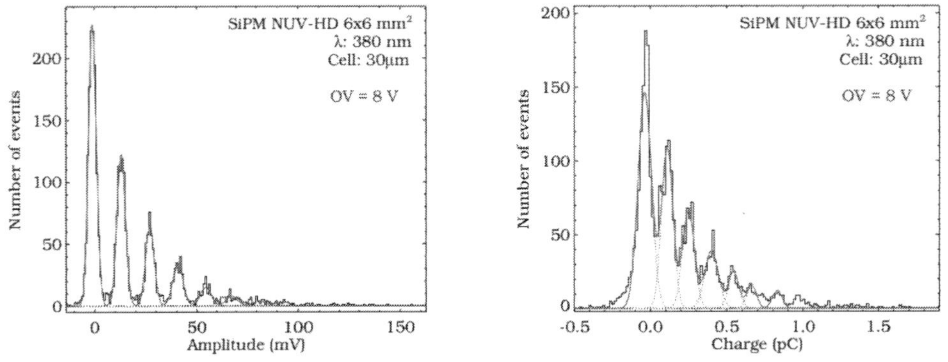

Figure 1. Distribution of signal events of the 6×6 mm^2 NUV–HD SiPM illuminated by a 380 nm LED and with OV = 8 V, expressed in terms of amplitude (*left panel*) and in terms of charge (*right panel*). The superimposed red line represents a Multi–Gaussian fit to the first seven photoelectron peaks.

a systematic study of the SiPM gain and of the first peak signal-to-noise ratio (SNR_1) as a function of the integration time. Looking at the results shown in the left panel of figure 2, we can conclude that integration times in the interval 50 ns – 75 ns represent a good compromise for obtaining high gains with a good SNR. Moreover, our tests reveal a uniformity in the behavior of the device in terms of gain. The right panel of figure 2 shows the gains versus OV calculated for different integration times and using all LEDs. The average of the slopes obtained for each wavelength is the gain g, expressed in fC/V. Small deviations among wavelengths are seen for long integration times and high biases. This effect is probably due to the fluctuations in the waveform tail.

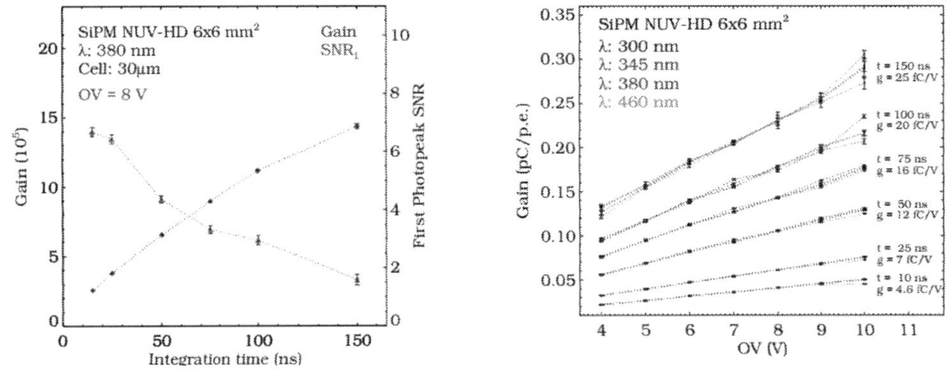

Figure 2. *Left panel*: SiPM gains (blue dots) and SNR$_1$ (red triangles) of the 6×6 mm^2 NUV–HD SiPM illuminated by a 380 nm LED and with OV = 8 V, displayed as a function of the integration time. The connecting lines do not represent fits and just connect subsequent points. *Right panel*: Gain versus OV, calculated considering different integration times and using four different LEDs.

DOI: 10.1051/epjconf/201713603022

2 Dark count rate analysis

In order to study the dark count rate, we performed a second set of test campaigns, which were conducted in a climate chamber located at the SITAEL laboratory [3]. The chamber was operated in complete darkness between -40 °C and +40 °C. The waveform analysis was performed first by filtering the signal and then by applying the differential leading edge discriminator (DLED) method [4]. In this way we obtained a baseline-compensated signal by calculating the difference between the measured signal and its delayed replica. We chose a delay Δt = 560 ps. The DLED analysis for T = 0 °C and OV = 6 V is shown in the top left panel of figure 3. We then selected those filtered peaks which rise above a threshold set at about half of the single dark photoelectron (PE) height. In the top right panel of figure 3 we show the density plot of the obtained peak distribution versus the time distance between successive peaks. The horizontal distributions correspond to the first, second, third, etc. dark PEs (from bottom to top). A histogram of the dark PE distribution is given in the bottom left panel of figure 3. The mean time between single dark PEs is about 2 μs (see bottom right panel of figure 3), which is negligible with respect to the single PE rates due to the night sky background. In order to perform a full characterization of the forward and reverse bias, we also tested

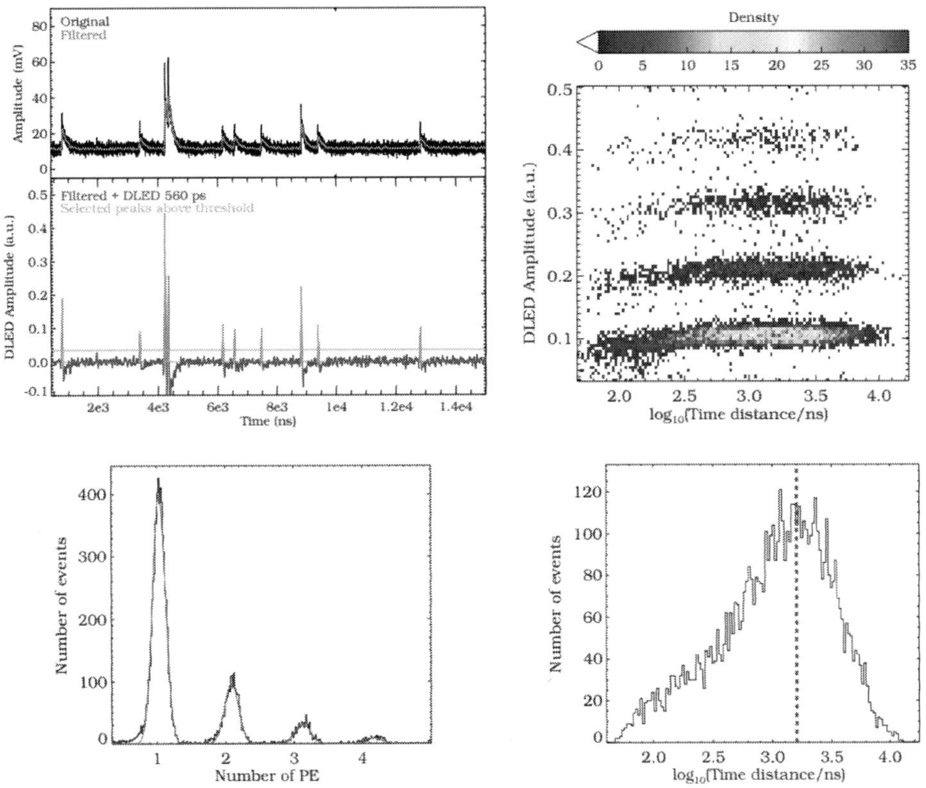

Figure 3. *Top left panel*: Waveform analysis of the dark signal for T = 0 °C and OV = 6 V before (top) and after (bottom) the application of the DLED method with a delay Δt = 560 ps. *Top right panel*: Density plot of the peak distribution versus the time distance between successive peaks. *Bottom left panel*: Histogram of the dark PE distribution. *Bottom right panel*: Temporal distribution of the single dark PEs. The dotted red line indicates a mean value of about 2 μs.

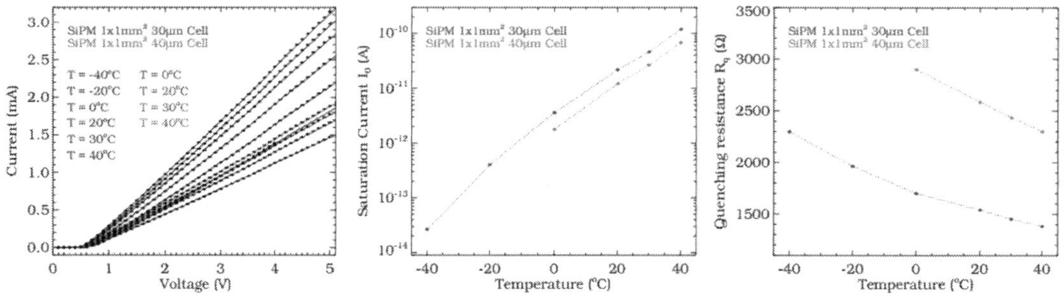

Figure 4. Forward IV-characteristics of 1×1 mm^2 SiPMs with 30 μm (blue curves) and 40 μm (red curves) cell sizes. Saturation current (middle panel) and quenching resistance (last panel) versus temperature (same color code).

in the climate chamber some SiPMs with a smaller area (1×1 mm^2) and two different cell sizes (30 μm and 40 μm). We derived the forward IV-characteristics, which are shown in the first panel of figure 4. From the data fitting we deduced the evolution of the saturation current (middle panel) and of the quenching resistance (last panel) as a function of the temperature for both devices (blue and red curves). This represents a crucial step towards a correct modeling of the SiPM equivalent circuit, in order to optimize the preamplifier design, i.e. getting a short signal within 10 ns.

3 SiPM in matrix configuration

Finally, we performed an analysis of our NUV-HD SiPMs arranged in an 8×8 matrix configuration. We tested the reverse IV-characteristics of all 64 matrix channels and demonstrated that the breakdown voltages have a uniform behavior [5]. Moreover, we measured waveforms at different OVs by covering all but one channel with a black mask. We found that the SNRs vary between 1.5 and 5.5, whereas amplitudes are characterized by a variation of a few percent. It is interesting to note that the amplitude is generally lower if compared with the previous technology of NUV SiPMs [6]. This fact makes the NUV-HD SiPMs with 6×6 mm^2 area suitable detectors for the pSCT, if coupled to electronics able to detect and process positive current signals of a few millivolts per photoelectron.

Acknowledgments

We gratefully acknowledge support from the agencies and organizations listed under Funding Agencies at this website: http://www.cta-observatory.org/. E.B. ackowledges the Italian "Fondo di Sviluppo e Coesione 2007-2013 - APQ Ricerca Regione Puglia - Future In Research".

References

[1] G. Zappalà et al., JINST **11** (2016)
[2] `http://advansid.com/products/product-detail/asd-ep-eb-n`
[3] `http://www.sitael.com`
[4] A. Gola et al., IEEE Transactions on Nuclear Science **59** Issue:2 (2012)
[5] G. Ambrosi et al., Nuclear Physics B - Proceedings Supplements (CRIS2016) (2016)
[6] G. Ambrosi et al., Proceedings of Science (ICRC2015) **992** (2015)

Status and prospects of the JUNO experiment

Gioacchino Ranucci[1 a], on behalf of the JUNO Collaboration
[1]*Istituto Nazionale di Fisica Nucleare, sezione di Milano, via Celoria 16, 20133 Milano, Italy*

Abstract. The JUNO Jiangmen Underground Neutrino Observatory, a 20 kton multi-purpose underground liquid scintillator detector, has been proposed and approved for realization in the south of China. In this work I describe first the broad physics capabilities of the experiment, which include the crucial measure of the neutrino mass hierarchy, the high precision determination of three oscillation parameters, and a rich astroparticle program. Then, I give the details of the mass hierarchy determination procedures and an outlook on the progress and schedule of the experiment.

1 Introduction

In the global context of the future neutrino oscillation studies, the JUNO detector [1] will play a central role on two aspects: the determination of mass hierarchy and the precise measurements of the solar oscillation parameters, i.e. Δm_{21}^2, $sin^2\theta_{12}$, as well as of the atmospheric squared mass difference Δm_{31}^2.

JUNO is designed and realized as a huge liquid scintillator detector, therefore exploiting a mature and well proved technology, which has already provided fundamental contributions to the neutrino oscillation study through several implementations (Borexino [2], KamLAND [3], Daya Bay [4], Reno [5] and Double Chooz [6] being the most recent examples). It will base its measurements on the detection of the global antineutrino flux coming from the cores of two nearby nuclear complexes, Yangjiang and Taishan, located at about 53 km from the experimental site.

The program will be complemented by an ensemble of astroparticle physics measurements, which will significantly enhance the physics potential of JUNO.

2 Summary of characteristics and of physics goals

JUNO will join the renowned, long tradition family of reactor neutrino experiments based on the scintillation technology, whose first well known example was the Savannah River experiment, with which Cowan and Reines revealed for the first time the (anti)neutrino particle.

In Fig. 1 there is the summary of reactors' results accumulated so far, expressed as ratio of observed over expected events, contrasted with the prediction from the oscillation survival probability function. On the horizontal axis the reactor-detector distance is reported; the plot shows the well-known fact that at small distance the impact of the oscillation phenomenon on the detector count rate

[a] Corresponding author: gioacchino.ranucci@mi.infn.it

is not visible, while it starts to manifest from roughly little less than 1 km baseline. At the special distance of 53 km the count rate suppression, mainly driven by the solar oscillation parameters, is maximal, therefore creating the best condition to study the interference effect governed in turn by the atmospheric mass squared difference, which is responsible for the ripple superimposed on the count rate suppressed profile. This is, therefore, the rationality beyond the choice of the optimum site and distance between JUNO and the emitting anti-neutrino cores.

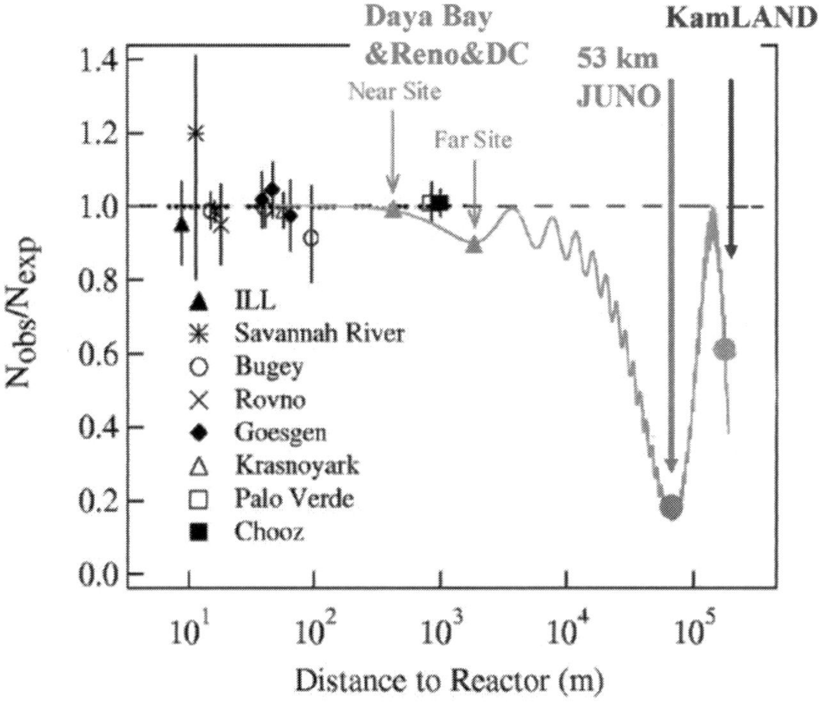

Figure 1. Summary of past reactors' results as ratio of observed to expected count rate, together with the predicted JUNO point.

To fully exploit this optimal baseline, in order to perform an effective and successful measurement of the mass hierarchy, the detector must be endowed with two essential characteristics: large mass to perform a high statistic measurements, and stringent energy resolution to clearly distinguish the ripple induced by the atmospheric mass squared term. The two key numbers in this respects are the total mass of 20 kton of liquid scintillator, and the energy resolution of 3% at 1 MeV, which represent, therefore, the major technical features which characterize the experiment.

In term of physics reach, JUNO thanks to its large mass can tackle a plurality of measurements. Beyond mass hierarchy and precision determination of neutrino oscillation parameters, it can provide fundamental results concerning many hot topics in the astroparticle field. An incomplete list comprises supernova burst neutrinos, diffuse supernova neutrinos, solar neutrinos, atmospheric neutrinos, geo-neutrinos, sterile neutrinos, nucleon decay, indirect dark matter search, as well as a number of additional exotic searches, as thoroughly illustrated in the physics program of the experiment (yellow book), published in [7].

3 Basic features of the program: detector structure, location and Collaboration

In term of implementation characteristics, JUNO is a spherical unsegmented liquid scintillator detector that will push such a technology beyond the present limit, as far as the mass (20 kton) and the resolution (3%) are concerned. Succinctly, the detector can be described as a large spherical acrylic vessel, which will hold the scintillator volume, contained in turn in a water pool, to ensure adequate shielding against the gamma radiation and neutrons from the rock.

The vessel will be surrounded by a stainless steel truss which will perform the twofold task to sustain the vessel, by relieving its internal stress, and to provide the anchor support for the 18000 20" photomultipliers observing the scintillation photons. The light detection system will comprise also an additional set of 3" PMTs, up to 36000, which will be used for calibration purpose and to cross check the performances of the main PMTs, with the scope to control and reduce the systematic effects of the measurements performed by the main 20" PMT system.

Moreover, the shielding water around the acrylic vessel will be converted into a Cherenkov detector, being instrumented with about 2000 phototubes, which will detect the muon induced Cherenkov light. Such an arrangement, together with the top tracker that will be deployed on the roof of the detector itself, will allow an efficient muon veto capability, an essential feature at the planned shallow depth of the experiment, i.e. 700 m.

JUNO has been approved in China at the beginning of 2013 and has been later joined by groups from all over the world. Currently the Collaboration encompasses 66 institutions from Asia, Europe and America, with more than 450 researchers, and it is still expanding.

The experiment is located in the South of China, Guangdong province, Jianmeng County, Kaiping city, at 53 km from the two sites of Yangjian and Taishan, where 6 and 4 nuclear cores are planned, respectively. By 2020 according to the construction schedule of the plants 26.6 GW will be installed (2 cores will be missing at Taishan), while eventually the total power of 35.8 GW will be available.

4 How to infer the mass hierarchy

The observable quantity from which the mass hierarchy will be inferred is the positron spectrum detected in the liquid scintillator, stemming from the Inverse Beta Decay reaction through which antineutrino detection will occur. Specifically, the determination of the mass hierarchy relies on the identification on such a spectrum of the "imprinting" of the anti-ν_e survival probability.

The Inverse Beta Decay Reaction a là Cowan Reines is the following

$$\bar{\nu}_e + p \rightarrow e^+ + n \tag{1}$$

The energy deposited by the positron in the scintillator, i.e. its kinetic energy plus the total 1.022 keV energy of the two annihilation gammas, reflects faithfully the energy of the incoming anti-neutrinos

$$E_{vis}(e^+) = E(\nu) - 0.8\, MeV \tag{2}$$

$E_{vis}(e^+)$ is, thus, the specific measurement output to be analyzed for the hierarchy evaluation.

The time coincidence (mean difference of the order of 250 μs) between the positron event and the γ ray from the subsequent neutron capture on protons allows to identify effectively the occurrence of neutrino detection and to pick up the positron scintillation signal, even in presence of uncorrelated background.

In order to describe the specific algorithm through which the MH can be unraveled, we resort to the electron (ant)neutrino survival probability, which in a full three flavor framework can be written as

$$P_{ee} = 1 - cos^4\theta_{13} sin^2 2\theta_{12} sin^2(\Delta_{21}) - sin^2\theta_{13} sin^2(|\Delta_{31}|) - sin^2\theta_{12} sin^2 2\theta_{13} sin^2(\Delta_{21}) cos(2|\Delta_{31}|) \pm \frac{sin^2\theta_{12}}{2} sin^2 2\theta_{13} sin(2\Delta_{21}) sin(2|\Delta_{31}|) \tag{3}$$

exploiting the approximation $\Delta m^2_{32} \approx \Delta m^2_{31}$, and assuming

$$\Delta_{ij} \equiv \frac{\Delta m_{ij}^2 L}{4E_\nu}$$

The sign flip in front of the last term of eq. (3) is due to the hierarchy: positive for direct hierarchy, negative for the inverse one. The presence in this term of the multiplicative factor $sin^2 2\theta_{13}$ questioned the effectiveness of this methodology, proposed for the first time in [8], until the θ_{13} experimental determination from Daya Bay, Reno and Double Chooz. Indeed, should θ_{13} have been resulted close to 0, the last term of the P_{ee} expression would have been vanishing small, making the proposed approach unfeasible. In reality, the discovery that θ_{13} is actually very close to the previous Chooz limit [9], opened the door to the actual implementation of the method.

The effect of P_{ee} on the reactor spectrum is shown in Fig. 2; the y axis is proportional to the event rate, while on the x axis the ratio L/E_nu is reported. The dashed line is the un-oscillated spectrum; the continuous black line is the spectrum distorted and suppressed as effect of the "solar" oscillation: it is this large effect the key for the very precise determination of the two "solar" mixing parameters Δm_{21}^2 and $sin^2\theta_{12}$.

The blue and red lines superimposed on the smooth black line, instead, display the effect of the interference term driven by the atmospheric mass squared difference. The frequency of the ripple depends on $|\Delta m_{31}^2|$ (which therefore can also be determined with high accuracy from the precise "tracking" of the ripple itself), while its phase is linked to the MH, as shown by the reciprocal shift of the blue and red lines in the figure. Unraveling the phase of the ripple, hence, is the clue for the MH determination. Quantitative assessments show that this method can lead in JUNO to a 4σ discrimination capability between the true and wrong hierarchy in 6 years of data taking.

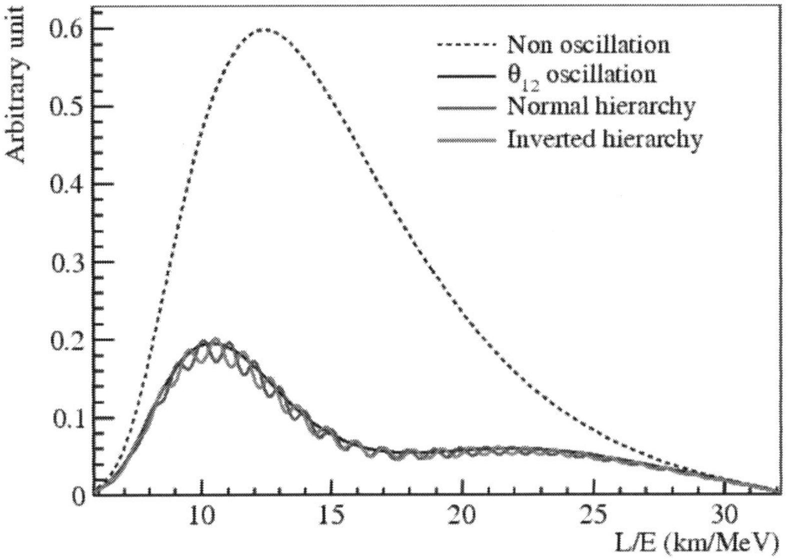

Figure 2. Effect of Pee electron neutrino survival probability on the reactor spectrum.

5 JUNO progress and schedule

The experiment is scheduled to start data taking in 2020. The ground breaking signalling the start-up of the excavation occurred in January 2015. So far, more than half of the slope tunnel (900 out of 1340 m) and about half of the vertical shaft (300 out of 611 m) have been excavated. The former will allow to bring the scintillator underground, the latter will enable access of personnel and construction materials.

The civil construction is foreseen to be completed by about middle of 2018, included the large experimental hall. The preparation of the detector components, e.g. phototubes, acrylic panels, etc., has started in the current year 2016 and will encompass the whole 2017 and part of 2018, while the global onsite installation will be completed by the end of 2019. All this is in line to ensure scintillator fill and start-up of data taking within the targeted 2020 year.

References

1. T. Adam et al. (JUNO Collaboration), "The JUNO Conceptual Design Report", arXiv:1508.07166 (2016)
2. G. Alimonti et al. (Borexino Collaboration), Nucl. Instr. and Meth A 600, 568 (2009)
3. S. Abe et al (KamLAND Collaboration), Phys. Rev. Lett. 100, 221803 (2008)
4. F. P. An et al. (Daya Bay Collaboration), Phys. Rev. Lett. 108, 171803 (2012)
5. J. K. Ahn et al., (RENO Collaboration), Phys. Rev. Lett. 108, 191802 (2012)
6. S. Abe et al., (DOUBLE CHOOZ Collaboration), Phys. Lett. B, 723, 66 (2013)
7. F. An et al. (JUNO Collaboration), J. of Phys. G (Nucl. and Part. Phys.) 43, id. 030401 (2016)
8. S.T. Petcov and M. Piai, Physics Letters B 553, 94-106 (2002)
9. M. Apollonio et al. (Chooz Collaboration), The European Physical Journal C 27, 331-374 (2003)

Results from the ANTARES neutrino telescope

Agustín Sánchez Losa[1,a] on behalf of the ANTARES Collaboration

[1] *INFN - Sezione di Bari, Via Edoardo Orabona 4, 70125 Bari, Italy*

Abstract. The ANTARES detector is an underwater neutrino telescope, the largest in the Northern Hemisphere and the first one ever built under the sea, located in the Mediterranean Sea 40 km off the Southern coast of France, at a depth of 2.5 km. It comprises 885 photomultiplier tubes distributed along twelve detection lines. The signal due to neutrinos is searched by reconstructing the tracks of secondary particles produced in the surroundings of the detector. The detector is in data taking with its final configuration since 2008. It is aimed at identifying the sources, either steady or flaring, of cosmic neutrinos, and is also suitable for detection of dark matter within the Sun and/or Galactic Centre. ANTARES can contribute in the confirmation of the cosmic neutrino flux observed by IceCube, being particularly competitive for the Galactic Centre, and in general for galactic sources, due its latitude and at lower energies and softer spectra due its configuration. Several multi-messenger analyses have been also attempted, including the search of coincidence signals of neutrinos with gravitational-waves. Additional topics include neutrino oscillations or the search of exotic particles, like nuclearites and magnetic monopoles. Results from the latest analyses are presented.

1 Introduction

ANTARES [1] is the largest underwater neutrino telescope on the Northern Hemisphere. It has been taking data smoothly in its final configuration since 2008 and is placed on the seabed at 40 km off the coast of Toulon (France) with an instrumental volume of approximately 0.01 km^3 distributed between 2.5 and 2.0 km depth on the Mediterranean Sea. It consists of 885 10$''$ photomultipliers distributed in a three-dimensional matrix in such volume that primarily reconstructs the muon tracks that cross its volume, in some occasions the result of a charged-current interaction of a ν_μ in the vicinity of the detector. Its main purpose is to discover the origin of high energy cosmic neutrinos while at the same time give an answer to other mysteries like the dark matter nature and other phenomena.

In the following, the latest highlighted ANTARES analysis are reported, which are point source, diffuse flux, multi-messenger (section 2 to 4 respectively) and dark matter analyses (see contribution [2] in this conference). Some analyses are explained in detail when not covered in other contributions of the conference, providing the latest upper limits of ANTARES.

2 Point sources

Point source searches is one of the main neutrino telescope analyses. They could resolve the source or sources responsible of the excess observed by IceCube [4].

[a] e-mail: agustin.sanchez@ba.infn.it

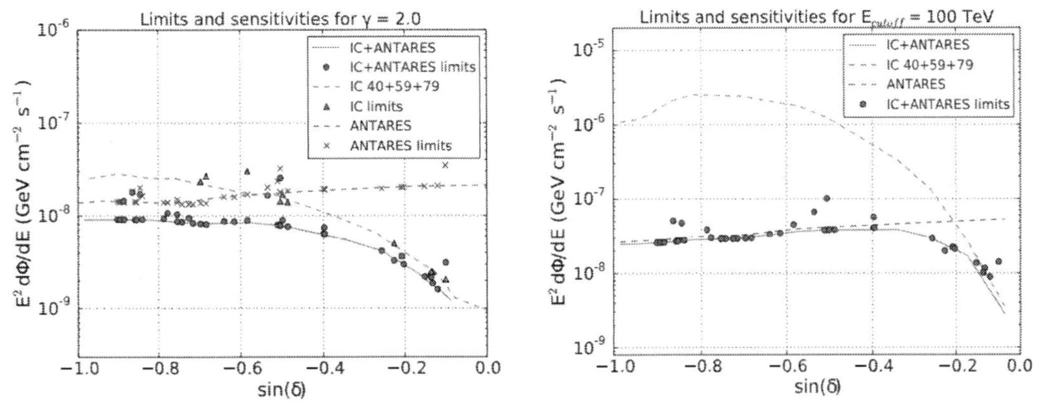

Figure 1. Combined upper limits for E^{-2} (left) and $E^{-2}\exp^{-\sqrt{E/100\text{TeV}}}$ (right) for the first combined IceCube-ANTARES point source analysis [3].

In a previous analysis [5], ANTARES constrained the extension of a possible source responsible of the IceCube cosmic neutrino flux observed close to the Galactic Centre, in particular a source with a Gaussian-like extension smaller than 0.5° for declinations below −11° and smaller than 1° for $\delta \lesssim -33°$ for a E^{-2} spectrum. In a recent update [6], these limits have been extended to softer spectral indexes up to a $E^{-2.5}$ spectrum, disfavouring an origin of the IceCube signal for a point-like source with these spectra.

The TANAMI collaboration reported observations of 6 bright blazars locally compatible with the 2 first PeV IceCube events IC14 (Bert) and IC20 (Ernie). ANTARES realised a point source analysis of these 6 sources in collaboration with TANAMI [7] in order to discover or constrain them as possible source of the Ice Cube events, excluding at a 90% confidence level such possibility unless the neutrino spectrum is softer than −2.4.

2.1 IceCube-ANTARES combined analysis

For the first time, a combined IceCube-ANTARES analysis has been performed [3] on the Southern Sky by using muon tracks detected by both experiments from Jan 29, 2007 to Dec 31, 2012 for ANTARES and from Apr 5, 2008 to May 13, 2011 for IceCube, including its IC-40, IC-59 and IC-79 configurations. This joint analysis summed up the great statistics of IceCube with the privileged ANTARES sensitivity for the Southern Sky (including the Galactic Centre) and lower energies (i.e. soft spectra) and its good angular resolution.

From the IceCube data 146,018 Southern-sky track-like events were selected, while 4,136 tracks were selected from ANTARES. An unbinned maximum likelihood ratio estimation has been used to evaluate them in the search of a possible signal excess, weighting each detector contribution as a function of their acceptances.

A full Southern-sky search and a candidate list search of 40 Galactic and extra-Galactic sources have been carried out finding no statistically significant excess in both searches. The most significant cluster is located at equatorial coordinates $\alpha = 332.8°$, $\delta = -46.1°$ with a post-trial significance of 24%. From the candidate list, HESS J1741-302 shows the lowest p-value, with a 11% post-trial probability of being produced by the background. In figure 1 the combined upper limits are shown,

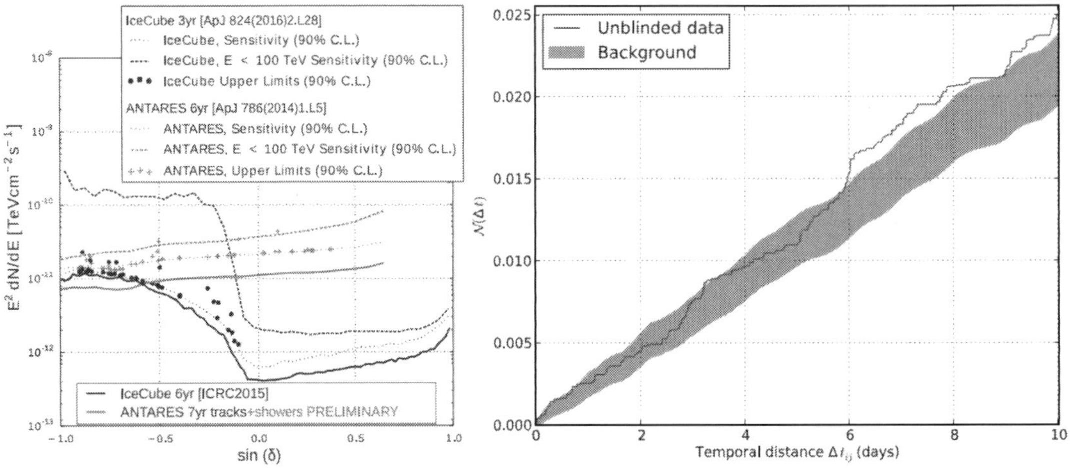

Figure 2. Left: ANTARES combined sensitivity for tracks and showers with data collected between 2007 and 2013. Right: Cumulative two-point distribution over 10 days for the time correlation analysis. The green area corresponds to the standard deviation of each bin of the cumulative distribution, in the background only hypothesis while the two-point cumulative distribution for the unblinded dataset is represented in blue.

where both detectors contributed to improve the limits that the respective datasets would have achieved separated.

2.2 Point source update

The first ANTARES point source analysis including the shower channel has been performed [8] with data between 2007 and 2013, comprising 1690 days of live-time. 6490 tracks and 172 showers with a median angular resolution of 0.3° and 3° respectively, have been selected for a full sky and candidate list search. It has been estimated a ~10% contamination of atmospheric muons for the shower channel. The candidate list included 54 Galactic and extra-Galactic sources, the Galactic Centre and 8 IceCube muon tracks [4].

The most significant cluster found is at equatorial coordinates $\alpha = 311.7°$, $\delta = -48.3°$ with a post-trial 1.3σ significance and the source with the largest excess was HESS J0632+057 with a 0.75σ post-trial significance. In the absence of a signal excess, upper limits on the neutrino fluxes have been set (see figure 2 left).

3 Diffuse fluxes

Diffuse flux searches look for cosmic neutrino fluxes that do not presume a point source origin, but an extensive region or even no location dependence at all. Therefore, they suffer of a larger background than the point source searches but are not constrained to any source origin.

For the first time, a search for an cosmic neutrino diffuse flux using both muon tracks and shower events in ANTARES has been carried out, using data between 2007 and 2015, where 7 events have been observed with 5 events expected from background and 2 from an IceCube flux. Results and used methods are covered in contribution [9] in this conference. Also there, an update is presented for a dedicated analysis of neutrinos coming from the Fermi bubbles [10], with ANTARES muon tracks reconstructed in the period 2007–2015 and a 1σ excess, and the search of a neutrino diffuse flux coming from the Galactic Ridge [11] with 2007–2013 data, where 2 events were observed with 3.7 expected from background.

3.1 Time correlation with IceCube HESE

A search of ANTARES neutrinos in temporal coincidence with IceCube High-Energy Starting Events (HESE) and space compatible has been carried out. The analysis [12] uses ANTARES tracks occurred between May 2010 and November 2012 for $E^{-2.5}$ to E^{-2} spectra in coincidence with the 8 IceCube HESE [4] that are within 45° from the Galactic Centre during that period. This is done by a two-point time-spatial correlation function without need of a prior on the burst timing structure nor on the electromagnetic emission, providing an effective way to acquire information on the possible origin of the IceCube astrophysical signal from a transient source.

The ANTARES event selection criteria have been optimized through Monte-Carlo simulations to reach a constant neutrino candidate rate over time. This has been done with a Model Discovery Potential optimization on the cut on the track quality parameter for each of the five sub-samples of optical module mean counting rates chosen for the analysis. The signal model used has been the IceCube flux as reported in [13]. These sub-samples reflect different data taking conditions regarding its data quality. The selection of data this way lead to a final sample consisting of 4337 events.

ANTARES results show a fluctuation with at least a 35% probability of being of background origin. Thus, no significant correlation has been found. Comparison between data and background (see figure 2 right) shows that the largest deviation between the two-point distributions corresponds to a time scale of 6.1 days, being this the most likely time scale of a possible transient signal responsible of the selected IceCube events.

4 Multi-messenger

The search for neutrinos in coincidence with other astronomical messengers allows to reduce the amount of signal required for a discovery under the assumption of a correlation in both messenger productions.

With the discovery of the first gravitational-wave, GW150914 [14], possible high energy neutrino events in coincidence have been searched in a joint analysis of ANTARES with the IceCube collaboration [15]. The neutrino follow-up of this gravitational-wave and of two IceCube HESE event alerts are described in detail in contribution [16] in this conference.

On a different approach, the Telescopes-ANTARES as Target of Opportunity [17, 18] (TAToO) multi-wavelength follow-up program has studied more than 200 high energy neutrino alerts with optical follow-ups (with the optical robotic telescopes TAROT, ROTSE and MASTER) since mid-2009 and 12 alerts with X-ray follow-ups with the Swift-XRT instrument since mid-2013. The results of these analyses are presented in contribution [16] in this conference, with a recent update of the Murchinson Widefield Array (MWA) radio follow-up of two neutrino alerts.

A new analysis evaluating two promising GRB neutrino emission models, internal shock [19] and photospheric [20] ones, has been carried out. On this analysis, neutrinos were searched for coincidence with four of the brightest GRBs visible by ANTARES between 2008 and 2013: GRB080916C, GRB110918A, GRB130427A and GRB130505A. The details of this analysis are extended in contributions [21] and [22] of this conference.

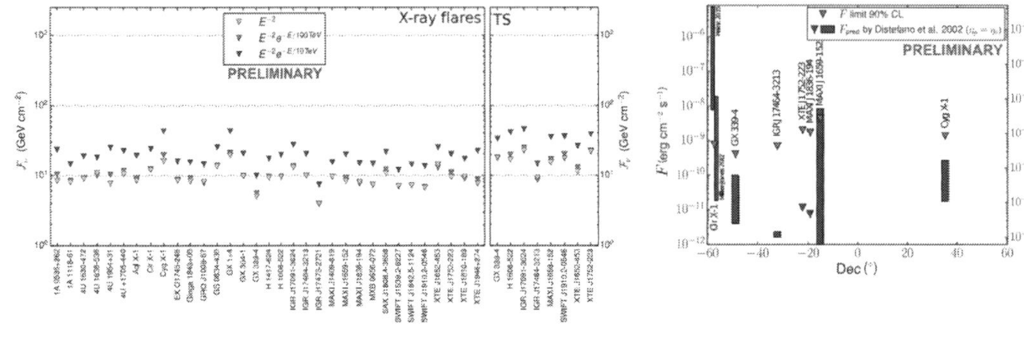

Figure 3. Upper limits at 90% C.L. on (left) the neutrino fluence for the 33 X-ray binary candidates and 8 hardness transition state sources and (right) on the neutrino energy flux for the microquasars studied considering a $E^{-2}\exp^{-\sqrt{E/100\mathrm{TeV}}}$ spectra, compared with the expectations from [23].

4.1 X-ray binaries

A time-dependent search looking for possible Galactic high energy neutrino sources has been performed, testing X-ray binaries as possible candidates, using ANTARES muon tracks reconstructed from 2008 to 2012 [24]. This analysis completes one carried out for blazars as extra-Galactic transient neutrino source candidates [25] with similar methodology and updates one precedent where six microquasars were studied [26].

The search for cosmic neutrinos has been carried out during flaring periods deduced with Bayesian block time-series analysis methods [27] from X-ray emission light curves measured on 33 X-ray binary candidates by the Swift/BAT [1] telescope, completed with data from RXTE/ASM [2] and MAXI [3] instruments. Additionally, hardness transition states on 8 X-ray binaries have been considered. The tested spectra are E^{-2} and $E^{-2}\exp^{-\sqrt{E/cut\text{-}off}}$ with 10 TeV and 100 TeV cut-offs. ANTARES data selection is based on a Model Discovery Potential optimization. The devised maximum likelihood ratio analysis considered also a possible lag between the neutrino and the electromagnetic signals.

The data unblinding provided no significant excess, with the most significant fluctuation corresponding to GX1+4 with a 74% post-trial probability of being produced by the background. Upper limits have been computed on the neutrino fluence and on different neutrino flux model parameters found in the literature (see figure 3).

5 Conclusions

The latest ANTARES analyses have provided numerous constrains to the possible origin of the IceCube signal, from Galactic to extra-Galactic candidates, point sources or extended ones, and even limits on a possible transient nature. Not to mention the latest dark matter analysis, which proved to be very competitive on the exclusion of certain dark matter candidate parameters. In the future, KM3NeT will take over of its contribution to the neutrino astronomy, as it has been presented on contribution [28] in this conference.

[1] http://swift.gsfc.nasa.gov/results/transients
[2] http://xte.mit.edu/ASM_lc.html
[3] http://maxi.riken.jp

References

[1] M. Ageron et al. (the ANTARES Coll.), Nucl. Instrum. Methods Phys. Res., Sect. A **656**, 11 (2011), 1104.1607 1
[2] C. Tönnis (on behalf of the ANTARES Coll.), NU/261 **RICAP16**, 4 (2016) 1
[3] S. Adrián-Martínez et al. (the ANTARES and IceCube Coll.), Astrophys. J. **823**, 65 (2016), 1511.02149 2
[4] M.G. Aartsen et al. (the IceCube Coll.), Phys. Rev. Lett. **113**, 8 (2014), 1405.5303 1, 3, 4
[5] S. Adrián-Martínez et al. (the ANTARES Coll.), Astrophys. J. Lett. **786**, 5 (2014), 1402.6182 2
[6] J. Barrios Martí (on behalf of the ANTARES Coll.), 1077 **ICRC2015**, 6 (2015) 2
[7] S. Adrián-Martínez et al. (the ANTARES and TANAMI Coll.), Astron. & Astrophys. **576**, 6 (2015), 1501.07843 2
[8] T. Michael (on behalf of the ANTARES Coll.), 1078 **ICRC2015**, 8 (2015) 3
[9] L. Fusco (on behalf of the ANTARES Coll.), Nu/102 **RICAP16**, 4 (2016) 3
[10] S. Hallmann (on behalf of the ANTARES Coll.), 1059 **ICRC2015**, 8 (2015) 3
[11] S. Adrián-Martínez et al. (the ANTARES Coll.), Phys. Lett. B **760**, 143 (2016), 1602.03036 4
[12] A. Coleiro (on behalf of the ANTARES Coll.), 1073 **ICRC2015**, 8 (2015) 4
[13] M.G. Aartsen et al. (the IceCube Coll.), Phys. Rev. D **91**, 16 (2015), 1410.1749 4
[14] B. Abbott et al. (the LIGO Scientific and Virgo Coll.), Phys. Rev. Lett. **116**, 16 (2016), 1602.03837 4
[15] S. Adrián-Martínez et al. (the ANTARES, IceCube, LIGO Scientific and Virgo Coll.), Phys. Rev. D **93**, 15 (2016), 1602.05411 4
[16] A. Coleiro (on behalf of the ANTARES Coll.), GW/82 **RICAP16**, 4 (2016) 4
[17] M. Ageron et al. (the ANTARES Coll.), Astropart. Phys. **35**, 530 (2012), 1103.4477 4
[18] S. Adrián-Martínez et al. (the ANTARES Coll.), J. Cosmol. Astropart. Phys. **02**, 25 (2016), 1508.01180 4
[19] T. Piran, Rev. Mod. Phys. **76**, 1143 (2004), astro-ph/0405503 4
[20] P. Kumar, B. Zhang, Phys. Rept. **561**, 1 (2015), 1410.0679 4
[21] M. Sanguineti (on behalf of the ANTARES Coll.), NU/101 **RICAP16**, 4 (2016) 4
[22] S. Celli (on behalf of the ANTARES Coll.), Porters/295 **RICAP16**, 4 (2016) 4
[23] C. Distefano, D. Guetta, E. Waxman, A. Levinson, Astrophys. J. **575**, 378 (2002), astro-ph/0202200 5
[24] S. Adrián-Martínez et al. (the ANTARES Coll.), Submitted to J. Cosmol. Astropart. Phys. (2016), 1609.07372 5
[25] S. Adrián-Martínez et al. (the ANTARES Coll.), J. Cosmol. Astropart. Phys. **12**, 26 (2015), 1506.07354 5
[26] S. Adrián-Martínez et al. (the ANTARES Coll.), J. High Energy Astrophys. **3–4**, 9 (2014), 1402.1600 5
[27] J.D. Scargle, J.P. Norris, B. Jackson, J. Chiang, Astrophys. J. **764**, 26 (2013), 1207.5578 5
[28] J. Brunner (on behalf of the KM3NeT Coll.), CR&γ/76 **RICAP16**, 6 (2016) 5

Follow-up of GW150914 and multi-messenger studies of transient astrophysical sources with the ANTARES neutrino telescope

Alexis Coleiro[1,a] on behalf of the ANTARES collaboration

[1] *APC, Université Paris Diderot, CNRS/IN2P3, CEA/Irfu, Observatoire de Paris, Sorbonne Paris Cité, 10 rue Alice Domon et Léonie Duquet, 75205 Paris Cedex 13, France.*

Abstract. By constantly monitoring at least one complete hemisphere of the sky, neutrino telescopes are well designed to detect neutrinos emitted by transient astrophysical sources. In particular, the ANTARES telescope is currently the largest high-energy neutrino detector in the Northern Hemisphere. Searches for ANTARES neutrino candidates coincident with multi-wavelength and multi-messenger transient phenomena are performed by triggering optical, X-ray and radio observations immediately after the detection of an interesting ANTARES event and also by looking for neutrino emission spatially and temporally coincident with transient astrophysical events detected across the electromagnetic spectrum or with new messengers as gravitational-wave signals. The latest results of the multi-messenger analyses performed with ANTARES will be presented in this contribution. In particular, we will focus on the neutrino follow-up performed after the detection of the first gravitation-wave event, GW150914.

1 Introduction

Time-domain astroparticle physics has entered an exciting period with the recent development of wide-field-of-view instruments, communication strategies and low latency alert triggering of gravitational wave and high-energy neutrino (HEN) signals, but also across the electromagnetic spectrum. In particular, neutrinos represent unique probes to study high-energy cosmic sources. They are neutral, stable and weakly interacting. Contrary to cosmic rays (CRs), they are not deflected by the magnetic fields and unlike high-energy photons, they are not absorbed by pair production via $\gamma\gamma$ interactions with cosmic microwave and infrared backgrounds. A HEN diffuse flux of cosmic origin has been identified by the IceCube telescope (see e.g. [1]), the sources of which have still to be identified. In this context, multi-messenger approaches consisting in simultaneously looking for the same sources with both neutrino telescopes, gravitational-wave interferometers and/or multi-wavelength facilities can constitute a viable mean of locating HEN/CR sources and thus further understanding the acceleration mechanisms at play in these sources.

The ANTARES neutrino telescope is currently the largest neutrino telescope in the Northern hemisphere. Located in the Mediterranean Sea, 20 km offshore Toulon (France), it is composed of 885 pho-

[a] e-mail: coleiro@apc.in2p3.fr

DOI: 10.1051/epjconf/201713604003

tomultipliers installed on 12 detection lines, sensitive to the Cherenkov light emitted by relativistic up-going muons produced by the interaction of HEN close to the detector.

In particular, search for transient sources of HEN is promising since the short timescale of emission drastically reduces the background level, mainly composed of atmospheric muons and neutrinos and consequently increases the sensitivity and discovery potential of neutrino telescopes. This contribution briefly presents the most recent results of the ANTARES multi-messenger program.

2 ANTARES neutrino alerts

A multi-wavelength follow-up program of ANTARES alerts, denoted TAToO (Telescopes-ANTARES Target of Opportunity) has been operating since 2009 [2]. It triggers optical and/or X-ray observations within a few seconds after the detection of selected high-energy neutrino events. In particular, more than 200 alerts have been sent to optical robotic telescopes (TAROT, ROTSE and MASTER) since mid-2009 while 12 X-ray targets of opportunity have been sent to the XRT instrument on board the Swift satellite since mid-2013. The angular resolution of the neutrino direction is better than 0.5° at high energy (> 1 TeV). Three online neutrino trigger criteria are currently used in TAToO: *(i)* detection of at least two neutrino candidates with similar directions (angular separation below 3°) within 15 minutes; *(ii)* detection of a single high-energy (> 7 TeV) neutrino candidate; *(iii)* detection of a neutrino candidate directionally consistent (< 0.5°) with a local galaxy (distance < 20 Mpc).

From January 2010 to January 2016, 93 alerts with early optical follow-up have been analyzed. No optical counterparts were found and upper limits on the R-band magnitude of a transient astrophysical source have been derived. By comparing these upper limits with optical afterglow light cuves of gamma-ray bursts (GRB), it becomes possible to reject a GRB association with each neutrino alert, in particular when the optical follow-up is performed within a few minutes after the neutrino trigger [3]. A similar analysis has been carried out with *Swift*-XRT follow-ups of 12 ANTARES alerts [3]. The probability to reject the GRB hypothesis reaches more than 70% if the X-ray follow-up occurs within 1.1 hour after the trigger.

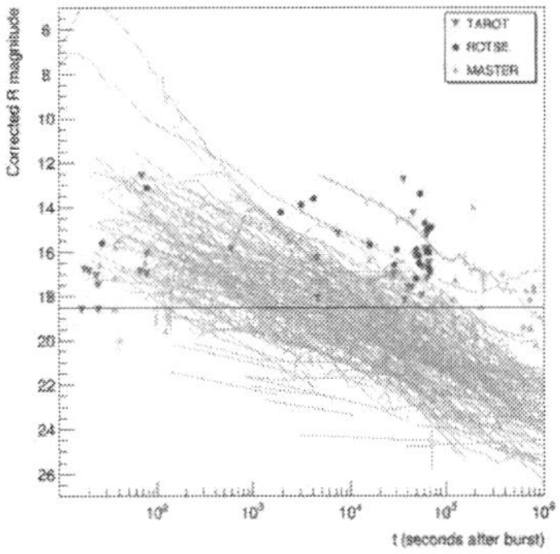

Figure 1. Corrected R-band magnitude as a function of time for 301 GRB afterglows. Red, blue and green dots indicate upper limits on GRB afterglow magnitudes for neutrino alerts observed by TAROT, ROTSE and MASTER respectively. The horizontal dashed line corresponds to the sensitivity of the optical telescopes.

Follow-up observations of ANTARES neutrino candidates are now performed over a broad range of the electromagnetic spectrum. Recently, the Murchinson Widefield Array (MWA), a low frequency (80 – 300 MHz) precursor of the Square Kilometre Array, searched for radio counterpart of two candidate high-energy neutrino events consistent with the locations of galaxies within 20 Mpc of Earth [4]. No counterparts were detected and upper limits for low-frequency radio luminosity have been derived. Likewise, two ANTARES alerts have been followed by H.E.S.S. and 36 by HAWC since November 2014.

3 Follow-up analyses

In addition to the follow-up observations described above, specific strategies are developed to look for neutrino events in both time and space coincidence with transient events announced by an alert distributed through the Gamma-ray Coordinated Network (GCN). Hereafter, we describe the follow-up analysis performed with ANTARES after the detection of the first gravitational-wave event by LIGO/Virgo in September 2015 (section 3.1) and following the detection of high-energy neutrino candidates by IceCube (section 3.2).

3.1 High-energy neutrino follow-up of the gravitational-wave event GW150914

The observation of two significant gravitational wave (GW) sources by Advanced LIGO on September 14th and December 26th, 2015 [5, 6] represents an important step forward in the era of multi-messenger astrophysics.

In a joint analysis with the IceCube and the LIGO/Virgo collaborations [7], we searched for directional and temporal coincidences between GW150914 and reconstructed HEN candidates. Relying on the methodology defined in [8], we looked for *(i)* temporal coincidences within a ±500 s time window around the GW alert and *(ii)* spatial overlap between the 90% probability contour of GW150914 and the neutrino point spread function. To this end, we used ANTARES's online reconstruction pipeline [3] which selects up-going neutrino candidates with atmospheric muon contamination less than 10%. An energy cut was also applied to reduce the background of atmospheric neutrinos which finally leads to an event rate of 1.2 events/day. Consequently, the expected number of neutrino candidates within 1000 s is 0.014. This corresponds to a Poisson probability of observing at least one background event of ~1.4%. No neutrino candidates temporally coincident with GW150914 were found with ANTARES while IceCube detected 3 events within the ±500 s time window. Both results are fully compatible with the background expectations. The absence of neutrino candidate both temporally and positionally coincident with GW150914 allowed us to derive an upper limit on the spectral fluence emitted in neutrinos by the source at 90% confidence level, as a function of the location of the source in equatorial coordinates. Two different spectral models were considered: a standard $dN/dE \propto E^{-2}$ model and a model with a spectral cutoff at 100 TeV expected for sources with exponential cutoff in the primary proton spectrum. Figure 2 shows in each direction of the sky the most stringent fluence upper limit (UL) provided either by ANTARES or IceCube (the white contour on figure 2 defines the region where ANTARES is the most sensitive) for the spectral model with cutoff.

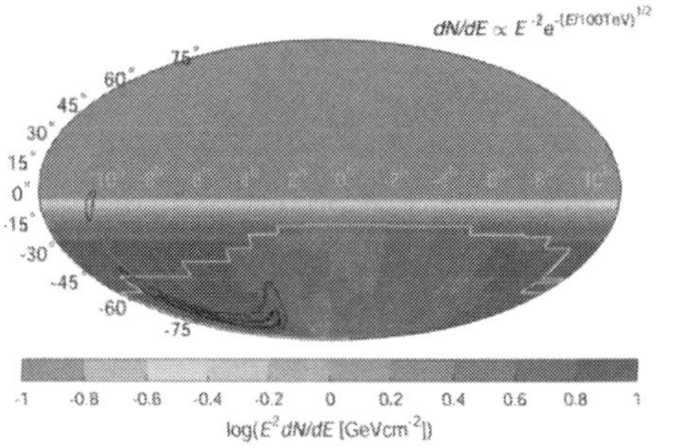

Figure 2. Upper limit on the HEN spectral fluence ($\nu_\mu + \bar{\nu}_\mu$) from GW150914 assuming the spectral model with cutoff at 100 TeV.

Using the constraints on the distance of the GW source and the neutrino fluence UL, we derived the ULs on the total energy emitted in neutrinos by this source. This was obtained by integrating the emission between 100 GeV and 100 PeV for each spectral model and each location in the sky map. The total energy UL depends on the source distance and equatorial coordinates. To account for these uncertainties, the lowest and highest total energy UL within the 90% confidence level interval are provided. The ULs on the total energy radiated in neutrinos are $5.4 \times 10^{51} - 1.3 \times 10^{54}$ erg and $6.6 \times 10^{51} - 3.7 \times 10^{54}$ erg respectively for the spectral model without and with cutoff. These ULs could be finally compared to the energy radiated in GW of $\sim 5 \times 10^{54}$ erg.

3.2 Follow-up of IceCube HEN events

IceCube is currently the largest neutrino telescope. Located at the geographic South Pole, it is composed of 86 detection lines distributed over a cubic-kilometer of ice. High-energy events starting into the detector (HESE, see e.g. [1]) and extremely high-energy ones (with energy above 1 PeV) are received by the Astrophysical Multi-messenger Observatory Network (AMON, [9]) and distributed to the community via an alert of the GCN. A coincident detection by both IceCube and ANTARES would be a significant proof of the astrophysical origin of these neutrino candidates and would point directly to the position of the source in the sky. In that context, the ANTARES collaboration is performing a follow-up analysis of each IceCube event whose position is below the horizon of ANTARES (which could consequently yield to an up-going event at the time of the alert). Up to now, ANTARES has followed three IceCube alerts [10–12]. No neutrino candidates were found compatible with one of the alerts within a time window of $\pm$ 1 day. We used these non-detections to derive preliminary 90% confidence level upper limits on the radiant neutrino fluence of these events of the order of $\sim$15 GeV $\cdot$ cm^{-2} and $\sim$30 GeV$\cdot$ cm^{-2} for the E^{-2} and the E$^{-2.5}$ spectral models respectively.

4 Conclusion

By simultaneously monitoring at least half of the sky, neutrino telescopes are well-suited to detect transient sources. In this context, multi-messenger approaches are destined for a bright future and will help to probe the physical processes at work in these objects. In particular, a multi-wavelength follow-up program has been operating in ANTARES since 2009 and enables to increase the sensitivity of the telescope by looking for a coincident electromagnetic detection both in time and space. Furthermore, the ANTARES collaboration is deploying specific strategies to search for joint detections of neutrinos and other messengers such as GWs. ANTARES will continue following-up future GW events during the second observing run of advanced LIGO, starting in December 2016. Because of the better angular accuracy of the neutrino telescopes compared to GW detectors with two interferometers, a coincident detection would drastically constrain the position of the GW source on the sky, bringing valuable information for subsequent electromagnetic follow-ups.

References

[1] Aartsen M G et al. (IceCube), *ApJ*, 809(1), 98 (2015).
[2] Ageron M et al., *Astroparticle Physics*, 35:530-536 (2012).
[3] Adrìan-Martínez S et al., (ANTARES), *JCAP*, 02:62 (2016).
[4] Croft S et al., *ApJL*, 820(2), 24 (2016).
[5] Abbott B P et al., *Phys. Rev. Lett.*, 116 061102 (2016).
[6] Abbott B P et al., *Phys. Rev. Lett.*, 116 241103 (2016).
[7] Adrìan-Martínez et al., *Phys. Rev. D*, **93** 122010 (2016).
[8] Baret B et al., *Astropart. Phys.* **35**, 1:7 (2011).
[9] Smith M W E et al., *Astroparticle Physics*, 45:56-70 (2013).
[10] Dornic D & Coleiro A, on behalf of ANTARES, *ATeL*, 9324 (2016).
[11] Dornic D & Coleiro A, on behalf of ANTARES, *ATeL*, 9440 (2016).
[12] Dornic D & Coleiro A, on behalf of ANTARES, *ATeL*, 9715 (2016).

Search for high energy neutrinos from bright GRBs with ANTARES

M. Sanguineti[1,2,a], S. Celli[3,4], and D. Turpin[5] on behalf of the ANTARES collaboration

[1] *Dipartimento di Fisica dell'Università, Via Dodecaneso 33, 16146, Genova, Italy*
[2] *INFN-Sezione di Genova, Via Dodecaneso 33, 16146, Genova, Italy*
[3] *Gran Sasso Science Institute, Viale Francesco Crispi 7, 00167, L'Aquila, Italy*
[4] *INFN-Sezione di Roma, P.le Aldo Moro 2, 00185, Roma, Italy*
[5] *Aix Marseille Université, CNRS/IN2P3, CPPM UMR 7346, 13288, Marseille, France*

Abstract. Gamma-ray bursts are a possible site of hadronic acceleration, thus neutrinos are expected in correspondence of a GRB event. The brightest GRB observed between 2008 and 2013 (GRB080916C, GRB110918A, GRB130427A and GRB130505A) have been investigated using the data of the ANTARES high energy neutrino telescope. In this paper two of most promising models of the GRB neutrino emission will be studied: the internal shock model and the photospheric model. No muons have been measured in space and time correlation with the selected GRBs and upper limits at 90% C.L. on the expected neutrino fluxes have been derived. This measure allows also setting constraints on some parameters used in the modeling of the neutrino flux: the bulk Lorentz factor of the jet Γ and the baryon loading f_p.

1 Introduction

The ANTARES detector (Astronomy with a Neutrino Telescope and Abyss environmental RESearch) is the largest neutrino telescope currently in operation in the Northern Hemisphere and the first operating in sea water. It is a three dimensional array of photo-multiplier tubes (PMTs), in which neutrinos are detected through the Cherenkov radiation induced by ultra-relativistic particles created in a neutrino interaction. The main goal of the ANTARES telescope is the detection of high energy cosmic neutrinos and in particular the identification of point-like sources.

In particular GRBs are one of most interesting types of celestial objects, they represent the most powerful outburst of energy in the Universe since the Big Bang itself. Gamma ray bursts are intense flashes of gamma rays, whose duration can vary from a fraction of a second up to a few minutes [1] [2]. Several gamma ray burst models have been proposed in the last few years, the most promising of which are the internal shock [1] and photospheric models [3] (see Sect. 2). Both scenarios predict a neutrino flux in correspondence of a GRB event. ANTARES ([4] and [5]) and IceCube ([6] and [7]) have already performed several research of a neutrino flux from GRBs, but they did not succeed in the identification of a significant excess of events over the expected background.

[a] e-mail: matteo.sanguineti@ge.infn.it

In Sec. 2 the two neutrino emission mechanism (internal shock and photospheric) are introduced and in Sec. 3 the analysis approaches will be described. Finally in Sec. 4 the derived upper limits on the neutrino flux and the constraints on the GRBs parameters will be presented (Sec. 5).

2 The internal shock and photospheric models

Both internal shock and photospheric models assume that the gamma ray emission is due to a relativistic jet of particle ejected by an inner engine, but the location of the interaction is different. In the case of the internal shock model the gamma rays are produced by the interaction of different shock waves inside the jet, on the other hand in the photospheric scenario the interaction takes place in the initial part of the expansion of the jet, when it is still opaque to photons.

The predicted neutrino energetic range is completely different in the two scenarios. In the case of the internal shock model neutrino are expected to be above 100 TeV [1], while in the photosphetic picture a low energy component is predicted [3]. For the internal shock model the GRB neutrino spectra have been computed thanks to the numerical code 'NeuCosmA' [8] and in the case of the photospheric scenario the analytical approach in [3] have been used. Both predictions rely on the measured parameters of the γ-ray emission light curve and spectrum, and assume as bulk Lorentz factor of the jet $\Gamma = 316$ and as baryon loading $f_p = 10$.

3 GRB selection and analysis

The brightest GRBs provide more probability of a neutrino flux discovery, because the per-burst neutrino fluence is directly scaled to the γ-ray fluence. The GRBs included in this analysis have a γ-ray fluence F_γ larger than 1×10^{-4}erg cm^{-2}. The selected GRBs were also below the horizon at the trigger time and information on the redshift of the bursts are available. Four bright GRBs fulfil all the requirements: GRB080916C, GRB110918A, GRB130427A and GRB130505A.

The data acquisition system of the ANTARES neutrino telescope is based on the unique "all-data-to-shore" concept: all photon signals recorded by the detector are transported to the shore station where filtering is performed. The filtering algorithms are also operating during GRB events, but in this case also raw data are saved on disks for a couple of minutes, so that data cover the majority of the burst duration. Because of the different energy ranges of the searches, in the internal shock model case the filtered data are used [9], while for the photospheric model study raw data are exploited. In both analysis the aperture of the search cone around the burst is set to 10°, while the search time window in the internal shock analysis is fixed to be equal to each burst duration with a symmetric extension of 2 seconds, while in the photospheric model case it depends on the raw data buffer duration. Since for GRB080916C and GRB110918A raw data are not available, the filtered data approach and its corresponding time window have been used to derive the photospheric model upper limits on the fluence for these two gamma ray bursts.

The MC signal simulations and the analysis optimisation are performed independently for each burst, while the background is estimated though data. The signal and background angular resolution are derived in order to compute pseudo-experiments, relying on an extended maximum likelihood ratio test statistic. This procedure is repeated requiring different track quality parameters: the set of cuts that maximize the model discovery potential is chosen.

 DOI: 10.1051/epjconf/201713604004

4 Results

No neutrino events were found spatially and temporally in coincidence with any of the four bright GRBs presented, so 90% C.L limits on the neutrino fluence have been computed for the internal shock model and the photospheric model (Figure 1).

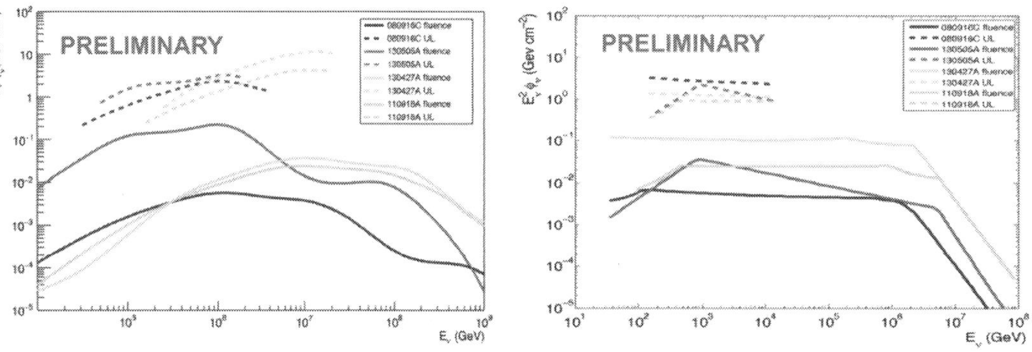

Figure 1. Expected $\nu_\mu + \overline{\nu}_\mu$ fluence (solid line) and ANTARES 90% C.L. upper limit (dashed line) on the selected GRBs, in the energy band where 90% of the signal is expected to be detected by ANTARES according to internal shock model (left) and photospheric model (right).

5 Constraints on GRB parameters

The 90% C.L limits on the neutrino fluence allows setting constraints on the free parameters that significantly impact the neutrino flux such as the baryonic loading factor f_p and the bulk Lorentz factor Γ. In Figure 2 the best 90% and 50% C.L exclusion limits in the $\Gamma - f_p$ plane for both the internal shock model and the photospheric scenario are shown: they are respectively on GRB130505A and on GRB130427A. The results are obtained assuming that $f_p \in [1 - 200]$ and $\Gamma \in [10 - 900]$ and that the two parameters are not correlated.

The constraints on GRB130505A start to significantly challenge the internal shock scenario up to $\Gamma \sim 200$, while constraints on GRB130427A according to the photospheric picture can rule out a high baryonic content ($f_p < 100$) in its jet.

6 Conclusions

The analysis of the four most promising gamma ray bursts did not yield a measurement of a neutrino flux in correspondence of the GRB events, but it allows setting an upper limit on the neutrino flux from the gamma ray burst for two different emission models (internal shock and photospheric). Two dedicated analysis methodology have been developed in order to maximize the detector sensitivity for each model and both analysis were optimized in order to provide the highest model discovery potential for each burst. This measurement allows setting constraints on the parameters that affect the neutrino yield: the bulk Lorentz factor of the jet Γ and the baryon loading f_p. A similar analysis has been performed by the IceCube collaboration [10] excluding small bulk lorentz boost factor and large baryonic loading. However the photospheric limits shown in Figure 1 (right) constrain emission at substantially lower energies than the IceCube results.

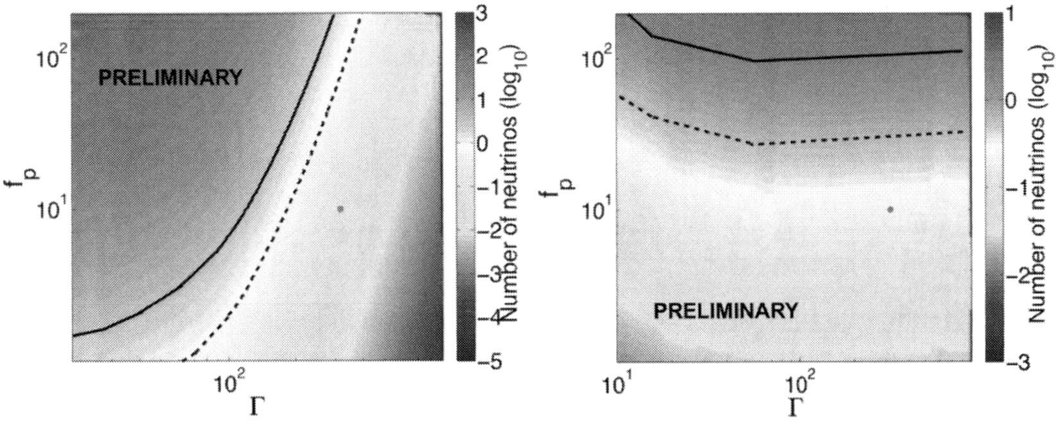

Figure 2. Constraints on the $\Gamma - f_p$ plane. The solid (dashed) black line corresponds to the exclusion limits at 90 (50)% C.L. The red dot shows the benchmark value $f_p = 10$ and $\Gamma = 316$. Internal shock scenario for GRB130505A (left) and photospheric scenario for GRB130427A (right).

The future KM3NeT-ARCA telescope will be the ideal instrument to constrain the neutrino flux from GRBs. It will be a kilometre scale detector composed of a two building blocks of 115 detection strings each. The detection of neutrinos from gamma ray bursts will provide important information of the process of emission of GRB, in particular it will give strong indications of the hadronic interactions occurring in the jets of the gamma ray bursts and it will help to distinguish between the different GRB model scenarios.

References

[1] T. Piran, Rev. Mod. Phys. **76**, 1143 (2004), astro-ph/0405503
[2] P. Meszaros, Rept. Prog. Phys. **69**, 2259 (2006), astro-ph/0605208
[3] P. Kumar, B. Zhang, Phys. Rept. **561**, 1 (2014), 1410.0679
[4] S. Adrian-Martinez et al. (ANTARES), JCAP **1303**, 006 (2013), 1302.6750
[5] S. Adrian-Martinez et al. (ANTARES), Astron. Astrophys. **559**, A9 (2013), 1307.0304
[6] M.G. Aartsen et al. (IceCube), Astrophys. J. **805**, L5 (2015), 1412.6510
[7] M.G. Aartsen et al. (IceCube), Astrophys. J. **824**, 115 (2016), 1601.06484
[8] S. Hummer, M. Ruger, F. Spanier, W. Winter, Astrophys. J. **721**, 630 (2010), 1002.1310
[9] S. Celli et al. (ANTARES), Proceeding of RICAP16, EPJ Web Conf. (2016)
[10] M.G. Aartsen et al. (IceCube), Astrophys. J. **824**, 115 (2016), 1601.06484

EPJ Web of Conferences **136**, 04005 (2017) DOI: 10.1051/epjconf/201713604005
RICAP16

Searches for diffuse fluxes of astrophysical neutrinos with the ANTARES telescope

Luigi Antonio Fusco[1,2,a], Steffen Hallmann[3], Sergio Navas[4], and Federico Versari[1,2]on behalf of the ANTARES Collaboration

[1] *Dipartimento di Fisica e Astronomia dell'Università, Viale Berti-Pichat 6/2, 40127 Bologna, Italy*
[2] *INFN - Sezione di Bologna, Viale Berti-Pichat 6/2, 40127 Bologna, Italy*
[3] *Friedrich–Alexander-Universität Erlangen–Nürnberg, Erlangen Centre for Astroparticle Physics, Erwin-Rommel-Str. 1, 91058 Erlangen, Germany*
[4] *Departamento de Fisica Teorica y del Cosmos, Universidad de Granada, Av. Severo Ochoa, 18071 Granada, Spain*

Abstract. In this proceedings we report on the status of searches for diffuse fluxes of cosmic neutrinos with the ANTARES neutrino telescope data. A complete overview of diffuse neutrino searches will be given, together with the search for a neutrino emission from regions such as the Fermi Bubbles or the Galactic Plane. A non-significant, though intriguing, excess of events above the atmospheric background is observed in all-sky analysis both for the track and shower channels.

1 Introduction

Cosmic Rays (CR) produced at astrophysical objects such as Supernova Remnants, Active Galactic Nuclei or Gamma Ray Bursts can interact in the interstellar medium and produce neutrinos. A diffuse flux of cosmic neutrinos is expected from unresolved individual sources or CR propagation in the Universe. The energy spectrum of these neutrinos should be similar to that of primary CRs and flatter than the observed atmospheric neutrino background.

The ANTARES detector [1] is at present the largest neutrino telescope in the Northern Hemisphere, located at a depth of 2475 m in the Mediterranean Sea, 40 km off-shore Toulon, France and has been continuously operated since 2007. It consists of 885 photomultiplier tubes (PMTs), distributed along 12 mooring lines anchored to the sea bed and kept taut by top buoys. It detects the products of all-flavour neutrino interactions by collecting the Cherenkov photons emitted by relativistic charged particles. Charged current (CC) ν_μ interactions, where a straight muon track is produced, can be reconstructed in direction with a good pointing accuracy [2]. All-flavour neutral current (NC) interactions, as well as CC ν_e and ν_τ interactions are as well reconstructed with good accuracy, thanks to the optical properties of water [3].

The main background in searches for diffuse fluxes of a cosmic signal is given by atmospheric neutrinos and muon bundles produced in CR air showers. To reduce the background from atmospheric

[a] e-mail: lfusco@bo.infn.it

muons, only upward-going events are selected since only neutrinos can traverse the Earth. Nonetheless, wrongly reconstructed atmospheric muons can survive this directional selection: a proper cut on the reconstruction quality parameter can reduce this background. The background from atmospheric neutrinos, whose spectrum is softer, is reduced by a cut on energy estimators [4].

The IceCube Collaboration has reported [5] the observation of an all-flavour diffuse and isotropic excess of high energy neutrinos, not compatible with atmospheric expectations. Different approaches are possible to analyse this signal and the efforts made by the ANTARES collaboration are presented in the following sections. Blinded analysis procedures have been followed, with only 10% of data available for tests and data/Monte Carlo comparison. The good agreement between data and Monte Carlo allowed for a final cut choice based on simulations.

2 Diffuse flux analyses

The analysis of diffuse ν_μ fluxes uses track-like events in the complete ANTARES data-sample (2007-2015). The equivalent livetime is 2451 days. Most of the down-going atmospheric muon events are rejected applying a preliminary cuts on the reconstructed zenith angle ($\theta_{zen} > 90°$) and on the angular error estimation β ($\beta < 0.5°$). To further reduce the atmospheric background combined cut on reconstruction quality parameter, the reduced log-likelihood Λ, and on the number of PMT-signals (hits) related to the reconstructed track, N_{hit}, has been applied, thus reducing the muon background level well below 1%.

Atmospheric neutrinos can be rejected by applying an energy-related cut, based on the estimation of the muon energy using an Artificial Neural Network [6]. The optimal cut on this variable is chosen by the Model Rejection Factor (MRF) procedure [7] to give the best sensitivity flux. After the final cut 13.5^{+3}_{-4} events are expected from the background and 3 events should be produced by an IceCube-like signal [5]. After unblinding and applying the energy selection, 19 events are found in data, as shown in figure 1 - left.

All events reconstructed with the shower algorithm of [3] are considered in the cascade analysis. Data from 2007 to 2013, with a total livetime of 1451 days, have been used. An event pre-selection is done using a containment cut on the vertex position and requiring events to be upward-going. Various quality parameters and track-distriminating quantities are used to reduce the background from atmospheric muons. Finally the MRF optimisation is done on the reconstructed shower energy. The optimal cut is found to be $E_{shower} > 30$ TeV. With this selection, 5±2 background events are expected, and an IceCube-like signal flux would correspond to about 1.5 cosmic events added to the atmospheric expectations. 7 events are observed in data after unblinding as shown in figure 1 - right.

By construction there is no overlap between the track and the shower samples, since an anti-track selection is applied in the shower selection chain. The samples can be combined and, overall, 26 events are observed in data over an expected background of 18±5. Since the excess is not statistically significant ($<1.5\sigma$), an upper limit at 90% confidence level (c.l.) can be computed according to [8] and is shown in figure 2. This estimation of the upper limit does not take into account yet systematic effects, mainly related to the uncertainties of the optical module efficiency and water properties.

3 Special regions

Southern-sky areas showing a diffuse and enhanced emission of γ-rays are suitable targets to search for a diffuse neutrinos emission, since ANTARES has large exposure to these parts of the sky. For this kind of analysis, an on/off-zone procedure has been followed: the background is estimated from regions of the sky with the same exposure as the expected signal regions and is then compared to the

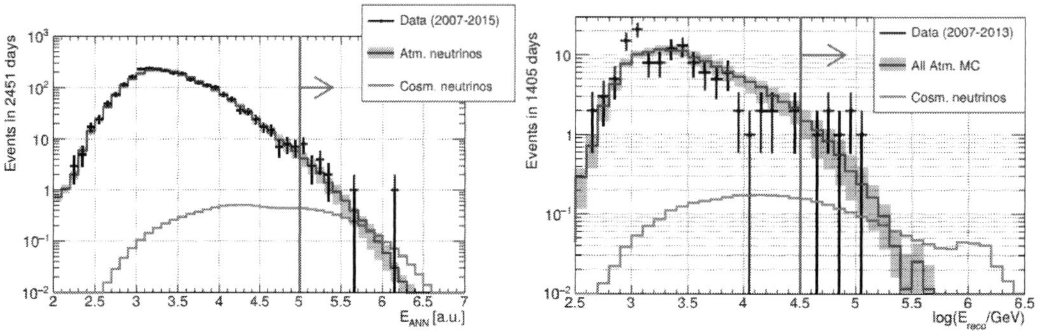

Figure 1: Left: energy estimator distribution for the track channel after the event selection chain. Right: same for the shower channel. Red lines indicate the cosmic neutrino expectation for the IceCube cosmic flux, while blue lines represent the contribution from atmospheric events, with the shaded areas providing an estimate of the related uncertainties. The gray lines and arrow show the chosen energy-related cut.

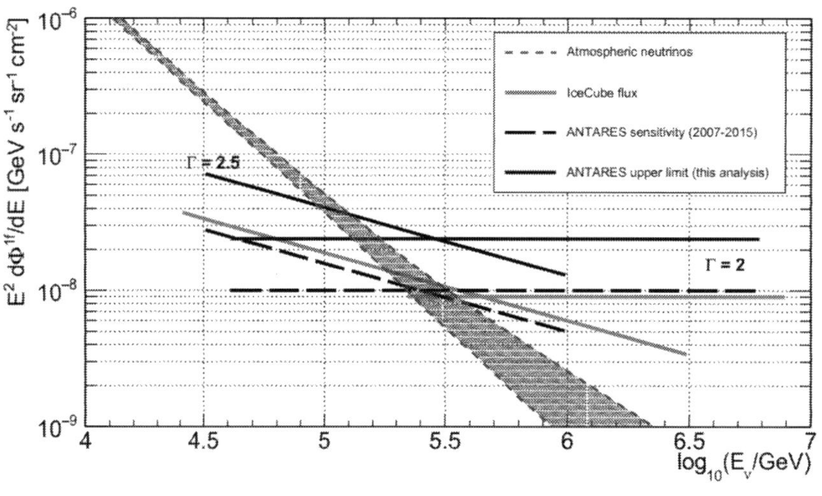

Figure 2: Upper limit for the combination of the two analyses (solid black lines) for the two spectral indices considered (Γ = 2 and 2.5) compared to the final sensitivity (black dashed lines) expected when including the whole available livetime. Also shown in red the best fit from IceCube analyses for the two spectral hypotheses and the contribution from atmospheric neutrinos (blue shaded area).

observation in the signal region. In the absence of signal, the observed number of events in on and off-zones should be statistically compatible.

3.1 Fermi Bubbles

Fermi telescope data [9, 10] have revealed the presence of two giant γ-ray emission regions above and below the Galactic Plane. If hadronic mechanisms are responsible for the production of such a signal,

 DOI: 10.1051/epjconf/201713604005

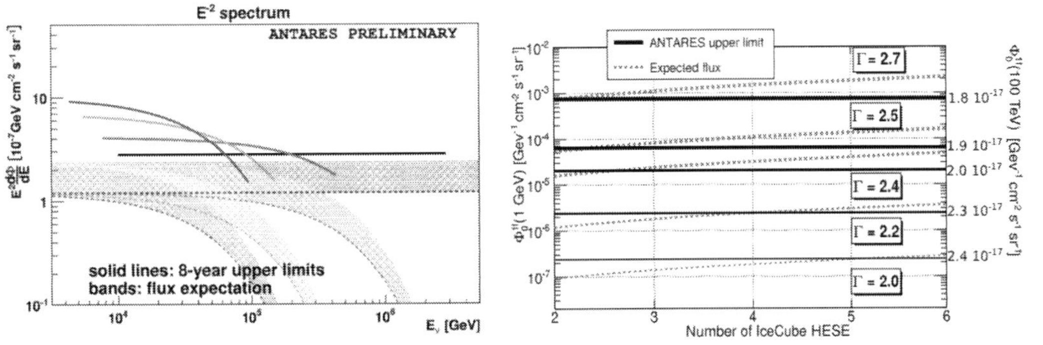

Figure 3: Left: upper limit calculation for neutrino fluxes from the Fermi Bubbles for different energy cut-offs (∞, 100 TeV, 50 TeV, 10 TeV - black, red, green and blue lines). Shaded areas are for theoretical predictions with the various cuf-offs. Right: upper limits and sensitivity for Galactic Plane analyses from ANTARES for different spectral indexes compared to a flux producing a certain number of IceCube High Energy Starting Events (3 years sample) [16].

diffuse neutrino emissions are expected from these regions with various possible energy cut-offs from few to some hundreds of TeV [11].

Data collected in the ν_{μ} CC channel with the ANTARES telescope from 2007 to 2015 are considered in this analysis. The event selection is based on cuts on the quality of up-going reconstructed tracks and, as in the previous analysis, on the energy estimator based on the ANN method. The final selection was obtained by minimizing the MRF and optimizing for a neutrinos flux with a 100 TeV cut-off. After the unblinding of the on-zone, 28 events are observed, while 19.7 are expected, on average, from the off-zones. The significance of this excess can be estimated, following the prescription of [12], as 1.5σ, and 90% c.l. upper limits, computed as in [8] are shown in figure 3.

3.2 Galactic Plane

A diffuse neutrino flux is expected from the decays of charged mesons produced in CR interactions in the interstellar medium in the Galactic Plane [13]. The corresponding emission from neutral mesons is clearly visible in γ-ray observation of the sky [14]. Different models for the CR propagation are proposed and can be tested using neutrino data.

In this case, as in the case of the diffuse analysis, only muon data from 2007 to 2013 have been analyzed [15] following the MRF procedure to set an optimal cut on the area of the on-zone and on the quality and energy of reconstructed tracks. The 9 selected off-zones show on average 3.5 events, and 2 are observed in the on-zone after unblinding. Correspondingly, 90% c.l. upper limits are put (equal to the sensitivity) on the possible contribution from the Galactic Plane to the IceCube signal from [16] as shown in figure 3.

4 Conclusions

Data collected by the ANTARES telescope from 2007 to 2015 have been analysed and various searches have produced interesting results regarding the diffuse neutrino signal observed by the IceCube collaboration. An intriguing excess of events is reported in both the channels of all-sky searches, even though still compatible with the expectations from atmospheric backgrounds. Stringent limits are also being put on the possible galactic origin of the IceCube signal, lowering the possible number of IceCube HESE events origintating from the Galactic Plane down to 2 for the three years sample.

More data are being analysed and the expected sensitivity for the whole data sample in all-sky searches is close to the observed IceCube flux for all the considered scenarios. Further improvements are possible especially in the shower channel, for which the event selection chain can be optimised in order to reduce the background from atmospheric events. This is also true for special region searches, since the angular resolution of ANTARES in the shower channel is of the order of few degree and an enhancement of the detector performance is expected when adding also cascade-like events in the analysis [17].

References

[1] M. Ageron et al., Nucl. Instr. and Meth. A **656**: 11 (2011).
[2] J. Barrios Martí et al., Proceedings of the 34th ICRC, The Hague, Netherlands, arXiv:1510.04508 (2015).
[3] T. Michael et al., Proceedings of the 34th ICRC, The Hague, Netherlands, arXiv:1510.04508 (2015).
[4] S. Adrián-Martínez et al. EPJ C **73**: 2606 (2013).
[5] M.G. Aartsen, et al., Astrophys. J. **809**: 98 (2015).
[6] J. Schnabel et al., Nucl. Instr. and Meth. A **725**: 106-109 (2013).
[7] G.C. Hill and K. Rawlins, Astropart. Phys. **19**: 393 (2003).
[8] G.J. Feldman and R.D. Cousins, Phys. Rev. D **57**: 3873-3889 (1998).
[9] M. Su et al., Astrophys. J. **724**: 1044 (2010).
[10] M. Ackermann et al., Astrophys. J. **793**: 64 (2014).
[11] R. M. Crocker and F. Aharonian, Phys. Rev. Lett. **106**: 101102 (2011).
[12] T. P. Li and Y. Q. Ma, Astrophys. J. **272**: 317 (1983).
[13] D. Gaggero et al., ApJ Letters **815**: L25 (2015).
[14] M. Ackermann et al., Astrophys. J. **750**: 3 (2012).
[15] S. Adrián-Martínez et al. Phys. Lett. B **760**: 143-148 (2016).
[16] M.G. Aartsen et al., Phys. Rev. Lett. **113**:101101 (2014).
[17] T. Grégoire et al. Proceedings of the 27th Neutrino Conference, London, UK (2016).

A time-dependent search for high-energy neutrinos from bright GRBs with ANTARES

Silvia Celli[1,2,3,a]
On behalf of the ANTARES Collaboration

[1] *Gran Sasso Science Institute, Viale Francesco Crispi 7, 67100 L'Aquila, Italy*
[2] *INFN-Sezione di Roma, Piazzale Aldo Moro 2, 00185 Roma, Italy*
[3] *Dipartimento di Fisica dell'Università La Sapienza, Piazzale Aldo Moro 2, 00185 Roma, Italy*

Abstract. Astrophysical point-like neutrino sources, like Gamma-Ray Bursts (GRBs), are one of the main targets for neutrino telescopes, since they are among the best candidates for Ultra-High-Energy Cosmic Ray (UHECR) acceleration. From the interaction between the accelerated protons and the intense radiation fields of the source jet, charged mesons are produced, which then decay into neutrinos. The methods and the results of a search for high-energy neutrinos in spatial and temporal correlation with the detected gamma-ray emission are presented for four bright GRBs observed between 2008 and 2013: a time-dependent analysis, optimised for each flare of the selected bursts, is performed to predict detailed neutrino spectra. The internal shock scenario of the fireball model is investigated, relying on the neutrino spectra computed through the numerical code NeuCosmA. The analysis is optimized on a per burst basis, through the maximization of the signal discovery probability. Since no events in ANTARES data passed the optimised cuts, 90% C.L. upper limits are derived on the expected neutrino fluences.

1 Introduction

The search for neutrinos from astrophysical sources is motivated by the still poor understanding of the origin of UHECRs. GRBs are huge explosive phenomena, that offer a promising environment for proton and heavier nuclei acceleration at shock fronts. Photo-hadronic interactions produce high-energy gamma-rays and neutrinos mainly through the following resonant channel:

$$p + \gamma \xrightarrow{\Delta^+} \left\{ \begin{array}{l} p + \pi^0 \\ n + \pi^+ \end{array} \right. \longrightarrow \begin{array}{l} \pi^0 \longrightarrow \gamma + \gamma \\ n \longrightarrow p + e^- + \overline{\nu}_e \\ \pi^+ \longrightarrow \mu^+ + \nu_\mu \\ \mu^+ \longrightarrow e^+ + \nu_e + \overline{\nu}_\mu \end{array} \quad (1)$$

In addition, kaon production leads to a high-energy component in neutrino spectra. The search for such high-energy neutrinos from bright sources is motivated by the fact that theoretical GRB-models, as the internal shock scenario of the fireball model [1], predict a neutrino flux linearly scaled with the detected γ-ray flux. Hence, after a brief description of the ANTARES detector in Sec. 2, the selection

[a] e-mail: silvia.celli@gssi.infn.it

of bright sources is introduced in Sec. 3. The methods of the time-dependent analysis are discussed in Sec. 4. Finally, in Sec. 5, results are presented and conclusions are derived.

2 The ANTARES neutrino telescope

The ANTARES detector [2] is the largest operational neutrino telescope of the Northern hemisphere, located in the Mediterranean sea, at a depth of about 2475 m: it is composed by 12 vertical strings, instrumented with 885 photo-multipliers, detecting the Cherenkov light induced by the passage in water of ultra-relativistic charged particles. The search for astrophysical point-like neutrino sources is among the ANTARES scientific goals: the signature of a point source is a cluster of events in the defined direction of the sky, where the source is located. In order to reduce the huge background of atmospheric muons, the search is performed by selecting up-going track-like events: GRBs from below ANTARES horizon are therefore considered in this search for muon neutrinos. Moreover, since GRBs are transient sources, a significant suppression of the background is further achieved by considering only the reduced time window corresponding to the bursting event. This constitutes the main difference with respect to steady source searches.

Figure 1: An artistic picture of the underwater ANTARES neutrino telescope, located in the Mediterranean sea, offshore Toulon (Fr), at a depth of about 2474 m.

3 Bright GRB selection and time-dependent neutrino spectra

Bright GRBs are here investigated since, according to the fireball model, neutrino fluxes are expected to scale with gamma-ray fluxes: GRBs between 2008 and 2013 are selected from the Fermi and Swift online catalogues and from the Konus-Wind GCNs with γ-ray fluence $F_\gamma > 10^{-4}$ erg/cm^2; it is also required that the sources have measured redshift and are below the ANTARES horizon at the trigger time (while ANTARES is taking physics data). Four GRBs satisfied such selection criteria: GRB 080916C, GRB 110918A, GRB 130427A and GRB 130505A. Their γ-ray light curves are represented in Fig. 2(a). In order to compute the neutrino spectra of each GRB according to the internal shock fireball model, the parameters measured by the γ-ray satellites are used: the detailed computation of neutrino emissions refers to single bursting episodes present in each GRB light curve. Such parameters are presented in Tab. 1. In this way, a time-dependent neutrino spectrum is predicted: the total expected fluence per burst is then obtained summing up the contributions from each time bin. The resulting neutrino expectation for GRB 080916C is presented in Fig. 2(b). The simulation through the numerical NeuCosmA [7] assumes $\Gamma = 316$ for the bulk Lorentz factor, $f_p = 10$ for the baryonic content, $\epsilon_e = \epsilon_B = 0.1$ as fraction of the jet kinetic energy transferred to electrons and to magnetic field, $\langle x_{p\to\pi} \rangle = 0.2$ as average fraction of energy transferred from proton to pion.

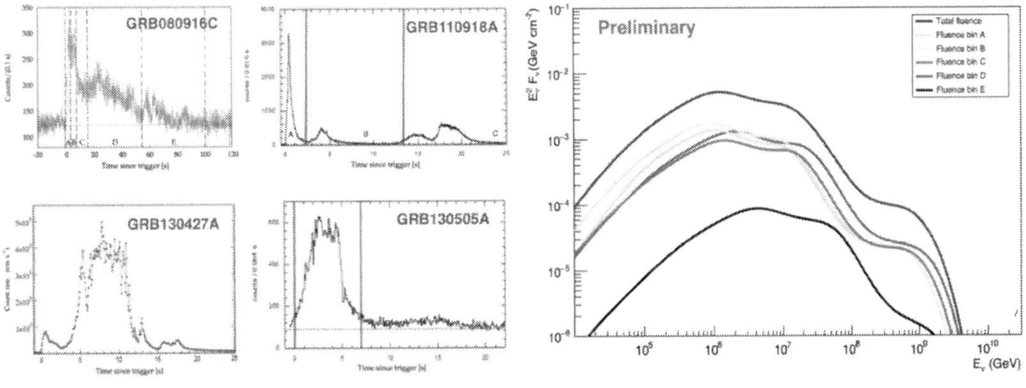

(a) (b)

Figure 2: (a) γ-ray lightcurves of the selected GRBs. *Top left*: GRB080916C (Fermi:GBM/LAT [3]); *Top right*: GRB 110918A (Konus-Wind [4]); *Bottom left*: GRB 130427A (Konus-Wind [5]); *Bottom right*: GRB 130505A (Konus-Wind [6]). (b) Time-dependent neutrino spectra of GRB 080916C: $\nu_\mu + \bar{\nu}_\mu$ fluxes are obtained according to NeuCosmA, for the five time bins reported in Tab. 1.

Table 1: Name of the burst and name of the detector reporting the γ-ray parameters used in the time-dependent analysis: time bin of the analysis, duration T, fluence F_γ (in the energy range from 20 keV to 2 MeV for GRB 080916C and from 20 keV to 10 MeV for the others), low-energy spectral index α, high-energy spectral index β, break energy of the spectrum, redshift z and variability time scale t_{var} (* is a default value).

NAME	Detector	BIN	T (s)	F_γ (10^{-4} erg/cm^2)	α	β	E_γ (keV)	z	t_{var} (s)
GRB 080916C	Fermi:GBM/LAT	A	3.6	0.15	-0.58	-2.63	440	4.35	0.23
"	"	B	4.1	0.21	-1.02	-2.21	1170	"	"
"	"	C	8.2	0.16	-1.02	-2.16	490	"	"
"	"	D	38.9	0.53	-0.92	-2.22	400	"	"
"	"	E	46.1	0.11	-1.05	-2.16	230	"	"
GRB 110918A	Konus-Wind	A	2.3	4.03	-1.95	-2.41	990	0.98	0.25
"	"	B	11.0	2.06	-1.00	-2.60	250	"	"
"	"	C	15.1	1.57	-1.20	-3.30	78	"	"
GRB 130427A	Konus-Wind	-	18.7	26.8	-0.96	-4.14	1028	0.34	0.04
GRB 130505A	Konus-Wind	-	7.0	3.13	-0.69	-2.03	631	2.27	0.01*

4 Analysis methods

Monte Carlo signal simulations have been performed for each burst, to account for the detector condition at the instrument trigger time. The number of background events μ_b at each GRB position in the angular and temporal search windows was evaluated from data, following the procedure outlined in [8]: the search time window is selected as the whole burst duration plus a symmetric extension of 2 seconds, while the angular window is a cone around the source with semi-aperture $\alpha = 10°$. In this way the normalized signal and background angular probability density function, $S(\alpha)$ and $B(\alpha)$, are obtained and pseudo-experiments relying on them are performed in both the background only and signal plus background hypothesis. At each pseudo-experiment with n_{tot} events, an extended maximum likelihood ratio test statistic, defined as [9]

$$Q = \max_{\mu'_s \in [0;n_{tot}]} \Big(\sum_{i=1}^{n_{tot}} \log \frac{\mu'_s S(\alpha_i) + \mu_b B(\alpha_i)}{\mu_b B(\alpha_i)} - \mu'_s \Big) \tag{2}$$

is evaluated in order to obtain the estimated number of signal events μ'_s. This technique is applied to event samples with different reconstructed track-quality parameter: such a cut is then optimised in order to yield the maximum probability of discovery assuming the NeuCosmA model.

5 Results and Conclusions

The expected signal and background event rates for each GRB at the optimal cuts and in the defined search windows are given in Tab. 2: for GRB 110918A, GRB 130427A and GRB 130505A, the expected background is negligible with respect to the predicted signal. After the data unblinding, no neutrino event was observed in temporal and spatial coincidence with any of the GRBs investigated, therefore 90% C.L. upper limits have been computed on the foreseen neutrino fluence. Defining the differential neutrino fluence ϕ_ν, ANTARES limits, expressed as $E^2\,\phi_\nu$, are in the interval $[10^{-1} - 10]$ GeV cm^{-2} (Fig. 3). The acceleration of UHECRs in GRBs cannot yet be ruled out. However, some parameters of the model tested here can be constrained [10].

Table 2: Expected number of signal μ_s and background μ_b events from each GRB at the optimal cuts.

	μ_s	μ_b
GRB 080916C	1.8×10^{-3}	8.6×10^{-3}
GRB 110918A	1.3×10^{-2}	7.2×10^{-3}
GRB 130427A	7.5×10^{-3}	4.1×10^{-3}
GRB 130505A	1.6×10^{-1}	2.4×10^{-3}

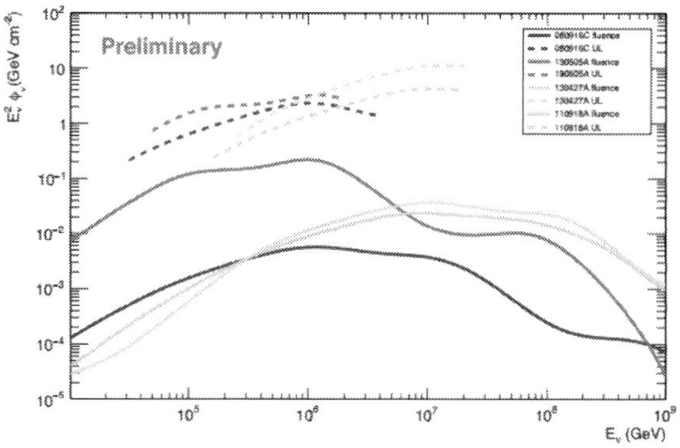

Figure 3: Expected neutrino fluences (solid lines) and ANTARES 90% C.L. upper limits (dashed lines).

References

[1] Guetta D et al., Astropart. Phys. **20**, 429-455 (2004)
[2] Ageron M at al., Nucl. Instrum. Meth. A **656**, 11-38 (2011)
[3] Abdo A et al., Science Mag. **323**, 1688 (2009)
[4] Frederiks D et al., Astrophys. J. **779**, 151 (2013)
[5] http://gcn.gsfc.nasa.gov/gcn3/14487.gcn3
[6] http://gcn.gsfc.nasa.gov/gcn3/14575.gcn3
[7] Hümmer S, Rüger M, Spanier F and Winter W, Astrophys. J. **721**, 630 (2010)
[8] Adrián-Martínez S et al., Astron. Astrophys. **559**, A9 (2013)
[9] Barlow R, Nucl. Instr. Meth. A **297**, 496 (1990)
[10] Sanguineti M et al., Proceeding of RICAP16, EPJ Web Conf

Baikal-GVD

A.D. Avrorin[1], A.V. Avrorin[1], V.M. Aynutdinov[1], R. Bannash[9], I.A. Belolaptikov[2], V.B. Brudanin[2], N.M. Budnev[3], I.A. Danilchenko[1], S.V. Demidov[1], G.V. Domogatsky[1], A.A. Doroshenko[1], R. Dvornicky[2,7], A.N. Dyachok[3], Zh.-A.M. Dzhilkibaev[1], L. Fajt[7,8], S.V. Fialkovsky[5], A.R. Gafarov[3], O.N. Gaponenko[1], K.V. Golubkov[1], T.I. Gress[3], Z. Honz[2], K.G. Kebkal[9], O.G. Kebkal[9], K.V. Konischev[2], A.V. Korobchenko[2], A.P. Koshechkin[1], F.K. Koshel[1], A.V. Kozhin[4], V.F. Kulepov[5], D.A. Kuleshov[1], M.B. Milenin[5], R.A. Mirgazov[3], E.R. Osipova[4], A.I. Panfilov[1], L.V. Pan'kov[3], E.N. Pliskovsky[2], M.I. Rozanov[6], E.V. Rjabov[3], F.A. Shamakhov[2], B.A. Shaybonov[2], A.A. Sheifler[1], M.D. Shelepov[1], F. Simkovic[7,2], A.V. Skurihin[4], A.A. Smagina[2], I. Stekl[2], O.V. Suvorova[1,a], V.A. Tabolenko[3], B.A. Tarashansky[3], S.A. Yakovlev[9], A.V. Zagorodnikov[3], and V.L. Zurbanov[3]

[1]*Institute for Nuclear Research, Moscow, 117312 Russia*
[2]*Joint Institute for Nuclear Research, Dubna, 141980 Russia*
[3]*Irkutsk State University, Irkutsk, 664003 Russia*
[4]*Institute of Nuclear Physics, Moscow State University, Moscow, 119991 Russia*
[5]*Nizhni Novgorod State Technical University, Nizhni Novgorod, 603950 Russia*
[6]*St. Petersburg State Marine Technical University, St. Petersburg, 190008 Russia*
[7]*Comenius University, Mlynská dolina F1, SK-842 48 Bratislava, Slovakia*
[8]*Czech Technical University in Prague, 12800 Prague, Czech Republic*
[9]*EvoLogics, Germany*

Abstract. We present the status of the Gigaton Volume Detector in Lake Baikal (Baikal-GVD) designed for the detection of high energy neutrinos of astrophysical origin. The telescope consists of functionally independent clusters, sub-arrays of optical modules (OMs), which are connected to shore by individual electro-optical cables. During 2015 the GVD demonstration cluster, comprising 192 OMs, has been successfully operated in Lake Baikal. In 2016 this array was upgraded to baseline configuration of GVD cluster with 288 OMs arranged on eight vertical strings. Thus the instrumented water volume has been increased up to about 5.9 Mtons. The array was commissioned in early April 2016 and takes data since then. We describe the configuration and design of the 2016 array. Preliminary results obtained with data recorded in 2015 are also discussed.

1 Introduction

The deep underwater neutrino telescope, Baikal Gigaton Volume Detector (GVD), is currently under construction in the southern basin of Lake Baikal [1]. The underwater depth is about 1360 m at a distance of more than three kilometres away from the shore. At the detector site the combination of hydrological, hydro-physical, and landscape factors were studied and found to be optimal for the deployment and the operation of the neutrino telescope [1]. The Baikal-GVD neutrino experiment is

[a]e-mail: suvorova@cpc.inr.ac.ru

targeting on high energy neutrinos that are expected from astrophysical sources including those visible in gamma-rays and from directions of accumulation sites of dark matter (DM) as the Galactic Centre or the Sun [2]. The Baikal-GVD is composed by a three-dimensional lattice of optical modules. These are photomultiplier tubes housed in transparent pressure resistent spheres, arranged in vertical load-carrying cables to form strings. The telescope has a modular structure and consists of functionally independent clusters, sub-arrays of 8 strings of OMs, which are connected to shore by individual electro-optical cables. The GVD demonstration cluster named "Dubna" has been deployed in 2015 in Lake Baikal. The array operation time spans 213 days from April 2015 till February 2016. In April 2016 the "Dubna" array was upgraded to the baseline configuration of the GVD-cluster which comprises 288 optical modules (OMs) attached to 8 strings at a depth ranging from 750 m to 1275 m. Here we describe the configuration and the basic parameters of the array deployed in 2016 in Lake Baikal and discuss the first preliminary results obtained from the data collected with the "Dubna" array in 2015.

2 Baikal-GVD cluster

The first configuration of the Baikal-GVD cluster was deployed in April 2015. It consisted of eight 345 m long strings of optical modules [4] spaced by 40 m (192 OMs in total). Each string comprises two sections of OMs. Each section consists of 12 optical modules and a central electronics module (CeM). One Hamamatsu R7081-100 photomultiplier tube with a 10-inch hemispherical photocathode and quantum efficiency up to 35% is used as light sensor of OM. The signals arriving from the OMs are digitized in ADC boards located at the CeM (12 channels with a sampling frequency of 200 MHz). When digitized, the signals from the ADC are transferred into a programmable gate array (FPGA). The FPGA memory buffer allows acquisition of the input signal waveform in a time interval of 5 μs. A FPGA logic has been configured to generate a trigger, read data from the ADC channels, and transmit them through Ethernet to the data acquisition center. The highest rate of data transfer from each section has been obtained with SHDSL modems — up to 10 Mbit/s. To reach the required event registration detector rate (100 Hz or higher) to be attained, the data arriving from the sections are processed in real time mode. Details on the data acquisition, the basic controls, the methods of calibrations and on the hardware and software triggers can be found in [3]. The cluster was operated from April 2015 to February 2016. About 1.6×10^9 events have been recorded during the cluster operation. The estimated operation efficiency was of about 72%. The first preliminary results obtained with the "Dubna" cluster are presented in section 3.

In winter expedition 2016 the cluster was extended with an additional a third top section at each string. Schematic view of cluster is shown in Fig. 1 (left). The upgraded cluster comprises 288 OMs operated at the active depths of 750–1275 m, while 7 side strings are distant at radius of 60 m from the central one, according to the baseline configuration of GVD cluster. Instrumented volume of cluster encloses 5.9 Megatons. The cumulative number of section records generated by the array in the period of April 10 – June 23 of 2016 is shown in Fig. 1 (right).

Basic trigger mode requires coincidences of any neighbouring OMs within a section (thresholds 1.5 p.e. and 4.5 p.e., trigger rate 60 -110 Hz). About 3×10^8 of triggered events have been recorded. The estimated operation efficiency was about 75%.

3 "Dubna" array: operation and preliminary results

The "Dubna" array has the potential to detect astrophysical neutrinos with the flux values measured by IceCube [5]. The search for high-energy neutrinos with the "Dubna" array is based on the selection of cascade events generated by neutrino interactions in the sensitive volume of the array. After

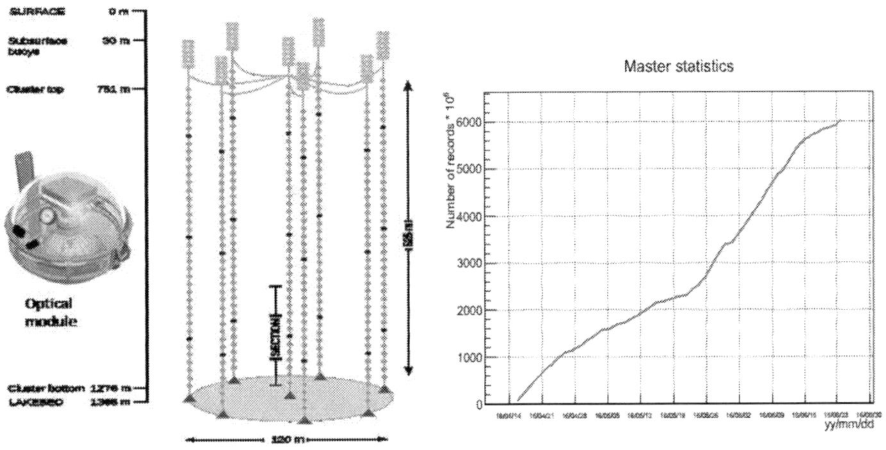

Figure 1. Left: Schematic drawing of the "Dubna" cluster operated in 2016. Right: Integrated number of events recorded by the cluster in period of April 10–June 23 of 2016.

the reconstruction of the cascade vertex, the energy and the direction and after applying quality cuts, events with a final multiplicity of hit OMs $N_{hit} > 20$ are selected as high-energy neutrino events. The accuracy of the cascade energy reconstruction is about 30%. The accuracy of the direction reconstruction is about 4° (median value) and the vertex resolution is of about 2 m [2]. For the IceCube astrophysical fluxes following the power laws of E^{-2} and $E^{-2.46}$ and for the single-flavour normalizations of $1.2 \cdot 10^{-8}$ GeV cm^{-2} s^{-1} sr^{-1} and $4.1 \cdot 10^{-6}$ GeV cm^{-2} s^{-1} sr^{-1} respectively [5–7] the expected number of events as function of the neutrino energy is reported in Fig. 2 (left panel). In the same plot the expected number events from the atmopheric neutrino flux is also reported. The expected number of background events from atmospheric neutrinos is strongly suppressed for energies higher than 100 TeV. About 0.5 and 0.4 cascade events per year with energies above 100 TeV and hit multiplicities $N_{hit} > 20$ from astrophysical fluxes with E^{-2} and $E^{-2.46}$ spectra respectively and 0.08 background events from atmospheric neutrinos are expected.

For the search of high energy neutrinos of astrophysical origin the data collected from 24 October till 17 December 2015 has been used. A data sample of 4.4×10^8 triggered events has been recorded by the array, which corresponds to 41.64 days of live time. The causality cuts and requirement of $N_{hit} > 3$, where N_{hit} is the hit OM multiplicity, reduce the number of events for following analysis at about 1.8×10^7 events.

After applying an iterative procedure of cascade vertex reconstruction followed by the rejection of hits contradicting the cascade hypothesis on each iteration stage, 12931 events survived as cascade-like events. Finally, after applying the cascade energy reconstruction and the event quality cuts, 1192 cascade-like events with energy above 100 TeV are selected. The hit OMs multiplicity distribution of these events is shown (top, black) in Fig. 2 (centre), as well as the expected event distribution (bottom, red) from the astrophysical flux with $E^{-2.46}$ spectrum and IceCube normalization. All events, but one, have hit OMs multiplicities less than 10 OMs and are consistent with the expected background events from atmospheric muons. The event with 17 hit OMs was reconstructed as downward moving cascade event with a zenith angle of 59° and an energy of 158 TeV energy. The neutrino telescope sensitivity to DM annihilation inside distant sources is estimated by the expected number of signal events in the search angular window ψ_0 for the life time T and astrophysical factor J of modeled DM profile, as follow

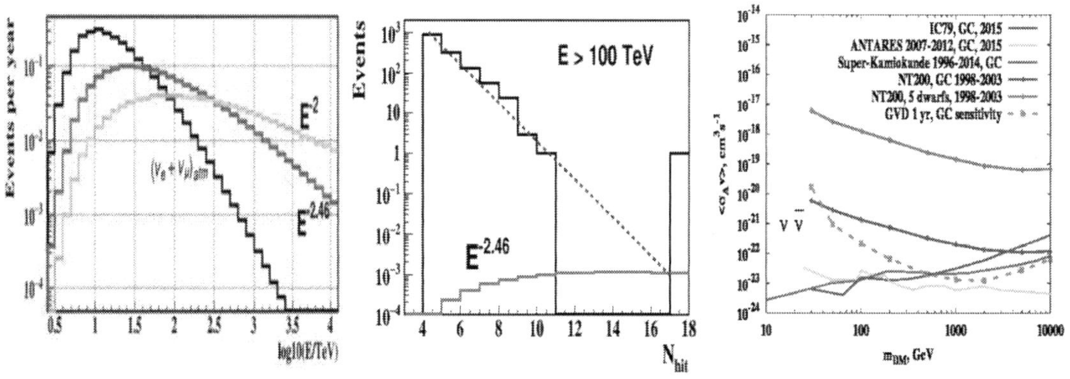

Figure 2. Left: Energy distributions of events expected for one year observation from astrophysical fluxes with E^{-2} and $E^{-2.46}$ spectra and IceCube normalization, and from atmospheric neutrinos (see text). Centre: Hit OMs multiplicity distribution of cascade-like events with energies above 100 TeV (histogram, see text). Right: Sensitivity of the GVD to neutrino-channel of the DM annihilations in the GC in comparison with the NT200 limits towards the GC direction and limits of joint dSphs analysis.

$$N(\psi_0) = T\frac{\langle\sigma_A v\rangle R_0 \rho^2_{local}}{8\pi m^2_{DM}} J_{\Delta\Omega}\int dE \cdot S(E)\frac{dN_\nu}{dE}, \tag{1}$$

where $S(E)$ is neutrino the effective area of the telescope averaged over the neutrino energy spectrum $\frac{dN_\nu}{dE}$ in the given annihilation channel. Here we present the GVD sensitivity based on the simulated muon-like events in the high energy neutrino interactions. Comparison of the Baikal results in most energetic $\nu\bar{\nu}$ channel of the DM annihilation is presented in Fig. 2 (right). The 90% c.l. sensitivity to DM annihilation in the Galactic Center (GC) as a function of DM mass for the GVD of 12 clusters [8] (red line) and the upper limits for the Baikal NT200 telescope [9] (blue line) are shown. In Fig. 2 (right) is also shown the new combined upper limit, performed with a likelihood analysis, for 5 dwarf spheroidal galaxies (dSphs),which have been chosen among 22 dwarfs with the NT200 data sample [10]. By 2020, GVD will consist of 2304 cumulative number of OMs on 8 clusters with a total volume of about 0.4 km^3, that aims at the discovery of non-atmospheric neutrinos.

The Baikal-GVD project is supported by the RFBR grants 14-02-00972, 16-29-13032. The work of S.V.Demidov and O.V.Suvorova was supported by RSCF grant 14-12-01430 in part where neutrino signal from dark matter annihilations has been studied.

References

[1] V.Aynutdinov et al., NIM A742 82-88 (2014).
[2] A.D.Avrorin et al., Phys.Part.Nucl. 46 no.2, 211-221 (2015).
[3] A.V.Avrorin et al., Instr. and Exp. Tech. 57, issue 3, 262-273 (2014).
[4] A.D.Avrorin et al., EPJ Web Conf. 116 , 01003 (2016).
[5] M.G.Aartsen et al. , Phys. Rev. Lett. 113 101101. Science 342 1242856 (2013).
[6] M.G.Aartsen et al., IceCube Coll., Phys. Rev. D 91, 022001 (2015).
[7] M.G.Aartsen et al., IceCube Coll., Astrophys. J. 809, 98 (2015).
[8] A.D.Avrorin et al., JETP Lett. 101 no.5, 289-294, (2015) .

DOI: 10.1051/epjconf/201713604007

[9] A.D.Avrorin et al., Astropart.Phys. 8, 12-20, (2016).
[10] A.D.Avrorin et al., in preparation for JETP

Design and production of the digital optical module of the KM3NeT project

Emanuele Leonora[1a], Valentina Giordano[1]
on behalf of the KM3NeT collaboration

[1]*INFN, sezione di Catania, Via S. Sofia, 64, 95123, Catania, Italy*

Abstract. The KM3NeT collaboration is building the ARCA and ORCA neutrino telescopes in the depths of the Mediterranean Sea. They will consist of 3-dimensional arrays of photodetectors, called digital optical modules, suspended in the sea by means of vertical string structures, called detection units. The optical modules are composed of a pressure-resistant 17-inch spherical glass vessel, which contains 31 small photomultiplier tubes and all the associated electronics. The multi- photomultiplier solution represents an innovative design with respect to optical modules of all currently operated neutrino telescopes comprising a single large photomultipliers.

1 Introduction

The KM3NeT Collaboration [1] started building a research infrastructure in the depths of the Mediterranean Sea hosting new generation neutrino telescopes located at the bottom of the Mediterranean Sea. The detectors of the KM3NeT telescopes can be described as a three dimensional matrix of photosensors that are sensitive to the Cherenkov radiation emitted by products of neutrino interactions in the deep water. The measured arrival time of the Cherenkov light at each sensor and the knowledge of their spatial position are used to reconstruct the trajectory of the electrically charged particle that produces the Cherenkov light. The amount of detected light provides information about the energy of the particle. The photosensors are called Digital Optical Modules (DOMs). In the sea, they are suspended in vertical structures, called Detection Units (DUs), supported by two Dyneema ® ropes, anchored to the seabed and kept taut with a system of buoys. Following its first phase of construction, KM3NeT will comprise two detectors with different density of the arrays of optical modules: KM3NeT/ARCA (Astroparticle Research with Cosmics in the Abyss) at the KM3NeT-It site, located at the South of the Sicilian coast, will consist of two detector building blocks of 115 detection units each and will be dedicated to the detection of high-energy neutrino astronomy. KM3NeT/ORCA (Oscillation Research with Cosmics in the Abyss), a single more densily instrumented building block located at the KM3NeT-Fr site, offshore Toulon, will be dedicated to a lower neutrino energy range for study of neutrino mass hierarchy. The DOMs are pressure-resistant 17-inch diameter transparent spherical glass vessels, each hosting 31 photomultiplier tubes (PMTs) with a 3-inch photocathode diameter and all associated front-end and readout electronics. The multi-PMT solution represents an innovative design with several advantages over traditional optical

[a] Corresponding author: emanuele.leonora@ct.infn.it

modules with single large PMTs. The DOM design and its physics potential have been proven *in-situ* by several prototypes and in several detection units [2][3]. The DOM design phase is thus concluded successfully and mass production of KM3NeT DOMs is currently on going.

2 The digital optical module design

The Multi-PMT design of the KM3NeT optical module [4] has several advantages over traditional optical modules of all the other currently operating neutrino telescopes- ANTARES, Baikal and IceCube - which have optical modules with a single large photomultiplier, typically with a photocathode diameter of 10 inch. The total photocathode area comprised in the module is almost three times of that of a 10" PMT. The influence of the Earth's magnetic field on small size PMT is negligible and thus magnetic shield is not required. The Multi-PMT design, in which each sensor works individually, allows also for an accurate photon counting, because the arrival of more than one photon at the DOM can be identified with high efficiency. In addition, the segmented layout provides an almost isotropic field of view and efficient rejection of optical background from the bioluminescence and ^{40}K in the sea just at the level of the DOM detection. In Fig. 1 a picture of a complete DOM is shown.

Figure 1. Picture of a DOM. The external collar, the break out box of the VEOC (Vertical Electrical-Optical Cable) backbone cable and the two parallel ropes of the detection unit are visible.

2.1 The main components of the DOM

The Digital Optical Module is composed of a transparent 17-inch diameter pressure resistant glass sphere housing 31 photomultiplier tubes and all the electronics. The transparent vessel is a Vitrovex® glass sphere, produced by Nautilus, composed of two separate hemispheres. The 31 PMTs are arranged in 5 rings of 6 PMTs plus a single PMT at the bottom pointing vertically downwards. The lower hemisphere contains 19 PMTs, the upper hemisphere 12 PMTs. The PMTs are kept in place by a 3D-printed support structure. The photon collection efficiency is increased by 20–40% by a reflector ring mounted around the face of each PMT. It allows increasing the detection area avoiding the use of a larger photocathode PMT.
The space between the support structure, the PMT windows and the glass sphere is filled with an optical silicone gel, in order to ensure an efficient mechanical and optical coupling between the different elements. It is a Silgel 612 A/B produced by Wacker. The support structure and the gel have to be sufficiently flexible to be resistant to vibration and shocks during handling and deployment and allow for the deformation of the glass sphere under the hydrostatic pressure in the deep sea.

In the Multi-PMT design, each PMT works as an individual optical sensor, with an individual low-power high-voltage base with integrated amplification and tunable discrimination. The power dissipated of each base is below of 5 mW. The photon arrival time and the time-over-threshold (ToT) of each PMT are recorded by an individual time-to-digital converter implemented in a Field Programmable Gate Array (FPGA). The threshold is set at the level of 1/3 of the mean single photon pulse height and the high voltage is set to provide a gain of $3x10^6$. The FPGA is mounted on the central logic board (CLB), which transfers the data to shore via an Ethernet network of optical fibers. The data provided by the PMT bases is collected and distributed to the CLB by means of two so-called Octopus Boards. The power board provides the DC power for all the systems inside the DOM. This board is monitored and controlled by the FPGA. It is powered with 12V DC supplied via a penetrator mounted in the upper hemisphere of the DOM and connected to a DC/DC converter in the break-out-box external to the glass sphere (see Fig.1). An inner view of the DOM is shown in Fig. 2.

Figure 2. Sketch of the inner components of the DOM. CLB, power board, octopus boards and PMTs are indicated.

Each DOM in a detection unit communicates at a dedicated wavelength, which is multiplexed with other DOM wavelengths for transfer via a single optical fiber to the shore. The broadcast of the onshore clock signal, needed for time stamping in each DOM, is embedded in the Gb-Ethernet protocol. The White Rabbit protocol [5] has been modified to implement the broadcast of the clock signal, and allows for synchronization of all the DOMs of the KM3NeT detector with 1 ns resolution.
An Aluminium structure provides heat conduction between the electronics inside and the environment outside of the sphere. Temperature and humidity sensors are mounted on the CLB and on the power board. The power consumption of a single DOM is about 7 W.
For the first phase of the project, a batch of 15000 3-inch Hamamatsu PMT R12199-02 was calibrated. The PMTs have a convex bialkali photo-cathode, with a diameter of 80 mm and a 10-stage dynode structure [6].
The DOM also contains three calibration sensors: a LED nano-beacon, which injects light in water to illuminate the neighbouring optical modules for time calibration; an acoustic piezo sensor is glued to the inner surface of the glass sphere to define the DOM position in water by means of an acoustic positioning system; an Attitude and Heading Reference System (AHRS) mezzanine board, mounted on the CLB, provides compass, tilt- and accelerometer data, to reconstruct the orientation of the DOM in the water.
A penetrator mounted in the upper hemisphere of the DOM contains two power cables and one optical fibre for data transmission. An external titanium collar holds the DOM to the two parallel Dyneema ropes of the string into the water. More details of the DOM design can be found in the Letter of Intent for KM3NeT 2.0 [7]

2.2 DOM main production phases

Each DOM is composed of over 80 different elements: mechanical objects, electronics and sensors. Therefore, the production of the DOM is a delicate process performed following a well-defined procedure. Many tools are used, and a lot of them have been custom designed by the KM3NeT collaboration.
Each DOM integration site in the Collaboration receives all the components, which meet acceptance criteria and have already passed characterization tests. During the assembly phases, all the components of each DOM are registered in database by means of a dedicated assembly software. As the glass vessel is composed of two halves, the two parts of the DOM, top and bottom, are constructed separately. A lot of the DOM integration steps can be performed in parallel to speed up the production. Also the PMT support structures are segmented in two halves, each one made of two parts. The white 3-D printing structures have to be black painted before being glued together. All the components are carefully cleaned before their integration following the properly method defined in the procedure. In the upper hemisphere an aluminium mushroom-shaped structure is glued to the inner glass sphere by means of a special tool that define its height and the distance with the inner surface of the glass. Power board and CLB are later mounted on the inner part of the aluminium mushroom. The penetrator is mounted in the top glass sphere, and its water tightness is tested by means of a helium leakage detector. The optical fiber coming from the penetrator is connected to the laser transceiver operating an optical splice on the fiber. The resulting optical power and attenuation are checked, and results are stored in the database. The piezo sensor is glued to the inner surface of the bottom glass hemisphere. All the PMTs are mounted in the PMT structures, and their positions are stored in the database. After the structures equipped with all the PMTs are placed into the glass hemispheres, all the bases are connected to the octopus boards, and a functional test of all systems is performed before applying the optical gel, which joins optically and mechanically all the components together.
At the end of the integration process, the DOM is closed connecting the two halves setting down one part upon the other by means of an electrical crane. After closure, 0.2 bar of under-pressure inside the DOM, followed by the application of a sealant and a special tape on the junction of the two halves, closes the DOM. The external titanium collars is later mounted on the already closed DOM.
Each produced optical module is submitted to a final acceptance test and calibration of all the sensors performed in a light-tight dark box. Test and calibration data are later stored in the database of the KM3NeT Collaboration.

3 Conclusions

An innovative Multi-PMT digital optical module was designed as the sensitive part of the cubic-kilometer sized KM3NeT neutrino telescope. The design has been demonstrated *in-situ* by several prototypes and the mass production of the KM3NeT DOMs is on-going in several integration sites. Installation of detection units in the deep Mediterranean Sea has already started; two fully equipped detection units are taking data at the KM3NeT-It site offshore Porto Palo di Capo Passero, Sicily.

References

1. KM3NeT TDR, ISBN 978-90-6488-033-9, http://km3net.org.
2. S. Adrián-Martínez et al., Eur. Phys. J. C, 74, 3056 (2014), arXiv:1405.0839 [astro-ph.IM]]
3. S. Adrián-Martínez et al., Eur. Phys. J. C, 76:54 (2016), DOI: 10.1140/epjc/s10052-015-3868-9
4. R. Bruijn and D. van Eijk for the KM3NeT Coll., in Proc. 34th Int. Cosmic Ray Conference, PoS (ICRC2015) 1157
5. Open Hardware Repository/CERN. http://www.ohwr.org/projects/white-rabbit
6. S. Aiello et al., AIP Conf. Proc. 1630 (2014), DOI: 10.1063/1.4902786
7. KM3NeT 2.0 Letter of Intent for ARCA and ORCA, arXiv:1601.07459 [astro-ph.IM]27 Jan 2016

Results and simulation of the prototype detection unit of KM3NeT-ARCA

C.M.F. Hugon[1,a] on behalf of the KM3NeT collaboration

[1] *INFN sezione di Genova*

Abstract. KM3NeT-ARCA is a deep sea high energy neutrino detector. A detection unit prototype was deployed in the future KM3NeT-ARCA deep-sea site, off of the Sicilian coast. This detection unit is composed of a line of 3 digital optical modules with 31 photomultiplier tubes on each one. The prototype detection unit was operated since its deployment in May 2014 until its decommissioning in July 2015. The results of the calibration of this detection unit and its simulation are presented and discussed.

1 Introduction

KM3NeT represents the next generation of Mediterranean undersea telescope, following the technical and scientific results of the ANTARES [1] neutrino telescope and the NEMO [2] and NESTOR [3] pilot projects. The two main goals of this detector are the detection of high energy neutrinos and their sources and the determination of the neutrino mass hierarchy with the detection of low energy neutrinos. The detection of neutrinos is possible thanks to the Cherenkov light emitted by the secondary charged particles. The secondary particles are produced by weak interaction in the vicinity of the detector. KM3NeT uses 17 inch digital optical modules (DOM) containing 31 3 inches photomultiplier tubes (PMT) for a total of photocathode area of 1400 cm^2 [4]. The final line design contains 18 DOM along a rope, anchored to the seabed and maintained close to vertical with a buoy. In May 2014, a prototype detection unit (Pre-Production Model DU, PPM-DU) with 3 DOMs was installed 80 km off the Sicily [5]. This prototype implements the mechanical structure, the electro-optical power supply and communication, allowing simultaneous data taking of the three DOMs. The PMT model installed in DOM 1 and DOM 2 is the D783KFLA produced by ETEL [6], while DOM 3 contains the R12199-02 Hamamatsu PMTs [7]. The current design for KM3NeT uses the R12199-02 PMT. The PMTs are surrounded by reflector rings at an angle of 45 degrees with respect to the PMT axis, 16 mm in width for ETEL and 17 mm for the Hamamatsu PMTs [4]. Figure 1 represents the prototype line and the DOM design and its PMT arrangement.

The simulation of the KM3NeT detector is based on the same principle as ANTARES [8]. Only two elements of the simulation chain had to be upgraded: The step-by-step high precision simulation of the optical module and the *km3* simulation software, briefly described in the next section. Then in section 3 the results will be presented and discussed.

[a]e-mail: chugon@ge.infn.it

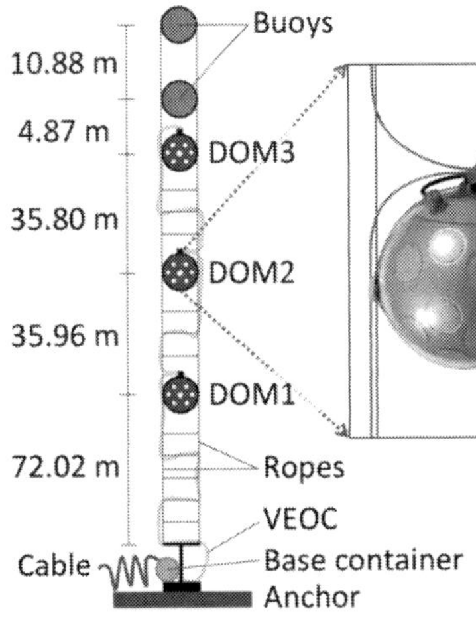

Figure 1: Schematic of the PPM-DU (not to scale). The DOMs are spaced by 36 m vertically. The buoys are made of 2 empty glass spheres. The vertical electro-optical cable (VEOC) connects the DOMs with the base. The base is connected to the submarine infrastructure, then to the shore station. The inset shows a DOM attached to the two ropes, which leaves the line free to move with the sea current.

2 The simulation

2.1 The step-by-step simulation

The step-by-step simulation is described in detail in [9, 10]. It is the very first step of the full simulation of ANTARES, NEMO and KM3NeT neutrino detectors. It manages the propagation of each photon, from their sources, through their different interactions in the matter, to the PMT photocathode. When these photons reach the photocathode, a complementary accurate calculation process is used in order to obtain a precise detection probability. This calculation takes into account the thickness and the complex index of the photocathode to provide the photoelectron production as a function of wavelength and incident angle, essential for underwater neutrino detectors. This simulation is used to produce the OM and DOM detection efficiency as a function of the angle and the wavelength of the incident photon (angular acceptance). This simulation has been validated by comparison with laboratory measurements. These measurements were done with an LED scanning of the ANTARES OMs and KM3NeT PMTs [11].

2.2 The full detector simulation

To simulate neutrino detectors, a full simulation would be far too slow. Therefore, the results of the step-by-step simulation are tabulated and used as inputs for the next simulation step, called *km3*. Thanks to this efficiency, the water properties and the number of photons produced by Cherenkov effect, the *km3* simulation is able to reproduce the full detector detection response. It is based on 4 main steps:The charged lepton directionality and energy are simulated a priori with the fast MUPAGE code [12], as a function of the water properties, to be propagated in the *km3* simulation can; the density of the Cherenkov photons production is calculated along the lepton (~100/cm); the photon probability density as a function on time is calculated along the Cherenkov front; at the position of each Optical module the probability to be detected is calculated as a function of the step-by-step simulation tabulated results. This method has been chosen for its good compromise between precision and calculation time, and is used for KM3NeT-ARCA and ORCA.

3 The prototype detection line simulation and calibration results

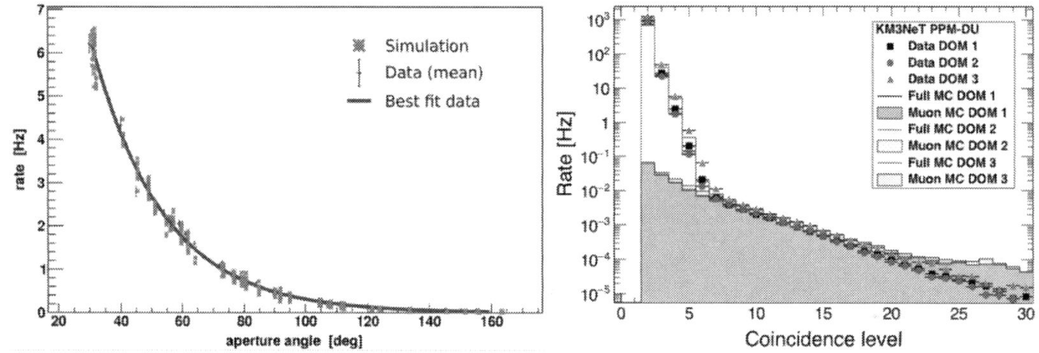

Figure 2: The upper figure shows the coincidence rates as a function to the angle separation between the pair of PMTs. The blue points represent the data from [5] (mean over multiple PMTs), the blue line represents the data best fit and the red points represent the step-by-step simulation results. The lower graphic shows the multi-fold coincidences for the data and the simulation. The step-by-step simulation is used for the ^{40}K coincidences and the global *km3* simulation is used for the muon simulation.

3.1 Intra-DOM calibration

Natural ^{40}K radioactive decays in sea water produce up to 150 Cherenkov photons. A decay close to the DOMs can be detected by one or multiple PMTs (single or multi-fold). Beyond the background it may represent, it's a calibration source for intra-DOM time offsets between the PMTs and their intrinsic transit time spread. The measured mean time spread of PMTs is 5 ns (FWHM). In addition, knowing precisely the water salinity [13], the coincidence rate can be used for the efficiency adjustment of the simulation. The two-fold coincidence rate is shown in Figure 2 as a function of the angular separation between pairs of PMTs in the DOM 3. The angular dependence has been fitted to an exponential function. The step-by-step simulation results are represented at their best fit, and show a good accordance between the measured rates and the simulation for all angles between the PMTs of the DOM.

3.2 The inter-DOM calibration

A LED nanobeacon system operating on the lowest DOM has been used to calculate the inter-DOM time offsets. The distribution of time differences, taking in account the travel time in water, of hits in coincidence with the LED emission time has been fitted with a Gaussian function. The resulting mean time offsets of DOMs 2 and 3 with respect to DOM 1 are, respectively, approximately 230 ns and 380 ns. Leaps of about ten ns have been observed, and are due to power cycles of the corresponding

DOI: 10.1051/epjconf/201713604009

module in the shore station. Otherwise no time offset shifts were observed within a few nanoseconds. In addition, thanks to its multi-DOM configuration, the prototype line was able to detect down-going muons. The signal from muons has been used to cross check the inter-DOM calibration by a χ^2 minimization algorithm applied to the data and the simulation. Figure 2 illustrate the agreement of simulation and data. This latest measurement was in agreement within 2 ns with the LED beacon calibration, and shows a good agreement with the full detector simulation *km3*.

Conclusion

The prototype project of the PPM-DU validated the KM3NeT structure at the depth of 3500 m, provided a test bench for the data taking and analysis and showed a reliable calibration procedure for the detection efficiency and the time offsets between PMTs and between DOMs. The simulation showed a good understanding of the detector, from the DOM with the step-by-step simulation to its line configuration with the *km3* full detector simulation. It provided the perspectives on the future KM3NeT-ORCA detector in this site, which is currently in the deployment stage.

References

[1] M. Ageron et al. (ANTARES), Nucl. Instrum. Meth. **A656**, 11 (2011), 1104.1607
[2] A. Capone et al. (NEMO), Nucl. Instrum. Meth. **A602**, 47 (2009)
[3] P.A. Rapidis (NESTOR), Nucl. Instrum. Meth. **A602**, 54 (2009)
[4] S. Adrian-Martinez et al. (KM3NeT), JINST **8**, T03006 (2013)
[5] S. Adrián-Martínez et al. (KM3NeT), Eur. Phys. J. **C76**, 54 (2016), 1510.01561
[6] R. Bormuth, L. Classen, O. Kalekin, H. Peek, J. Reubelt, D. Samtleben, E. Visser (KM3NeT), AIP Conf. Proc. **1630**, 114 (2014)
[7] S. Aiello, L. Classen, V. Giordano, O. Kalekin, E. Leonora, H. Peek, J. Reubelt, D. Samtleben, E. Visser (KM3NeT), AIP Conf. Proc. **1630**, 118 (2014)
[8] J. Brunner (ANTARES), *Antares simulation tools*, in *Technical aspects of a Very Large Volume Neutrino Telescope in the Mediterranean Sea. Proceedings, Workshop, VLVnuT Workshop, Amsterdam, Netherlands, October 5-8, 2003* (2003), pp. 109–113, http://www.vlvnt.nl/proceedings/
[9] C.M.F. Hugon (ANTARES, KM3NeT), Nucl. Instrum. Meth. **A787**, 189 (2015)
[10] C.M.F. Hugon, PoS **ICRC2015**, 1106 (2016)
[11] T. Avgitas, A. Creusot, A. Kouchner (KM3NeT), EPJ Web Conf. **116**, 06007 (2016)
[12] G. Carminati, M. Bazzotti, A. Margiotta, M. Spurio, Comp. Phys. Comm. **179**, 915 (2008)
[13] G. Riccobene, A. Capone (NEMO), Astropart. Phys. **27**, 1 (2007), astro-ph/0603701

DAMA/LIBRA results and perspectives

R. Bernabei[1,2], P. Belli[1,2], A. Di Marco[1,2], F. Montecchia[2a], A. d'Angelo[3,4], A. Incicchitti[3,4], F. Cappella[5], V. Caracciolo[5], R. Cerulli[5], C.J. Dai[6], H.L. He[6], H.H. Kuang[6], X.H. Ma[6], X.D. Sheng[6], R.G. Wang[6], and Z.P. Ye[6b]

[1]*Dip. di Fisica, Università di Roma "Tor Vergata", I-00133 Rome, Italy*
[2]*INFN sez. Roma "Tor Vergata", I-00133 Rome, Italy*
[3]*Dip. di Fisica, Università di Roma "La Sapienza", I-00185 Rome, Italy*
[4]*INFN sez. Roma, I-00185 Rome, Italy*
[5]*INFN, Laboratori Nazionali del Gran Sasso, I-67100 Assergi (AQ), Italy*
[6]*Key Laboratory of Particle Astrophysics, IHEP, 100049 Beijing, P. R. China*

Abstract. The DAMA/LIBRA experiment ($\sim$ 250 kg sensitive mass composed by highly radio-pure NaI(Tl)) is in data taking in the underground Laboratory of Gran Sasso (LNGS). Its first phase (DAMA/LIBRA–phase1) and the former DAMA/NaI experiment ($\sim$ 100 kg of highly radio-pure NaI(Tl)) collected data for 14 independent annual cycles, exploiting the model-independent Dark Matter (DM) annual modulation signature (total exposure 1.33 ton $\times$ yr). A DM annual modulation effect has been observed at 9.3 σ C.L., supporting the presence of DM particles in the galactic halo. No systematic or side reaction able to mimic the observed DM annual modulation has been found or suggested by anyone. Recent analyses on possible diurnal effects, on the Earth shadowing effect and on possible interpretation in terms of Asymmetric Mirror DM will be mentioned. At present DAMA/LIBRA is running in its phase2 with increased sensitivity.

1 Introduction

The DAMA project is dedicated to the development and use of low background scintillators for underground physics. The main experiment is DAMA/LIBRA [1–15] that, after the pioneering DAMA/NaI [16, 17], is further investigating the presence of DM particles in the galactic halo by exploiting the model independent DM annual modulation signature [18, 19].

Because of the Earth's revolution around the Sun, which is moving in the Galaxy, the flux of DM particles impinging a terrestrial detector is expected to be maximum around $\simeq 2^{nd}$ June when the projection of the Earth orbital velocity on the Sun velocity with respect to the Galactic frame is maximum, and minimum around $\simeq$ December 2nd when the two velocities are opposite. This effect, known as DM annual modulation signature, is very effective because the signal induced by DM particles must simultaneously satisfy many requirements: the rate must contain a component modulated according to a cosine function (1) with one year period (2) and a phase peaked roughly at $\simeq 2^{nd}$ June (3); the modulation must only be present in a well-defined low energy range (4); it must

[a]also Dip. di Ingegneria Civile e Ingegneria Informatica, Università di Roma "Tor Vergata", I-00133 Rome, Italy
[b]also University of Jinggangshan, Jiangxi, China

apply only to those events in which just one detector among many actually "fires" (*single-hit* events), since the DM particle multi-interaction probability is negligible (5); the modulation amplitude in the region of maximal sensitivity must be $\simeq 7\%$ for usually adopted halo distributions (6), but it can be larger (even up to $\simeq 30\%$) in case of some possible scenarios (see for example Ref. [20–24]). This signature is model independent and might be mimicked only by systematic effects or side reactions able to simultaneously satisfy all the requirements given above; no one is available [1–4, 7, 8, 12, 13, 16, 17, 25].

The full description of the DAMA/LIBRA set-up and performance during the phase1 and phase2 (presently running) and other related arguments have been discussed in details in Refs. [1–4, 6–8, 13] and references therein.

2 DM annual modulation results

The data of DAMA/LIBRA–phase1 and of DAMA/NaI experiment correspond to an exposure of 1.33 ton $\times$ yr collected in 14 annual cycles. In order to investigate the presence of annual modulation with proper features, many analyses have been carried out. All these analyses point out the presence of annual modulation satisfying all the requirements of the signature [2–4, 8]. In Fig. 1, as example, the time behaviour of the experimental residual rate of the *single-hit* scintillation events for DAMA/LIBRA–phase1 in the (2–6) keV energy interva is plotted. When fitting the *single-hit* residual rate of DAMA/LIBRA–phase1 together with the DAMA/NaI ones, with the function: $A\cos\omega(t-t_0)$, considering a period $T = \frac{2\pi}{\omega} = 1$ yr and a phase $t_0 = 152.5$ day (2nd June) as expected by the DM annual modulation signature, the following modulation amplitude in NaI(Tl) is obtained: $A = (0.0110 \pm 0.0012)$ cpd/kg/keV, corresponding to 9.2 σ C.L.. When the period, and the phase are kept free in the fitting procedure, the modulation amplitude is (0.0112 ± 0.0012) cpd/kg/keV (9.3 σ C.L.), the period $T = (0.998 \pm 0.002)$ year and the phase $t_0 = (144 \pm 7)$ day, values well in agreement with expectations for a DM annual modulation signal. In particular, the phase is consistent with about 2nd June and is fully consistent with the value independently determined by Maximum Likelihood analysis [4]. The run test and the χ^2 test on the data have shown that the modulation amplitudes singularly calculated for each annual cycle of DAMA/NaI and DAMA/LIBRA–phase1 are normally fluctuating around their best fit values [2–4].

No modulation was found in any possible source of systematics or side reactions; thus, cautious upper limits on possible contributions to the DAMA/LIBRA–phase1 measured modulation amplitude were obtained (see Refs. [2–4]). It is worth noting that they do not quantitatively account for the measured modulation amplitudes, and are even not able to simultaneously satisfy all the many requirements of the signature. Similar analyses were also performed for the DAMA/NaI data [16, 17].

In particular, in Ref. [7, 13] a simple and intuitive way why the neutrons, the muons and the solar neutrinos cannot give any significant contribution to the DAMA annual modulation results is outlined. Other arguments can be found in Refs. [1–4, 7, 8, 12, 13, 16, 17, 25]. In conclusion, DAMA gives model-independent evidence (at 9.3σ C.L. over 14 independent annual cycles) for the presence of DM particles in the galactic halo.

As regards comparisons, we recall that no direct model independent comparison is possible in the field when different target materials and/or approaches are used; the same is for the strongly model dependent indirect searches. In particular, the DAMA model independent evidence is compatible with a wide set of scenarios regarding the nature of the DM candidate and related astrophysical, nuclear and particle Physics; for examples some given scenarios and parameters are discussed e.g. in Refs. [2, 5, 8, 12, 14–17, 26–40] and references therein. Further large literature is available on the topics (see for example Ref. [41]). In conclusion, both negative results and possible positive hints

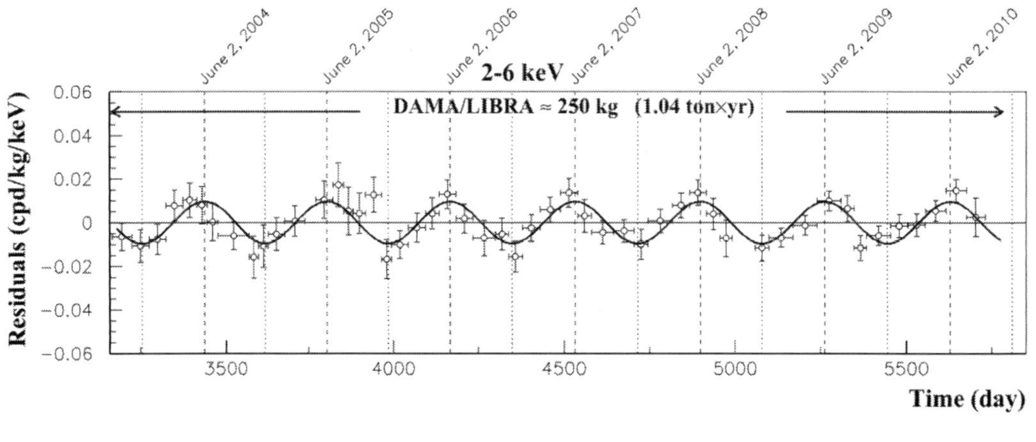

Figure 1. Experimental residual rate of the *single-hit* scintillation events measured by DAMA/LIBRA–phase1 in the (2–6) keV energy interval as a function of the time. The superimposed curve is the cosinusoidal function behaviour $A\cos\omega(t-t_0)$ with a period $T = \frac{2\pi}{\omega} = 1$ yr, a phase $t_0 = 152.5$ day (2nd June) and modulation amplitude, A, equal to the central values obtained by best fit on the data points of the entire DAMA/LIBRA–phase1. The dashed vertical lines correspond to the maximum expected for the DM signal (2nd June), while the dotted vertical lines correspond to the minimum.

reported in literature can be compatible with the DAMA model-independent DM annual modulation results in various scenarios considering also the existing experimental and theoretical uncertainties. Moreover, scenarios also exist for which the DAMA approach is favoured.

Three analyses on the DAMA/LIBRA–phase1 data are summarized in the following. In the first one the presence of possible diurnal effects has been investigated [12], while in the second one the so called "Earth Shadow Effect" has been considered [14]. Finally, the annual modulation result has been interpreted in terms of Asymmetric Mirror DM [15]. These items were extensively studied in the quoted references, here the results are shortly addressed.

3 Diurnal modulation

The results obtained by investigating the presence of possible diurnal variation in the low-energy *single-hit* scintillation events collected by DAMA/LIBRA–phase1 have been analysed in terms of a DM second order model-independent effect due to the Earth diurnal rotation around its axis [12].

Also this daily modulation of the rate on the sidereal time – expected when taking into account the contribution of the Earth rotation – presents some specific peculiarities. In particular the interest in this signature is also that the ratio R_{dy} of this DM diurnal modulation amplitude over the DM annual modulation amplitude is a model independent constant at given latitude; considering the LNGS latitude, $R_{dy} = \frac{S_d}{S_m} \simeq 0.016$. Thus, taking into account the DM annual modulation effect pointed out by DAMA/LIBRA–phase1 for *single-hit* events in the low energy region, the expected value of the diurnal modulation amplitude for the (2–6) keV energy interval is $\simeq 1.5\times 10^{-4}$ cpd/kg/keV. Fig. 2 shows the time behaviour of the experimental residual rates of *single-hit* events both as a function of solar (*left*) and of sidereal (*right*) time, in the (2–6) keV interval [12].

No diurnal variation with a significance of 95% C.L. is found at the reached level of sensitivity [12]. In conclusion the presence of any significant diurnal variation and of time structures can be

DOI: 10.1051/epjconf/201713605001

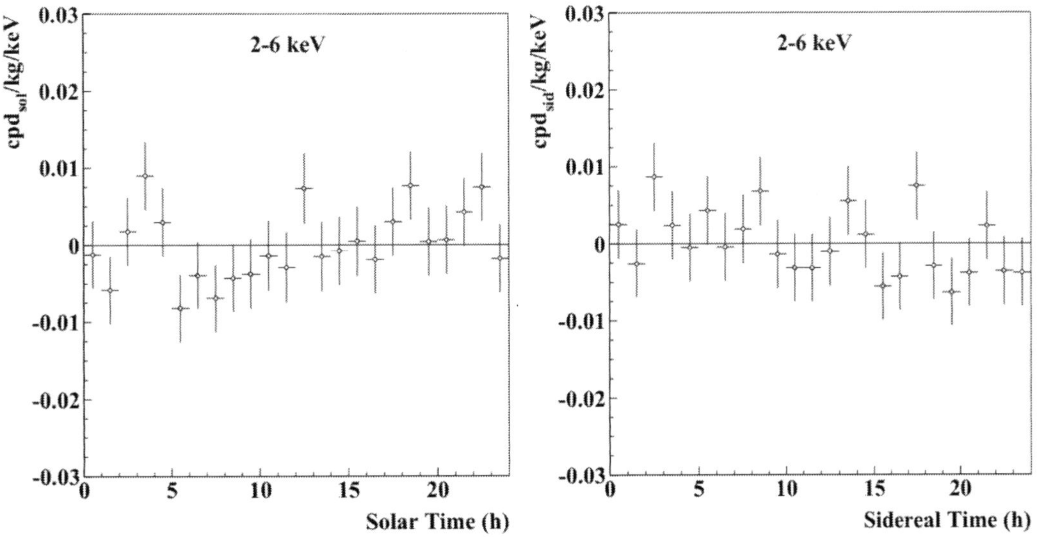

Figure 2. Experimental model-independent diurnal residual rate of the *single-hit* scintillation events, measured by DAMA/LIBRA–phase1 in the (2–6) keV energy intervals as a function of the hour of the solar (*left*) and sidereal (*right*) day. See Ref. [12] for details.

excluded at the reached level of sensitivity. In addition, considering the (2–6) keV energy interval the obtained upper limit on the DM diurnal modulation amplitude is 1.2×10^{-3} cpd/kg/keV (90% C.L.) [12]; thus, the effect of DM diurnal modulation, expected because of the Earth diurnal motion on the basis of the DAMA DM annual modulation results, is out the present sensitivity [12]. DAMA/LIBRA–phase2, presently running, with a lower software energy threshold [6] can also offer the possibility to increase sensitivity to such an effect.

4 Daily effect on the sidereal time due to the shadow of the Earth

The results obtained in the investigation of possible diurnal effects for low-energy *single-hit* scintillation events of DAMA/LIBRA–phase1 have been analysed in terms of Earth Shadow Effect, a model-dependent effect that is expected for DM candidates inducing only nuclear recoils and having high cross-section (σ_n) with ordinary matter [14].

In fact a diurnal variation of the low energy rate could be expected for these specific candidates, because of the different thickness of the shield due to the Earth during the sidereal day, eclipsing the wind of DM particles. The induced effect should be a daily variation of their velocity distribution, and therefore of the signal rate measured deep underground. However, this effect is very small and would be appreciable only in case of high cross-section spin independent coupled candidates. Such candidates must constitute a little fraction (ξ) of the Galactic dark halo in order to fulfil the positive DAMA result on annual modulation. By the fact, this analysis decouples ξ from σ_n. Considering the measured DM annual modulation effect and the absence – at the present level of sensitivity – of diurnal effects, the analysis selects allowed regions in the three-dimensional space: ξ, σ_n and DM particle mass in some model scenarios; for details see Ref. [14].

5 Asymmetric Mirror DM

The model independent annual modulation effect observed by the DAMA experiments has also been investigated in terms of a mirror-type dark matter candidates in some scenarios [15].

In the framework of asymmetric mirror matter, the DM originates from hidden (or shadow) gauge sectors which have particles and interaction content similar to that of ordinary particles. It is assumed that the mirror parity is spontaneously broken and the electroweak symmetry breaking scale v' in the mirror sector is much larger than that in the Standard Model, $v = 174$ GeV. In this case, the mirror world becomes a heavier and deformed copy of our world, with mirror particle masses scaled in different ways with respect to the masses of the ordinary particles. Then, in this scenario dark matter would exist in the form of mirror hydrogen composed of mirror proton and electron, with mass of about 5 GeV which is a rather interesting mass range for dark matter particles.

The data analysis in the Mirror DM model framework allows the determination of the $\sqrt{f}\epsilon$ parameter (where f is the fraction of DM in the Galaxy in form of mirror atoms and ϵ is the coupling constant). In the analysis several uncertainties on the astrophysical, particle physics and nuclear physics models have been taken into account in the calculation. The obtained values of the $\sqrt{f}\epsilon$ parameter in the case of mirror hydrogen atom ranges between 7.7×10^{-10} to 1.1×10^{-7}; they are well compatible with cosmological bounds [15].

In addition, releasing the assumption $M_{A'} \simeq 5m_p$, allowed regions for the $\sqrt{f}\epsilon$ parameter as function of $M_{A'}$, mirror hydrogen mass, obtained by marginalizing all the models for each considered scenario, have been obtained [15].

6 Conclusions and Perspectives

The cumulative exposure with ultra low background NaI(Tl) target by the former DAMA/NaI and DAMA/LIBRA–phase1 is 1.33 ton × yr (orders of magnitude larger than those available in the field) giving a model-independent positive evidence at 9.3 σ C.L. for the presence of DM candidates in the galactic halo with full sensitivity to many kinds of astrophysical, nuclear and particle Physics scenarios. Other rare processes have also been searched for by DAMA/LIBRA-phase1; see for details Refs. [9–11].

After the phase1, an important upgrade has been performed when all the PMTs have been replaced with new ones having higher Quantum Efficiency (QE). In this new configuration a software energy threshold below 2 keV has been reached [6]. DAMA/LIBRA is thus in its phase2, and after optimization periods it is continuously running with higher sensitivity.

DAMA/LIBRA–phase2 is continuously running in order: (1) to increase the experimental sensitivity thanks to the lower software energy threshold of the experiment; (2) to improve the corollary investigation on the nature of the DM particle and related astrophysical, nuclear and particle physics arguments; (3) to investigate other signal features; (4) to investigate rare processes other than DM with high sensitivity.

Future improvements to increase the sensitivity of the set-up can be considered by using high QE and ultra-low background PMTs directly coupled to the NaI(Tl) crystals. In this way a further large improvement in the light collection and a further lowering of the software energy threshold would be obtained.

Finally, for completeness, we also mention that low background $ZnWO_4$ crystal scintillators have recently been proposed within the DAMA collaboration for the study of the directionality of DM candidates inducing just nuclear recoils. The features and performances of such anisotropic scintillators are very promising [42] and are under exploration.

References

[1] R. Bernabei et al., Nucl. Instr. and Meth. **A 592**, 297 (2008).
[2] R. Bernabei et al., Eur. Phys. J. **C 56**, 333 (2008).
[3] R. Bernabei et al., Eur. Phys. J. **C 67**, 39 (2010).
[4] R. Bernabei et al., Eur. Phys. J. **C 73**, 2648 (2013).
[5] P. Belli et al., Phys. Rev. **D 84**, 055014 (2011).
[6] R. Bernabei et al., J. of Instr. **7**, P03009 (2012).
[7] R. Bernabei et al., Eur. Phys. J. **C 72**, 2064 (2012).
[8] R. Bernabei et al., Int. J. of Mod. Phys. **A 28**, 1330022 (2013).
[9] R. Bernabei et al., Eur. Phys. J. **C 62**, 327 (2009).
[10] R. Bernabei et al., Eur. Phys. J. **C 72**, 1920 (2012).
[11] R. Bernabei et al., Eur. Phys. J. **A 49**, 64 (2013).
[12] R. Bernabei et al., Eur. Phys. J. **C 74**, 2827 (2014).
[13] R. Bernabei et al., Eur. Phys. J. **C 74**, 3196 (2014).
[14] R. Bernabei et al., Eur. Phys. J. **C 75**, 239 (2015).
[15] A. Addazi et al., Eur. Phys. J. **C 75**, 400 (2015).
[16] R. Bernabei el al., La Rivista del Nuovo Cimento **26 n.1**, 1-73 (2003).
[17] R. Bernabei et al., Int. J. Mod. Phys. **D 13**, 2127 (2004).
[18] A.K. Drukier, K. Freese, D.N. Spergel, Phys. Rev. **D 33**, 3495 (1986).
[19] K. Freese, J.A. Frieman, A. Gould, Phys. Rev. **D 37**, 3388 (1988).
[20] D. Smith and N. Weiner, Phys. Rev. **D 64**, 043502 (2001).
[21] D. Tucker-Smith and N. Weiner, Phys. Rev. **D 72**, 063509 (2005).
[22] D. P. Finkbeiner et al, Phys. Rev. **D 80**, 115008 (2009).
[23] K. Freese et al., Phys. Rev. **D 71**, 043516 (2005).
[24] K. Freese et al., Phys. Rev. Lett. **92**, 11301 (2004).
[25] R. Bernabei et al., Eur. Phys. J. **C 18**, 283 (2000).
[26] R. Bernabei et al., Int. J. of Mod. Phys. A **31**, 1642009 (2016).
[27] R. Bernabei et al., Phys. Lett. B **389**, 757 (1996).
[28] R. Bernabei et al., Phys. Lett. B **424**, 195 (1998).
[29] R. Bernabei et al., Phys. Lett. B **450**, 448 (1999).
[30] P. Belli et al., Phys. Rev. D **61**, 023512 (2000).
[31] R. Bernabei et al., Phys. Lett. B **480**, 23 (2000).
[32] R. Bernabei et al., Phys. Lett. B **509**, 197 (2001).
[33] R. Bernabei et al., Eur. Phys. J. C **23**, 61 (2002).
[34] P. Belli et al., Phys. Rev. D **66**, 043503 (2002).
[35] R. Bernabei et al., Int. J. Mod. Phys. A **21**, 1445 (2006).
[36] R. Bernabei et al., Eur. Phys. J. C **47**, 263 (2006).
[37] R. Bernabei et al., Int. J. Mod. Phys. A **22**, 3155 (2007).
[38] R. Bernabei et al., Phys. Rev. D **77**, 023506 (2008).
[39] R. Bernabei et al., Mod. Phys. Lett. A **23**, 2125 (2008).
[40] R. Bernabei et al., Eur. Phys. J. C **53**, 205 (2008).
[41] S. Scopel, K.H. Yoon, J.H. Yoon, arXiv:1512.08577
[42] F. Cappella et al., Eur. Phys. J. C **73**, 2276 (2013).

DOI: 10.1051/epjconf/201713605002

$ZnWO_4$ anisotropic scintillator for Dark Matter investigation with the directionality technique

R. Bernabei[1,2], P. Belli[2], F. Cappella[3], V. Caracciolo[3], R. Cerulli[3], F.A. Danevich[4], A. d'Angelo[5], A. Di Marco[1], A. Incicchitti[5], V. M. Mokina[4,5], D.V. Poda[4,6], O.G. Polischuk[4], C. Taruggi[1,3], and V.I. Tretyak[4,5]

[1]*Dip. di Fisica, Università di Roma* Tor Vergata, *I-00133 Rome, Italy*
[2]*INFN, sez. Roma* Tor Vergata, *I-00133 Rome, Italy*
[3]*INFN, Laboratori Nazionali del Gran Sasso, I-67100 Assergi (AQ), Italy*
[4]*Institute for Nuclear Research, MSP 03680 Kyiv, Ukraine*
[5]*INFN, sez. Roma, I-00185 Rome, Italy*
[6]*CSNSM, Univ. Paris-Sud, CNRS/IN2P3, Université Paris-Saclay, 91405 Orsay, France*

Abstract. The $ZnWO_4$ crystal scintillator has unique features that make it very promising to realize a pioneering experiment to pursue Dark Matter investigation with the directionality technique. In particular in this detector the light output and the scintillation pulse shape for heavy particles (p, α, nuclear recoils) depend on the direction of the impinging particle with respect to the crystal axes, while the response to γ/β radiation is isotropic. The anisotropy of the light output can be considered to point out the presence in the diurnal counting rate of a Dark Matter signal produced by candidate particle inducing just nuclear recoils. In addition this crystal detector has also other important characteristics for a Dark Matter experiment: high light output, high level of radiopurity. In this paper the present performances of the developed $ZnWO_4$ crystal scintillator will be summarized together with the possible future improvements. Some reachable sensitivities – under given assumptions – in the investigation of DM candidate particles with the directionality technique will also be addressed.

1 Introduction

In Dark Matter (DM) direct experiments, in order to point out the presence of a signal of Galactic origin, it is mandatory to pursue a signature for signal identification. The most important one is the model independent DM annual modulation that has been successfully exploited by the DAMA/NaI and DAMA/LIBRA experiments obtaining, cumulatively, an annual modulation effect at 9.3σ C.L. over 14 annual cycles [1–3].

An independent evidence can be obtained by pursuing a different approach, but effective for those DM candidate particles able to induce just nuclear recoils: the directionality [4]. This strategy is based on the correlation between the arrival direction of the DM particles (and thus of the induced nuclear recoils) and the Earth motion in the Galactic rest frame. Because of the rotation of the Milky Way, the Galactic disc passes through the halo of DM and the Earth is crossed by a wind of DM particles apparently flowing along a direction opposite to that of solar motion. Since the Earth rotates around its axis, the average direction of DM particles with respect to an observer fixed on the Earth changes during the sidereal day. Thus, the directions of the induced nuclear recoils are expected to be strongly correlated with the impinging direction of the considered DM candidates while the background events are not.

In principle, an experiment able to measure the nuclear track might be suitable to investigate the directionality. One possibility is to use low pressure gas detector (such as Time Projection Chambers, TPC) where the range of recoiling nuclei is of the order of mm. However, a realistic experiment with low pressure TPC can be limited e.g. by the necessity of an extreme operational stability, of large detector size and of a great spatial resolution in order to reach a significant sensitivity. The limitations affecting experiments aiming to measure recoil tracks, can be overcome by using the anisotropic scintillation detectors [5, 6]. In this case there is no necessity of a track detection and recognition (in solid detectors the range of recoiling nuclei is typically of the order of μm). In these detectors the detector response for heavy particles and scintillation pulse shape depend on the incoming direction of the heavy particles relatively to the crystal axes and the information on the presence of DM induced recoils is given by a peculiar variation of the measured counting rate during the sidereal day [7].

2 The main features of the $ZnWO_4$ anisotropic scintillator

Recently, measurements and R&D works have shown that the $ZnWO_4$ scintillators can offer suitable features for a DM experiment based on the directionality. In this crystal scintillator the light output for heavy particles (p, α, nuclear recoils) depends on the direction of such particles with respect to the crystal axes while the response to γ/β radiation is isotropic; the scintillation decay time also shows the same property. In addition to the anisotropy, the recently developed $ZnWO_4$ scintillators have very good level of radiopurity [8], and can work at energy threshold of few keV [9]. The $ZnWO_4$ offers also a high atomic weight and the possibility to realize crystals with masses of some kg [10]. Moreover, three target nuclei with very different masses are present in this detector (Zn, W and O), giving sensitivity to both small and large mass for the considered DM candidates.

The luminescence of $ZnWO_4$ was studied sixty years ago [11]. Large volume $ZnWO_4$ single crystals of reasonable quality were grown [12] and studied as scintillators in the eighties [13]. Further developments for high quality radiopure $ZnWO_4$ have been performed and described in [10, 14, 15].

The first low background measurement with a small $ZnWO_4$ sample (mass of 4.5 g) was performed in the Solotvina underground laboratory (Ukraine) at a depth of $\approx$ 1000 m of water equivalent in order to study its radioactive contamination, and to search for double beta decay of zinc and tungsten isotopes [17]. A possibility to search for diurnal modulation of WIMP direction with $ZnWO_4$ scintillators was also pointed out in [17]. More recently, radiopurity and double beta decay processes of zinc and tungsten have been further studied also at LNGS using new developed $ZnWO_4$ detectors with masses 0.1 – 0.7 kg [8, 9, 18, 19]. The growth of the crystals, the scintillation properties, the pulse shape discrimination capability, the anisotropic properties, the residual radioactive contamination and the possible applications have been deeply studied [8, 10, 14, 15, 17–20]. The obtained results are very promising and an R&D to produce $ZnWO_4$ crystals having higher radiopurity is still ongoing. In particular, an R&D to improve $ZnWO_4$ crystals radiopurity by re-crystallization (recently demonstrated for $CdWO_4$ crystal [20]) is in progress.

In the measured $ZnWO_4$ scintillators the radioactive contamination is: < 0.002 mBq/kg for ^{228}Th and ^{226}Ra ($\sim$ 0.5 ppt for ^{232}Th and $\sim$ 0.2 ppt for ^{238}U, assuming the secular equilibrium of the ^{232}Th and ^{238}U chains), < 0.02 mBq/kg for ^{40}K; the total α activity is 0.18 mBq/kg [8].

As previously mentioned, the study of the directionality with the $ZnWO_4$ detectors is based on the anisotropic properties of these scintillators. In particular the light output of the detector for α and β

 DOI: 10.1051/epjconf/201713605002

particles of the same energy is different; in addition the response of the detector to α's depends also on the direction of the particels with respect to its crystallographic axes. In Fig. 1 the behaviuor of the light output ratio for α and β (α/β ratio) is reported as a function of the energy and the direction of the α beam in a $ZnWO_4$ crystal [17]. As shown in Fig. 1, the response of the detector for α particles

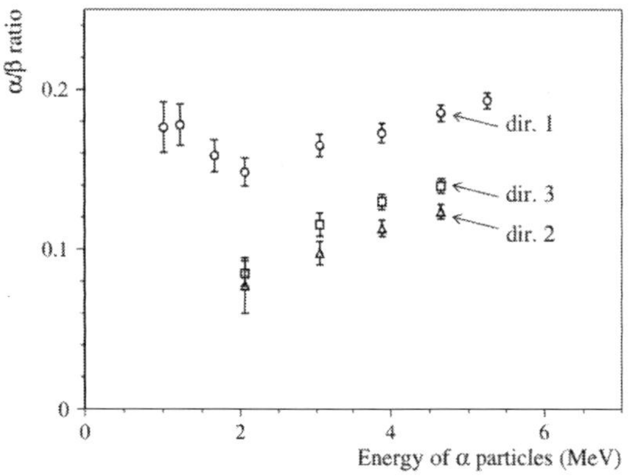

Figure 1. Dependence of the α/β ratio (ratio of the light output measured in the crystal for α and β particles) on the energy of α particles measured with $ZnWO_4$ scintillator. The crystal was irradiated in the directions perpendicular to (010), (001) and (100) crystal planes (directions 1, 2 and 3, respectively). The anisotropic behaviour of the crystal is evident [17].

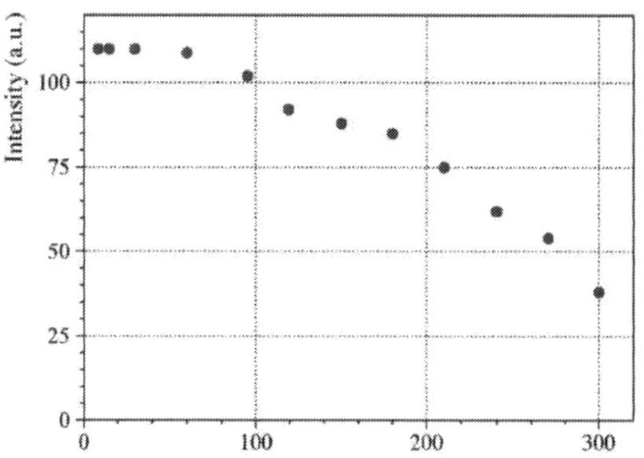

Figure 2. Dependence of the light output of the $ZnWO_4$ as a function of the temperature, for excitation with ^{241}Am α particles [15].

measured along direction 1 is about 1.5 times larger than that measured along direction 2, and about 1.4 times larger than that measured along direction 3. On the contrary, the anisotropy of the light response of the $ZnWO_4$ scintillator disappears in case of electron excitation. The same behaviour have been recently reported also for the anthracene in Ref. [16]. Moreover for $ZnWO_4$, as reported in Ref. [17], also the shape of the scintillation pulse depends on the type of irradiation; this feature allows one to discriminate $\gamma(\beta)$ events from those induced by α particles. This pulse shape discrimination capability can be of interest not only for a DM experiment but also for double beta decay searches. Measurements with a neutron beam to study the anisotropy response of the crystal for recoils at keV energy range will be performed in near future [21].

Another feature of this scintillator, important for a DM experiment, is the relatively high light output which is about 13-20% of the Na(Tl) scintillator. It has been observed that the light output largely increase when the crystal scintillator working temperature is decreased [15] (see Fig. 2).

3 Estimated sensitivity in some given scenarios

In case of a DM candidate interacting via elastic scattering with ordinary nuclei, by pursuing the directionality technique with anisotropic scintillators, the DM signal can be pointed out by studying the diurnal variation of the counting rate as a function of the sidereal time. In a laboratory on the Earth, the DM particles are expected to arrive preferentially from the opposite direction of the Sun velocity in the Galaxy. Due to the Earth rotation around its axis, the orientation of the crystallographic axes of the scintillator with respect to the arrival direction of the particle (and thus of the nuclear recoils)

 DOI: 10.1051/epjconf/201713605002

changes during the day. Because of the anisotropy of the scintillator, the response of the detector varies during the day. Thus the DM signal is expected to have a peculiar variation during the sidereal day [7]. On the contrary, the light response of the detector to e/γ background will be isotropic and its counting rate will show a flat behaviour as a function of the time. For this reason the observation of a peculiar diurnal variation can offer indication for the presence – in the Galactic halo – of DM candidates inducing just nuclear recoils.

To estimate the reachable sensitivity – in some given scenarios – of an experiment exploiting the directionality technique with the help of $ZnWO_4$ scintillators, we can consider a set-up composed by a matrix of detectors (as DAMA/LIBRA set-up) installed deep underground in a sealed low radioactive copper box, continuously flushed with high purity N_2 gas and placed in the center of a multi-ton, multi-component low radioactive passive shield. By considering 200 kg of $ZnWO_4$, in 5 years of data taking and with an energy resolution FWHM = $2.4\sqrt{E[keV]}$), the sensitivity reported in Fig. 3 can be reached [7]. In particular, two software energy thresholds have been considered: 2 keVee for Fig. 3-left and 6 keVee for Fig. 3-right. The sensitivity curves have been calculated considering four possible time independent background levels in the low energy region: 10^{-4} cpd/kg/keV (solid black lines), 10^{-3} cpd/kg/keV (dashed lines), 10^{-2} cpd/kg/keV (dotted lines) and 0.1 cpd/kg/keV (dotted-dashed lines). The response of the detector to Zn, W and O recoils for the three crystal axes have been estimated by considering the method described in Ref. [22], taking into account the data on the anisotropy obtained with the α particles (see Fig. 1). As shown in Fig. 3 the directionality approach can reach – for DM candidates inducing just nuclear recoils and the given scenarios – a sensitivity to spin-independent cross sections at level of $10^{-5} - 10^{-7}$ pb, depending on the candidate mass between few GeV and hundreds GeV. However, it is worth noting that these plots are model dependent and, thus, always affected by several uncertainties; to obtain this sensitivity plot the scenario described in details in Ref. [7] has been considered. In Fig. 3 the allowed regions (7.5σ from the null hypothesis) obtained by performing a corollary analysis of the DAMA model independent result in term of the scenarios described in Ref. [23] are also reported.

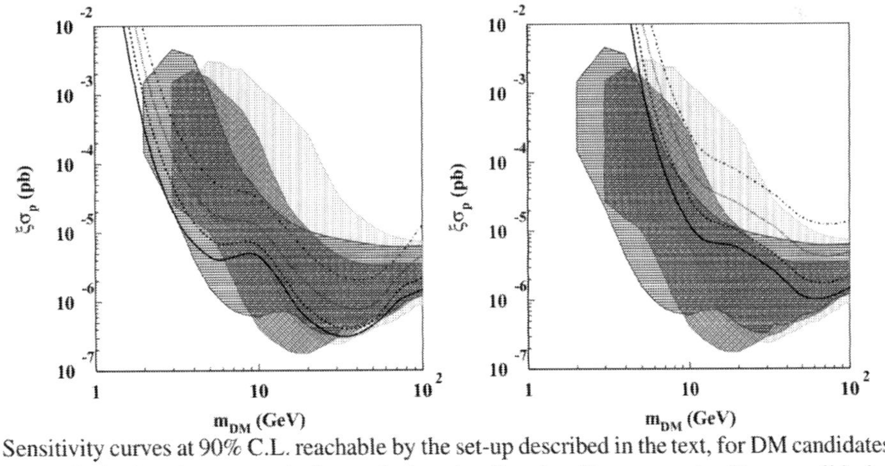

Figure 3. Sensitivity curves at 90% C.L. reachable by the set-up described in the text, for DM candidates inducing just nuclear recoils in the given scenario by exploring the directionality approach. Four possible background levels in the low energy region are considered (see text) and two software energy thresholds: 2 keVee in (a) and 6 keVee in (b). There m_{DM} is the particle mass, σ_p is the spin-independent elastic scattering cross section on nucleon and ξ is the fraction of the DM local density of the considered candidate. In the figures there are also shown (green, red and blue online) allowed regions obtained in Ref. [23] by a corollary analysis of the 9σ C.L. DAMA model independent result in terms of scenarios for DM candidates inducing just nuclear recoils.

4 Conclusions

The perspectives of a pioneering experiment with anisotropic $ZnWO_4$ detectors to further explore, with the directionality approach, those DM candidate particles inducing just nuclear recoils have been addressed. The features of these detectors can permit to reach, in some of the many possible scenarios, sensitivities not far from that of the DAMA/LIBRA positive result [2, 3]. In case of success the experiment can obtain an evidence for the presence of such DM candidate particles in the galactic halo with a new approach and will provide complementary information on the nature and interaction type of the DM candidate(s). In case of negative results the experiment would favor other kinds of astrophysical, nuclear and particle physics scenarios or other DM candidate particles, interaction types which can account as well for the 9.3σ C.L. DM model independent annual modulation effect already observed by the DAMA experiments. In all cases such an experiment would represent a first realistic attempt to investigate the directionality through the use of anisotropic scintillators and it could also represent a further activity in the application of highly radiopure $ZnWO_4$ detector in the field of rare processes.

References

[1] R. Bernabei *et al.* 2003 *La Rivista del Nuovo Cimento* **26** 1.
[2] R. Bernabei *et al.* 2008 *Eur. Phys. J. C* **56** 333.
[3] R. Bernabei *et al.* 2010 *Eur. Phys. J. C* **67** 39; P. Belli *et al.* 2011 *Phys. Rev. D* **84** 055014; R. Bernabei *et al.* 2012 *J. Instrum.* **7** P03009, R. Bernabei *et al.* 2012 *Eur. Phys. J. C* **72** 2064; R. Bernabei *et al.* 2013 *Eur. Phys. J. A* **49** 64; R. Bernabei *et al.* 2013 *Eur. Phys. J. C* **73** 2648; R. Bernabei *et al.* 2013 *Int. J. Mod. Phys., A* **28** 1330022; R. Bernabei *et al.* 2014 *Eur. Phys. J. C* **74** 2827.
[4] D.N. Spergel 1988 *Phys. Rev. D* **37** 1353.
[5] P. Belli *et al.* 1992 *Il Nuovo Cim. C* **15** 475.
[6] R. Bernabei *et al.* 2003 *Eur. Phys. J. C* **28** 203.
[7] F. Cappella *et al.* 2013 *Eur. Phys. J. C* **73** 2276.
[8] P. Belli *et al.* 2011 *Nucl. Instrum. Meth. A* **626-627** 31.
[9] P. Belli *et al.* 2011 *J. Phys. G* **38** 115107.
[10] E.N. Galashov *et al.* 2009 *Functional Materials* **16** 63.
[11] F.A. Kroger, Some Aspects of the Luminescence in Solids, Elsevier Pub. Co, Amsterdam 1948, p.109.
[12] B.C. Grabmaier 1984 *IEEE Trans. Nucl. Sci.* **31** 372.
[13] Y.C. Zhu *et al.* 1986 *Nucl. Instrum. Meth. A* **244** 579; F.A. Danevich *et al.* 1989 *Prib. Tekh. Eksp.* **5** 80 [1989 *Instrum. Exp. Tech.* **32** 1059].
[14] L.L. Nagornaya *et al.* 2008 *IEEE Trans. Nucl. Sci.* **55** 1469.
[15] L.L. Nagornaya *et al.* 2009 *IEEE Trans. Nucl. Sci.* **56** 994.
[16] P. Schuster, 2016 *IEEE Trans. Nucl. Sci.* **63** 1942.
[17] F.A. Danevich *et al.* 2005 *Nucl. Instrum. Meth. A* **544** 553.
[18] P. Belli *et al.* 2008 *Phys. Lett. B* **658** 193.
[19] P. Belli *et al.* 2009 *Nucl. Phys. A* **826** 256.
[20] A.S. Barabash *et al.* 2016 *Nucl. Instrum. Meth. A* **833** 77.
[21] C. Taruggi, ROM2F/2016/06.
[22] V.I. Tretyak 2010 *Astropart. Phys.* **33** 40.
[23] P. Belli *et al.* 2011 *Phys. Rev. D* **84** 055014.

Impact of next future Direct Detection experiments on Dark Portals and beyond

Giorgio Arcadi[1,a]

[1] *Max Planck Institüt für Kernphysik, Saupfercheckweg 1, 69117 Heidelberg, Germany*

Abstract. WIMPs are attractive DM candidates. The simplest model realizations, dubbed Dark Portals are in strong tensions with detection constraints. Two component DM setups can relax these tensions.

1 Introduction

Weakly interacting massive particles (WIMPs) are among the most popular DM candidates. Their production mechanism predicts the existence, in thermal equilibrium, of the DM in the primordial thermal bath and a subsequent decoupling (freeze-out). The value of the DM abundance is determined by a single particle physics input, consisting in the thermally averaged pair annihilation cross-section; more specifically $\Omega_{\rm DM} h^2 \propto 1/\langle \sigma v \rangle$ so that $\Omega_{\rm DM} h^2 \simeq 0.12 \rightarrow \langle \sigma v \rangle \sim 3 \times 10^{-26} {\rm cm}^3 {\rm s}^{-1}$ [1]. This value of the pair annihilation cross section implies sizeable enough interactions between the DM and the SM states to be in the reach of present experimental facilities. A simple framework for unveiling the potential correlations between the DM relic density and experimental searches is represented by the 'Dark Portals'; simple extensions of the SM in which the DM annihilations into SM states (mostly SM fermions) are induced by a s-channel mediator. In these kind of setups the DM annihilation cross-section and experimentally testable observables, like the scattering cross-section on nucleons, are related by simple crossing-symmetries. Because of these strong correlations, however, Dark Portals are in increasing tension with constraints from Direct Detection and many realizations will be substantially ruled-out in case of absence of signals from next generations of 1TON detectors, like XENON1T. A possible way-out for preserving the WIMP paradigm, compatibly with experimental constraints, is represented by considering two-component Dark portals. We will briefly illustrate a specific example in the second part of the text.

2 Single Component Dark Portals

Dark Portals are simple extension of the Standard model with the DM candidate and, possibly, a mediator of its interactions with the SM (fermions). Being these models built on purely phenomenological basis one can consider all the possible spin combinations between the DM and the mediator which are allowed by Lorentz invariance. For definiteness we will just focus on the cases of a scalar DM candidate χ or a vector state V^μ interacting with a scalar mediator. This kind of interaction does not

[a] e-mail: giorgio.arcadi@mpi-hd.mpg.de

require, in general, the presence of extra states, with respect to the SM, since it is realized, at the renormalizable level and in a Lorentz and gauge invariant way, by the four-field interactions $\lambda_\chi |\chi|^2 H^\dagger H$ or $\lambda_V V^\mu V_\mu H^\dagger H$, with H being the SM Higgs doublet and $\lambda_{\chi,V}$ adimensional couplings, which lead, after EW symmetry breaking, to trilinear interactions between the SM Higgs and a DM pair. These kind of interactions can be schematically written as (for simplicity we report just the interactions with SM fermions):

$$\mathcal{L}_\chi = \left(\mu_\chi |\chi|^2 + \lambda_f \bar{f} f\right) h, \quad \mathcal{L}_V = \left(\mu_V V^\mu V_\mu + \lambda_f \bar{f} f\right) h \tag{1}$$

where $\mu_{\chi,V} = \lambda_{\chi,V} v_h$ and $\lambda_f = m_f / v_h$, with v_h being the Higgs vev. The scalar or vectorial DM annihilates into SM fermions (as well as gauge bosons and higgs pairs) through s-channel exchange of h. The corresponding cross sections can be analytically approximated as [1]:

$$\langle \sigma v \rangle_\chi \approx \sum_f \frac{n_c^f |\mu_\chi|^2}{4\pi (m_h^2 - 4m_\chi^2)^2}, \quad \langle \sigma v \rangle_V \approx \sum_f \frac{n_c^f |\mu_V|^2}{4\pi (m_h^2 - 4m_V^2)^2} \tag{2}$$

At the same time, t-channel exchange of h induces Spin Independent (SI) scattering cross-sections on nucleons of the form:

$$\sigma_\chi^{\rm SI} = \frac{|\mu_\chi|^2 \mu_{N\chi}^2}{4\pi m_h^4 m_\chi^2} \frac{|Z f_p + (A-Z) f_n|^2}{A^2}, \quad \sigma_\chi^{\rm SI} = \frac{|\mu_V|^2 \mu_{NV}^2}{4\pi m_h^4 m_V^2} \frac{|Z f_p + (A-Z) f_n|^2}{A^2} \tag{3}$$

where $\mu_{(\chi,V)N}$ is the DM-nucleon reduced mass and $f_{p,n}$ are the couplings of the DM with protons and neutrons. A straightforward comparison between eq. (2) and (3) evidences similar parametric dependences for the two classes of cross-section. We can then compare, the constraints from the correct DM relic density and from DM DD in the bidimensional plane $(m_{\chi,V}, \mu_{\chi,V})$. An example of this kind of comparison is shown, in fig. (1), in the case of Scalar and Vector Higgs portal (as already mentioned above we can trade, in these kind of scenarios, the dimensional parameters $\mu_{\chi,V}$ with the dimensionless quantities $\lambda_{\chi,V}$.). By imposing the constraint $\Omega_{\chi,V} h^2 = 0.12 \pm 10\%$, it is possible to identify curves (represented in black in the two panels of fig. (1)) in the plane $(m_{\chi,V}, \mu_{\chi,V})$. These isocontours have been overlapped to the exclusion regions obtained by imposing that the DM scattering cross-sections (3) lie below current LUX sensitivity [3] (blue regions), and a 2 year projected sensitivity of XENON1T [4] (cyan regions), and that the decay rate of the Higgs into a DM pair, when kinematically open, corresponds to an invisible branching fraction below present limits [5].
As evident, current direct detection constraints already rule-out thermal DM ad exception of the 'pole' regions, i.e. $m_{\chi,V} \simeq m_h/2$ (low values of the DM mass, even outside the reach of DD experiments, are excluded since induce a too high invisible width of the Higgs). The WIMP hyphothesis would be completely ruled-out by absence of signals in XENON1T.

3 Two component Dark Portals

In order to keep relying on the WIMP paradigm, compatibly with DD constraints, it is necessary to decorrelate (at least partially) the DM annihilation cross-section and the scattering cross-section on nucleons. This makes necessary to go beyond the too simplistic structure of Dark Portals. An interesting possibility would be to consider a two-component DM scenario. In such a case the interplay between DM relic density and DD is more complicated. Indeed, a DD signal would arise from a sum of two, possibly different, contributions, weighted by the relative abundances of the two components. The latters are not necessarily dominated by pair annihilations into SM states, which can be directly

[1] These expressions are not valid close to the pole regions [2], i.e $m_{\chi,V} = m_h/2$.

DOI: 10.1051/epjconf/201713605003

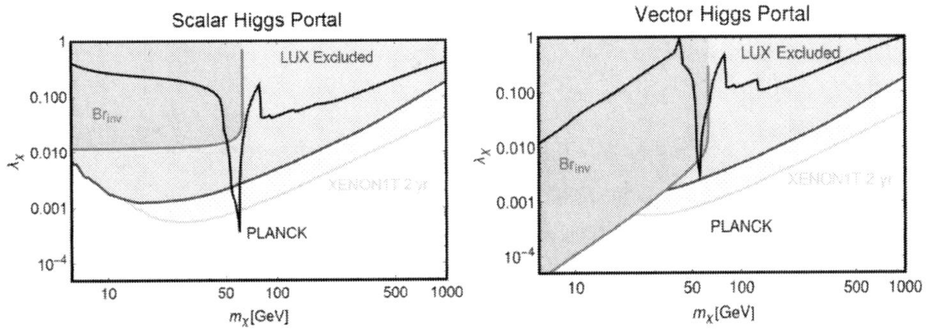

Figure 1. Combined constraints for scalar (left panel) and vector (right panel) Higgs portals. The black lines represent the isocontours of the correct DM relic density. The blue regions are excluded by current constraints by LUX [3] while the cyan region will be excluded in case of absence of signals at XENON1T after two years of excluded. In the red regions the invisible branching fraction, originated by decays into DM pairs, exceeds the experimental bounds [5].

related to the scattering rates, but processes like pair annihilations of one DM component to the other, can play a relevant role.

An interesting realization of a two-component DM scenario has been proposed in [6]. Here the SM is extended by a dark sector characterized by a new $SU(3)$ gauge symmetry, spontaneosly broken, by the vevs of two fields, ϕ_1 and ϕ_2, belonging to its fundamental representation, into a discrete global $\mathcal{Z}_2 \times \mathcal{Z}_2^{'}$ symmetry. The only states, apart the higgs bosons, present in the new sector are the 8 (massive) SU(3) gauge bosons. The lightest new states odd under the discrete symmetry are automatically DM candidates. The contact between the dark and the visible (i.e. the SM) sector is established by portal like interactions induced by operators of the type $\lambda_{H11}|H|^2|\phi_1|^2 + \lambda_{H22}|H|^2|\phi_2|^2$. The fields of the dark sector are described by the following Lagrangian:

$$\mathcal{L} = (D_\mu \phi_1)^\dagger (D^\mu \phi_1)^\dagger + (D_\mu \phi_2)^\dagger (D^\mu \phi_2)^\dagger - \frac{1}{4} G^a_{\mu\nu} G^{a\mu\nu}, \quad G^a_{\mu\nu} = \partial_\mu A^a_\nu - \partial_\nu A^a_\mu - \tilde{g} f^{abc} A^b_\mu A^c_\nu \tag{4}$$

where the Higgs triplets ϕ_1 and ϕ_2 are expressed, in the unitary gauge, as:

$$\phi_1 = \frac{1}{\sqrt{2}} \begin{pmatrix} 0 \\ 0 \\ v_1 + \varphi_1 \end{pmatrix}, \quad \phi_2 = \frac{1}{\sqrt{2}} \begin{pmatrix} 0 \\ v_2 + \varphi_2 \\ v_3 + \varphi_3 + i(v_4 + \chi) \end{pmatrix} \tag{5}$$

Under the assumption of CP-conservation the mixing with the Higgs doublet $H = (0 \ \frac{v+h}{\sqrt{2}})^T$ gives rise to 4 CP-even scalars $h_{i=1,4}$ and on CP-odd χ. The latter is odd also under the residual symmetry $\mathcal{Z}_2 \times \mathcal{Z}_2^{'}$ and can then represent a DM candidate.

Despite the general model is rather complicated (see [6] for more details), its most relevant features can be caught by the following simplified limit. Assuming a direct coupling of only ϕ_2 with the Higgs, i.e. $\lambda_{H11} = 0$ and setting $v_3 \sim 0, v_2 \ll v_1$, the relevant particle spectrum of the dark sector reduces to a two component DM, consisting into CP-odd dark Higgs χ and the mass degenerate vector pair $A^\mu_{1,2}$, interacting with the SM through the two lightest CP-even mass eigenstates $h_{1,2}$ (h_1 is identified with the SM Higgs). The DM interaction Lagrangian is thus simply written as:

$$\mathcal{L} = \frac{\tilde{g} m_A}{2}(-h_1 \sin\theta + h_2 \cos\theta) \sum_{a=1,2} A^a_\mu A^{a\mu} + \lambda_2 v_2 (-\sin\theta h_1 + \cos\theta h_2) + \lambda_{H22} v_h (\cos\theta h_1 + \sin\theta h_2) \chi^2 \tag{6}$$

 DOI: 10.1051/epjconf/201713605003

with $\tilde{g}$ and θ being the new gauge coupling and the mixing angle between SM and dark Higgs, and substantially resembles a two-component dark portal Lagrangian. The coupling of the scalar DM component χ depends on the λ_{H22} and on the quartic self-coupling of ϕ_2, λ_2. These can be rexpressed as:

$$\lambda_2 = \tilde{g}^2 \frac{\cos^2\theta m_{h_2}^2 + \sin^2\theta m_{h_1}^2}{4 m_A^2}, \quad \lambda_{H22} = \tilde{g} \frac{(m_{h_1}^2 - m_{h_2}^2)\sin\theta\cos\theta}{2 v_h m_A} \tag{7}$$

Despite the similarity with the Lagrangians (1) the DM phenomenology is sensitively different. We have indeed that the annihilation of the vectorial component is enhanced by the presence of the 'dark' channel $AA \to \chi\chi$ which is dominant, at least in some regions of the parameter space, because does on depend, contrary to the ones into SM states, on $\sin\theta$, which is constrained to be small because of the constraints on not standard decay branching fractions of the 125 GeV Higgs. For what regards the DM direct detection the scalar component features, instead, a peculiar behaviour. In the non relativistic limit, its effective coupling with the quarks is given by:

$$g_{\chi\chi qq} \propto \left[\frac{1}{m_{h_1}^2}\left(\cos^2\theta\lambda_{H22}v_h - \sin\theta\cos\theta\lambda_2 v_2\right) - \frac{1}{m_{h_2}^2}\left(\sin^2\theta\lambda_{H22}v_h + \sin\theta\cos\theta\lambda_2 v_2\right)\right] \tag{8}$$

It can be straightforwardly verified, by substituting eq. (7), that this coupling is null.
An interesting setup thus emerges: only one dark matter component is actually capable of scattering with nucleons but his abundance is reduced by a new annihilation channel, so that the total event rate at a DD experiment would be reduced. Moreover the annihilation and scattering cross-section are not fully correlated since the latter depends on $\sin\theta$ while $AA \to \chi\chi$ process not. The other DM component, with an unsuppressed abundance, is instead not capable of scattering off-nuclei.

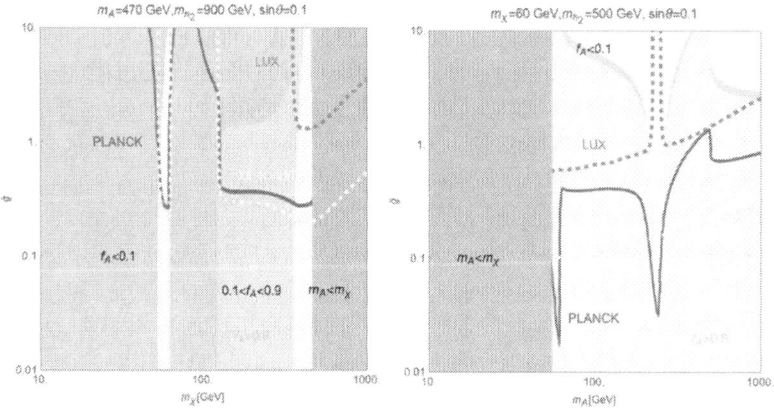

Figure 2. Combined constraints for two two-component DM benchmarks. The red contours represent the experimentally favoured value of the (total) DM relic density. The orange and yellow lines represent, respectively, the current limit by LUX [3] and a 2 yr exposure projected sensitivity by XENON1T [4]. The colored regions represent three ranges, < 0.1, $0.1 - 0.9$ and > 0.9, of the density fraction f_A of the vectorial DM component.

The phenomenological impact of these features is shown on fig. (2). Here for two benchmarks we have compared, similarly to what done in fig. 1, the constraints from DM relic density and DD as function of the DM coupling (it is $\tilde{g}$ for both components) and DM mass (one mass is varied and the other kept fixed, alternatively, in the two panels.). As evident bounds from LUX do not affect the region at the correct DM relic density, even when this is dominated by the vector component (see right panel of fig. 2). On the contrary this scenario will be probed, at least partially by XENON1T, although an exposure time of two years will be needed.

4 Conclusions

WIMPs are an attractive class of DM candidates. The simplest model realizations, dubbed dark portals, are in increasing tension with limits from Direct Detection. We have shown in an explicit example that this kind of bounds can be easily relaxed in multi-component DM constructions.
Acknowledgements: The author warmly thanks the organizers of the RICAP conference for the chance of giving this contribution.

References

[1] P.A.R. Ade et al. (Planck), Astron. Astrophys. **594**, A13 (2016), 1502.01589
[2] K. Griest, D. Seckel, Phys. Rev. **D43**, 3191 (1991)
[3] D.S. Akerib et al. (LUX), Phys. Rev. Lett. **116**, 161301 (2016), 1512.03506
[4] E. Aprile et al. (XENON), JCAP **1604**, 027 (2016), 1512.07501
[5] G. Aad et al. (ATLAS, CMS), JHEP **08**, 045 (2016), 1606.02266
[6] C. Gross, O. Lebedev, Y. Mambrini, JHEP **08**, 158 (2015), 1505.07480

Seasonal variations of the rate of multiple-muons in the Gran Sasso underground laboratory

F. Ronga[1,a]

[1] *Istituto Nazionale di Fisica Nucleare, Laboratori Nazionali di Frascati, via E.Fermi 40 - 00044 Frascati, Italy*

Abstract. It is well known that the rate of cosmic ray muons depends on the atmospheric temperature, and that for events with a single muon the peak of the rate is in summer, in underground laboratories in the northern hemisphere. In 2015 the MINOS experiment, in USA, found that, for small distances between the multiple-muons, the rate of multiple-muons peaks in the winter and that the amplitude of the modulation is smaller than in the case of a single muon. I have done a re-analysis of data of the past MACRO experiment. The result is that under Gran Sasso the rate of multiple-muons at small distances peaks in the summer. This difference with MINOS could be explained by differences in the atmospheric temperature due to latitude. This results could be of interest for dark matter experiments looking to dark matter seasonal modulation due to the Earth's motion.

1 Introduction

Underground muons originate primarily from the decay of mesons produced in high energy interactions between primary cosmic ray particles and atmospheric nuclei [1]. Fluctuations in atmospheric temperature lead to variations in the muon rate observed at ground level and underground.

While the temperature of the troposphere varies considerably within the day, the temperature of the stratosphere remains nearly constant during the day, usually changing on the time scale of seasons. An increase in temperature of the stratosphere causes a decrease in density. This reduces the chance of meson interactions, resulting in a greater number of mesons that decay and produce muons and increases the muon rate observed by several experiments located deep underground. For the a recent summary of the data see [2]. The majority of muons detected in an underground detector are produced in the decay of pions. All the underground experiments until 2015 analyzed the muon rate of events with a single track, or with specific cuts to select those events as for example in [3] or simply because in small detectors the rate is largely dominated by the single muons.

In 2015 the MINOS experiment published the seasonal modulation of the multiple-muons events [4]. We expect different seasonal oscillation between multi muon events and single muon events. One difference is due to the fact that multiple-muons are produced by primary cosmic rays with an energy higher than the one needed for single muons. Another difference is due to the different cosmic ray primary, multiple-muons production is preferred by heavy primaries. Heavy primaries have a cross section bigger than proton and therefore interact higher in the atmosphere. Another effect is geometrical: large size multiple-muons are produced at a bigger heigth than single muons or multiple

[a] e-mail: francesco.ronga@lnf.infn.it

muons of a small size. Of course this kind of geometrical effect depends also by the size of the detector.

MINOS reports results of two detector: a far detector (FD) with an overburden of 225 m.w,e, and a near detector (ND) at 2100 m.w.e. MINOS has observed a multiple-muon phase inconsistent with the summer maximum observed in the ND and the FD single-muon data. Data collected by the MINOS FD were used to show that there is a transition from a summer maximum in multiple-muon events with a large track separation to a winter maximum in multiple-muon events with a small track separation. This transition occurs at track separations of about 5-8 m. The amplitude of the oscillation was also dependent on the tracks separation.

The dependence of the muon rate variations on the atmospheric temperature at the first order can be expressed as [3] :

$$\frac{\Delta I_\mu}{I_\mu^0} = \alpha_T \frac{\Delta_{Teff}}{< T_{eff} >} \quad (1)$$

where α_T is a constant and T_{eff} is an $effective$ temperature. The effective temperature can be computed dividing the atmosphere in layers for which the temperature measurements exist. The effective temperature is a weighted mean of the layer temperatures. To compute T_{eff} I have used the formula given in the MACRO paper [3]. The T_{eff} used by MINOS [4] looks different, but numerically gives similar results to the one used by MACRO. For the atmospheric temperature I have used the temperature measurements at 37 atmospheric pressures provided by the ECWMF, the European Centre for Medium-Range Weather Forecasts [6].

MACRO [5] has been the largest acceptance cosmic ray detector located in the Gran Sasso underground laboratory in Italy. The large acceptance of MACRO allowed a large number statistics for multiple-muons. MACRO ended data taking in December 2000. The data presented on the single muon seasonal variation in the paper [3] were collected during the MACRO construction in the period December 1992-December 1994. The data presented in this paper have been collected in the period December 1995 -December 2000 with the full MACRO, corresponding to about 34.5×10^6 single muons and and 2.6×10^6 events having at least 2 tracks.

2 The data selection

The tracking package used in this analysis requires at least 4 horizontal streamer chamber planes or at least 2 horizontal streamer planes and 2 vertical streamer planes [5]. In this analysis I have used only one view: the streamer tube "wires view"; this in order to avoid the problems due to the possible wrong association of the two different views to define tracks in multiple-muons events.

This analysis is looking for effect at the 1% level. This means that is important to select runs having good efficiencies. MACRO was divided in 6 "supermodules" and the data acquisition was done with 3 separate data acquisitions, each collecting data from two supermodules. Sometimes a couple of supermodules was removed from the acquisition for the maintenance of the detector. Since the rate of single muons with the full MACRO is about 860 events/hour it is possible to select the runs with full MACRO applying a cut on the rate of the single muons. So I have analyzed only the runs with rate of the single muons between 710 and 1010 events/hour corresponding to a cut of more than 5 σ, a factor 9 larger than the 1.9% single muon variation due to the seasonal modulation.

I have not tried more sophisticated selection based on the logbook of the detector performance because of technical problems. I recall that the MACRO data analysis was designed around 1990 and it was based on alfaVAX Digital Equipment computers with VMS operatinfg system. A fraction of the analysis code, but not all, was ported to UNIX before the end of the experiment. So in this analysis

 DOI: 10.1051/epjconf/201713605004

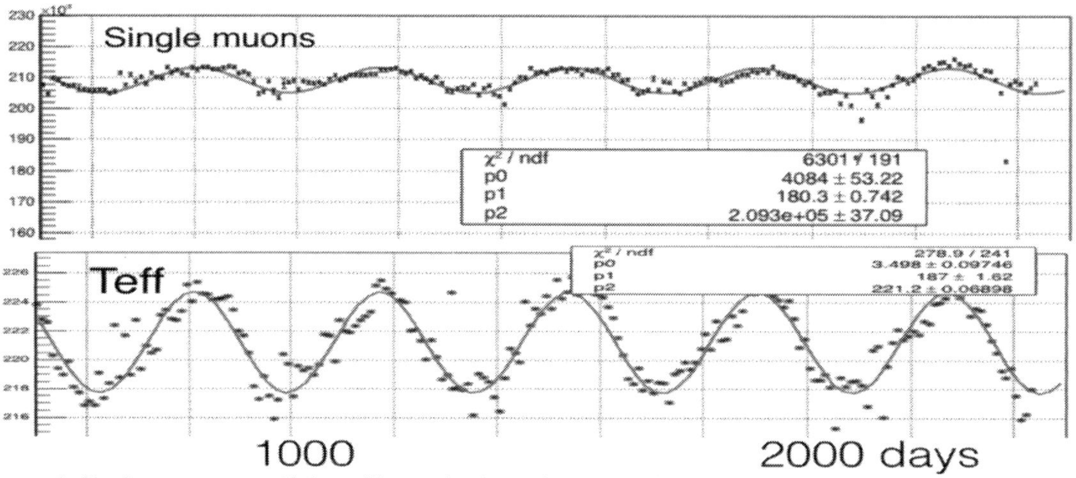

Figure 1. Single muons every 10 days. The x axis shows the day number starting from April 20th 1994. On the top are the single muons every 10 days. The bottom plot shows the effective temperature T_{eff} in Kelvin. The fit is done with a sinusoid + a constant (p2). The parameters p0/p2 = 1.9% give the oscillation amplitude, p1 gives the day of the peak, starting from day 0 =January 1th. p1 is 180.3 for single muons, The periodicity is fixed at 365.2 days

it was difficult to access to all the information on the data quality and to the detailed data base of the detector efficiencies. I have used only the information contained on data summary files containing track and hit informations.

However it is important to note the requirement on the very small number of hits to define a track (4 over a maximum possible of 14 for the horizontal chambers, or 2 horizontal and 2 vertical). Since the typical streamer tube efficiency is of the order of 97%, reasonable variations of this efficiency doesn't change very much the tracking efficiencies. MACRO started to take data with the full detector around April 20 1994, but in this analysis I have used only data starting from December 1995, to select data collected in stable conditions.

3 Results

The number of the single muon events every 10 days is shown at the top of Fig. 1. The x axis of the plot is the day number (day 0 is April 20 1994). The fit is done with a sinusoid + a constant value (parameter p2). The periodicity is fixed at the value of 365.2 days. The fit parameters ratio p0/p2 gives the fractional oscillation amplitude, p1 gives the day of the peak. p1 is 180.3 for single muons (Jun 30th-July 1th). From the daily correlation of the single muon rate with T_{eff} of Eq 1 is obtained $\alpha_T = 1.03 \pm 0.01(\textit{statistical})$ higher than the value $\alpha_T = 0.83 \pm 0.13$ reported in the MACRO seasonal modulation paper[5],to be compared the theoretical value $\alpha_T = \sim 0.92$ [2].

In case of multiple-muons the distance between couples of tracks has been evaluated using only the wire view, and the average value of the distances has been evaluated for each event. Fig. 2 shows the rate of the multiple muons with different cuts on the average distance: distance between 5 and 30 cm, 30 and 220 cm, 220 and 1000 cm and average distance bigger than 1000 cm. In the case of the multiple muons the peak of the rate varies from day 181, for distances bigger than 1000 cm, to day 198, for distances between 5 and 30 cm. The peak of the rate of multiple muons is delayed of a few days respect to the one of single muons, but is far from the winter months as observed in MINOS.

DOI: 10.1051/epjconf/201713605004

Also the amplitude varies from a minimum of 0.4% to a maximum of 5.8% for distances bigger than 1000 cm.

The results are seen in a better way in the polar graph of figure Fig. 3. On the left are the MACRO results and on the right are the MINOS results. The MACRO and MINOS data looks quite similar: in both there is a small amplitude at short distances and the oscillation amplitude is larger then the one of single muons at large distance (this is expected if every muon of multi-muons events is produced as "single" muon). In both experiments there are changes of the peak position correlated with the amplitude, but in MACRO the variations of the peaks are much smaller that in MINOS, while the variations of the amplitude are similar.

A comparison of the ECWMF atmospheric temperature in 37 layers shows different seasonal variations in the two experimental sites. In fact in MINOS there are atmospheric layers with a peak temperature in the winter season, while in MACRO the peak temperature is always in the summer months. This differences occurs at a height of about 13 km, where atmospheric jet streams are important. This difference is probably due to the different latitudes (42^0 for Gran Sasso and 48^0 for MINOS FD).

4 Conclusion

The peak of the rate of the multiple muons under Gran Sasso with different cuts on the distance, are in July, delayed respect to the one of single muons, while the expected peak of the dark matter

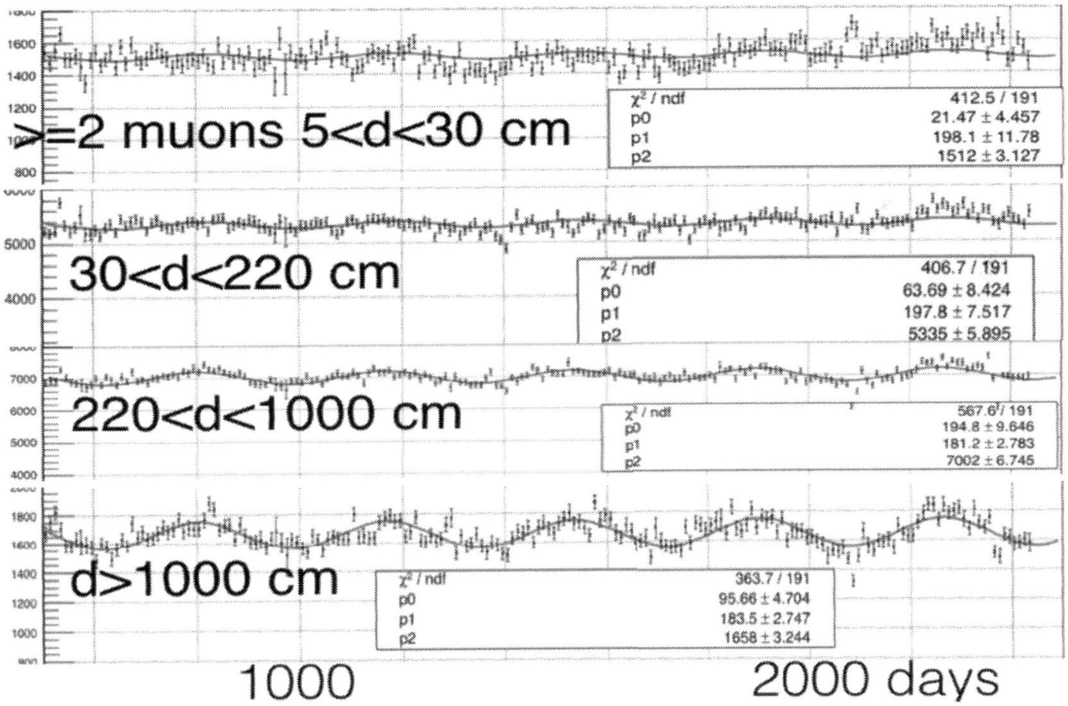

Figure 2. Multiple-muons every 10 days. The x axis shows the day number starting from April 20th 1994. Starting from the top: muons every 10 events with at least 2 tracks and with average distance between 5 and 30 cm, 30 and 220 cm, 220 and 1000 cm and average distance bigger than 1000 cm. The day of the peak of the rate (parameter p1) varies from 181 to 198 (July 1th- July 18th.)

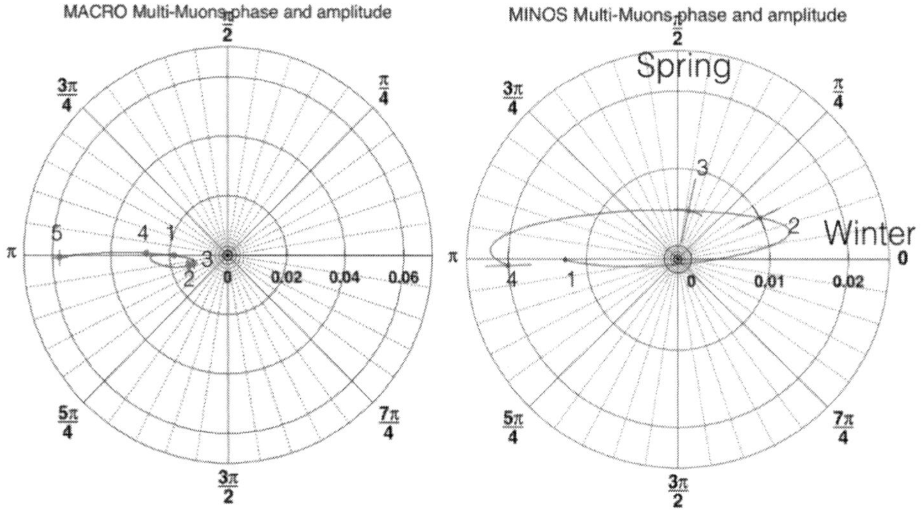

Figure 3. The amplitude and the phase of the first harmonic of the multiple muon rate in MACRO and MINOS. Phase 0 is near January 1th. Phase π is near July 1th. The line is only to guide the eye. The points labeled as 1 show the value for the single muons. The points after 1 are for multiple-muons. They are in increasing values of the separation between muons. In MACRO the points 2-5 corresponds to cuts in the average muon distance 5-30, 30-220,220-1000 cm , $\geq$1000. In MINOS the cut is in the minimum distance and the points 2-4 corresponds to cuts 60-450, 450-800,$\geq$800 cm. In MINOS there are big changes in the phase, but not in MACRO. In both experiments, the last point (cut at large distance) has practically the same phases of the one of single muons, while the amplitude is larger. Note that the radial scale of the two plots is different.

signal is expected around June 2; therefore multiple-muons originated backgrounds should not be a problem for the DAMA [7] dark matter experiment. This result is different in the MINOS FD site, the difference could be due to the different depths and to the different latitudes.

The author thanks A. Longhin, for giving me the ECWMF temperatures, A. Paoloni and A.Marini for useful suggestions and discussions, and all the MACRO past collaborators, listed in ref [5], contributing to the success of this experiment. This work also shows the importance to save data and the software of past experiments.

References

[1] T. Gaisser Cosmic Rays and Particle Physics Cambridge University Press, Cambridge (1990)
[2] M. Agostini *et al.* [GERDA Collaboration], Astropart. Phys. **84**, 29 (2016)
[3] M. Ambrosio *et al.* [MACRO Collaboration], Astropart. Phys. **7**, 109 (1997).
[4] P. Adamson *et al.* [MINOS Collaboration], Phys. Rev. D **91** (2015) no.11, 112006
[5] M. Ambrosio *et al.* [MACRO Collaboration], Nucl. Instrum. Meth. A **486** (2002) 663.
[6] ECWMF, European Centre for Medium-Range Weather Forecasts, (http://www.ecmwf.int/).
[7] R. Bernabei *et al.*, Eur. Phys. J. C **73** (2013) 2648